Endorsements for *Solar Power in Building Design*

Dr. Peter Gevorkian's *Solar Power in Building Design* is the third book in a sequence of comprehensive surveys in the field of modern solar energy theory and practice. The technical title does little to betray to the reader (including the lay reader) the wonderful and uniquely entertaining immersion into the world of solar energy.

It is apparent to the reader, from the very first page, that the author is a master of the field and is weaving a story with a carefully designed plot. The author is a great storyteller and begins the book with a romantic yet rigorous historical perspective that includes the contribution of modern physics. A description of Einstein's photoelectric effect, which forms one of the foundations of current photovoltaic devices, sets the tone. We are then invited to witness the tense dialogue (the ac versus dc debate) between two giants in the field of electric energy, Edison and Tesla. The issues, though a century old, seem astonishingly fresh and relevant.

In the smoothest possible way Dr. Gevorkian escorts us in a well-rehearsed manner through a fascinating tour of the field of solar energy making stops to discuss the basic physics of the technology, manufacturing process, and detailed system design. Occasionally there is a delightful excursion into subjects such as energy conservation, building codes, and the practical side of project implementation.

All this would have been more than enough to satisfy the versed and unversed in the field of renewable energy. But as all masters, Dr. Gevorkian wraps up his textbook in relevance by including a thorough discussion of the current solar initiatives (California being a prototype) and the spectrum of programs and financial incentives that are being created.

Solar Power in Building Design has the seductive quality of being at once an overview and course in solar energy for anyone with or without a technical background. I suspect that this book will likely become a standard reference for all who engage in the emerging renewable energy field.

DR. DANNY PETRASEK, M.D., PH.D.
California Institute of Technology

Dr. Gevorkian's *Solar Power in Building Design* is a great read. If you are able to envision a relatively arcane subject such as solar energy and photovoltaic applications as a compelling, page-turning read, this is your book.

Dr. Gevorkian is a very lucid writer. A dedicated grammarian as well as a master of a multiplicity of scientific disciplines, Dr. Gevorkian has crafted a text that broadens even the most jaded reader's perspective on the subject of solar power. He ranges from storytelling, as with his brief characterization of the controversy between early innovators Nikolai Tesla and Thomas Edison in the pioneering years of the modern Energy Age, to a full-blown historical tracing of the rapid advances of the expansive diversification of energy applications within the past 50 years.

Throughout the book, Dr. Gevorkian espouses a didactic approach that is thoroughly inductive. A wide swath of information, integral to the exposition of the text, is presented in survey fashion. Many questions are both asked and answered, which provides any level of audience with a great deal of satisfaction as they pour over the wide variety of data needed to complete this intriguing story.

Very little of this complex domain is left to speculation. The layperson drawn to this subject matter should be appreciative, as Dr. Gevorkian prepares any reader to be able to intelligently communicate with a consultant hired to install a solar power system, or to even install a system themselves.

Myriad examples of the range of installations, replete with visual support in the form of graphs, charts, renderings, and color photographs, serve to provide even the most technically inept audience with the certainty that they can adeptly navigate their way around a solar power system. That is the true beauty of *Solar Power in Building Design*: a gem of a book for even armchair wannabe experts.

DR. LANCE WILLIAMS
Executive Director
US Green Building Council
Los Angeles chapter

I think the best way to summarize *Solar Power in Building Design* is "more than you thought you needed to know." Dr. Gevorkian has covered many areas of the solar power design in a manner that allows the reader to really have a remarkable, practical understanding of solar renewable energy as it now exists in North America. Anyone getting into the business should really consider this book as a "must read," a super primer for a thorough understanding of the industry. In addition to covering the historical perspective, the book reflects a fresh perspective of the future direction of where the industry will be in 25 years.

In addition to magnificent coverage of technology issues, the reader is introduced to the critical underlying intricacies of capital assets management such as long functioning life and specificity of unique factors that must be taken into account for financial analysis. In addition the book skillfully covers asset profiling economics for the present and future and reflects a unique perspective as to how solar power cogeneration fits into the energy needs of the future.

Congratulations for another fine scholarly work!

GENE BECK
President
EnviroTech Financial, Inc.

I have never before enjoyed reading a technology book that could deliver so much information down to the core without once becoming boring and complicated.

The book takes the reader through the challenges of solar energy technology, engineering, and design and its applications, providing every detail and an in-depth perspective from the basic to the most complex issues.

Although it has been written for solar energy professionals, this book is a resource for anyone interested in solar energy technologies as Dr. Gevorkian has described the concepts of solar power engineering and design precisely and clearly.

The highly informative diagrams and illustrations with each chapter help in understanding the dynamics of solar power.

I also strongly recommend *Solar Power in Building Design* as a standard curriculum and guide on renewable energy for architectural and engineering schools and colleges as a resource for students.

DR. FARHAT IQBAL
President
Silica Solar LLC

The passion of the author in preservation of our ecological system, and in renewable energy in particular, is obvious in the pages of this scholarly text. His knowledge in the subject matter of solar energy is evidenced in the clear presentation that can be understood by the general populace, yet technological experts will appreciate the in-depth discussions.

The wealth of information and the abundance of resources contained in this volume will establish it as an encyclopedia for solar energy design.

Congratulation for a great landmark work! Save our planet! Keep it green!

EMILY LUK, M.D.

Dr. Gevorkian's *Solar Power in Building Design* is a very well written, scientifically sound, and technologically proficient book that should appeal to scientists, engineers, technologists, technicians, and laypersons alike. It is written in a simple, easy-to-read, and straightforward style that introduces the theory and then delves, immediately, into the practicalities. It is also replete with suitable illustrations. Energy is crucial for economic growth, but, as it now exists, extracts a heavy price in terms of environmental degradation. Solar energy offers an attractive mitigation here, albeit partially. But then, in an age of almost universal concerns about energy and environment, the fact is that a wide gulf separates our awareness of the need to use solar power and our familiarity with the nuts and bolts of harnessing this power. Dr. Gevorkian's book bridges that and can therefore be as much a resource for the solar energy professionals as for the solar energy enthusiasts as well as curious onlookers.

DR. PAL POORNA, M.S., M.B.A., PH.D.
Chair
Physical Science
Glendale Community College
Glendale, CA

It is an honor to be asked to review your latest book *Solar Power in Building Design.*

This book is systematically and simply presented to acquaint any student with the procedures necessary to incorporate solar power in the design of buildings.

The book covers straightforward delineation from basic physics through technologies, design, and implementation and culminates in an appreciation of our environment, and diminishing our dependence on fossil fuels, while punctuating the ultimate economics involved in preserving our planet.

The capturing of solar energy has been around for a long time, and educating the people of this planet as to how to utilize this natural resource remains a paramount task. This book is essentially a comprehensive educational resource and a design reference manual that offers the reader an opportunity to learn about the entire spectrum of solar power technology with remarkable ease. The book is a magnificent accomplishment.

WILLIAM NONA
Architect
National Council of Architectural Registration Boards

The author has provided very comprehensive material in the field of photovoltaic and solar systems in the text, starting from the basic knowledge on silicon technology, solar cell processing, and module manufacturing to the cutting edge and most aesthetically pleasing BIPV (building-integration photovoltaic) systems and installations. The content given has all the necessary tools and information to optimize solar system designs and integration. To all professionals interested in having a thorough knowledge in photovoltaic systems design, integration, and engineering, this book is a must.

FRANK C. PAO
Chairman Atlantis Energy Systems, Inc.

Solar Power in Building Design is a must-read primer for any professional or professional-to-be, who wants to learn about challenges and opportunities associated with design implementation or economics of solar systems.

The author emphasizes how engineering design is impacted by economics, environment, and local government policies.

The book precisely shows the current state of solar and renewable energy technology, its challenges, and its up-to-date successes.

Trivia facts and history of solar technology make this book fun to read.

ANDRZEJ KROL, P.E.
President
California Electrical Services
Glendale, CA

Producing electricity from the sun using photovoltaic (PV) systems or solar thermal systems for heating and cooling has become a major industry worldwide along with many helpful multilingual solar system simulation software tools. But engineering, installing, monitoring, and maintaining such systems requires constant knowledge update and ongoing training.

Dr. Gevorkian's *Solar Power in Building Design* makes a superb reference guide on solar electricity and offers a unique combination of technical and holistic discussion on building rating systems such as LEED with practical advice for students, professionals, and investors.

Well-illustrated chapter sequences with built U.S. examples offer step-by-step insights on the theory and reality of installed renewable energy systems, solar site analysis, component specifications, and U.S.-specific system costs and economics, performance, and monitoring.

THOMAS SPIEGELHALTER
Professor
School of Architecture
University of Southern California
R.A. EU, ISES, LEED
Freiburg, Germany, and Los Angeles

Solar Power in Building Design is a comprehensive book that is an indispensable reference for students and professionals.

Each of the topics is presented completely, with clear and concise text. A history of each subject is followed by both a global and a detailed view. The invaluable historical background amazingly spans topics as diverse as the centuries-old Baghdad battery to the photoelectric effect.

The figures in the text are excellent: the diagrams and illustrations, with the accompanying text in the book, walk the reader through each section, resulting in a better understanding of the concepts presented. The photographs are also excellent in that they clearly show their intended subject. In addition, the author has clearly thought through each topic, ensuring that there are no surprises for the professional embarking on incorporating a solar power system into a building's design.

DR. VAHE PEROOMIAN, PH.D.
Professor of Physics
University of California (UCLA)

Solar Power in Building Design is a comprehensive solar power design reference resource and timely educational book for our planet's troubled times. With global warming, pollution, and the waste of energy showing their irreparable damage, Dr. Peter Gevorkian's book is pioneering in the field of solar power cogeneration and fills a significant technology educational void sorely needed to mitigate our present environmental pollution. Dr. Gevorkian explains difficult technical concepts with such ease that it becomes a pleasure to read the entire book. It is truly remarkable how the detailed explanation of each process facilitates the conveyance of knowledge from a scientific source to a nontechnical reader.

EDWARD ALEXANIANS, S.E.
Sr. Engineer
Los Angeles County Research Engineering Department

Solar Power in Building Design is *the* comprehensive reference manual on solar power systems. This is a must read for anyone who will be implementing a solar power system from small residential applications to large commercial or industrial applications. This book has everything from the theory of solar power generation, to design guidelines, to the economics of solar systems. The field of solar power generation is advancing very quickly; however, this book includes the most current technology available and also includes emerging research and trends that provide a view of the future in solar power systems.

STEVE S. HIRAI, P.E.
Principal Engineer
Montgomery Watson Americas

Solar Power in Building Design is a remarkably comprehensive and easy-to-read book on solar power technology. The book as a design reference manual is very timely, relevant, and informative and exposes the reader to the entire spectrum of solar power technology as a whole. The most remarkable attribute of the book is that it can be read and understood by anyone without previous knowledge of the subject.

GUADALUPE FLORES, AIA
President
Pasadena and Foothill AIA Chapter

Dr. Peter Gevorkian's latest in a series of three books on sustainable energy truly hits the mark as the *ultimate* go-to guide for anything and everything quite literally "under the sun" relating to making use of this vital resource—solar energy.

In a text that is immediately engaging and understandable to anyone with the need or desire to expand knowledge in this vital technology, Dr Gevorkian reveals that harnessing sunlight into a practical energy source is doable in the here and now. Charts, graphs, photos, and illustrations are deftly used to support the vital concepts he explains so clearly. Everything you could possibly need, from conceptual presentation to a client, to site evaluation, to the nuts and bolts of components and installation methods are found here. Additionally, the text explains the historical development of the technologies and components from which today's products are derived.

MARY KANIAN
Environmentalist

Eons ago humans cowered from the sun fearing its power as a malevolent God.

Those who learned its rhythms became shaman, bringing the knowledge of the seasons to their people. By the end of the twentieth century humankind had harnessed much energy originally derived from the sun, and buildings sheltered rather than celebrated natural forces.

The twenty-first century brings new opportunities to choose alternative energy sources—or actually those that were always there. Dr. Gevorkian is a modern-day shaman. In a world where wars are fought for dinosaur remains, located deep in the earth, Dr. Gevorkian enlightens us on how to pluck energy from the sky.

Like a magician who has revealed all his secrets, *Solar Power in Building Design* is the definitive Manual for just that. Between its covers are all the answers for the application of solar power. If one had to grab just one book on solar energy prior to ducking into the bomb shelter, I would go for this one.

MARK GANGI, AIA
Principle
Gangi Architects

SOLAR POWER
IN BUILDING
DESIGN

About the Author

Peter Gevorkian, Ph.D., P.E., is President of Vector Delta Design Group, Inc., an electrical engineering and solar power design consulting firm, specializing in industrial, commercial, and residential projects. Since 1971, he has been an active member of the Canadian and California Boards of Professional Engineers. Dr. Gevorkian is also the author of *Sustainable Energy Systems in Architectural Design* and *Sustainable Energy Systems Engineering,* both published by McGraw-Hill.

SOLAR POWER IN BUILDING DESIGN

THE ENGINEER'S COMPLETE DESIGN RESOURCE

PETER GEVORKIAN, Ph.D., P.E.

New York   Chicago   San Francisco   Lisbon   London   Madrid
Mexico City   Milan   New Delhi   San Juan   Seoul
Singapore   Sydney   Toronto

1 2 3 4 5 6 7 8 9 0   DOC/DOC   0 1 3 2 1 0 9 8 7

ISBN 978-0-07-148563-0
MHID 0-07-148563-5

Sponsoring Editor: Cary Sullivan
Production Supervisor: Richard C. Ruzycka
Editing Supervisor: Stephen M. Smith
Project Manager: Arushi Chawla
Copy Editor: Lucy Mullins
Proofreader: Nigel Peter O'Brien
Indexer: Kevin Broccoli
Art Director, Cover: Jeff Weeks
Composition: International Typesetting and Composition

Printed and bound by RR Donnelley.

McGraw-Hill books are available at special quantity discounts to use as premiums and sales promotions, or for use in corporate training programs. For more information, please write to the Director of Special Sales, McGraw-Hill Professional, Two Penn Plaza, New York, NY 10121-2298. Or contact your local bookstore.

 This book is printed on recycled, acid-free paper made from 100% postconsumer waste.

CONTENTS

FOREWORD

Dr. Peter Gevorkian's *Solar Power in Building Design: The Engineer's Complete Design Resource* is the third book in a sequence of comprehensive surveys in the field of modern solar energy theory and practice. The technical title does little to betray to the reader (including the lay reader) the wonderful and uniquely entertaining immersion into the world of solar energy.

It is apparent to the reader, from the very first page, that the author is a master of the field and is weaving a story with a carefully designed plot. The author is a great storyteller and begins the book with a romantic yet rigorous historical perspective that includes the contribution of modern physics. A description of Einstein's photoelectric effect, which forms one of the foundations of current photovoltaic devices, sets the tone. We are then invited to witness the tense dialogue (the ac versus dc debate) between two giants in the field of electric energy, Edison and Tesla. The issues, though a century old, seem astonishingly fresh and relevant.

In the smoothest possible way Dr. Gevorkian escorts us in a well-rehearsed manner through a fascinating tour of the field of solar energy—making stops to discuss the basic physics of the technology, the manufacturing process, and detailed system design. Occasionally there is a delightful excursion into subjects such as energy conservation, building codes, and the practical side of project implementation.

All this would have been more than enough to satisfy the versed and unversed in the field of renewable energy. But as all masters, Dr. Gevorkian wraps up his textbook in relevance by including a thorough discussion of the current solar initiatives (California being a prototype), and the spectrum of programs and financial incentives that are being created.

Solar Power in Building Design: The Engineer's Complete Design Resource has the seductive quality of being at once an overview and course in solar energy for anyone with or without a technical background. I suspect that this book will likely become a standard reference for all who engage in the emerging renewable energy field.

DR. DANNY PETRASEK, M.D., PH.D.
California Institute of Technology

INTRODUCTION

Since the dawn of agriculture and civilization, human beings have hastened deforestation, impacting climatic and ecological conditions. Deforestation and the use of fossil fuel energy diminish the natural recycling of carbon dioxide gases. This accelerates and increases the inversion layer that traps the reflected energy of the sun. The augmented inversion layer has an elevated atmospheric temperature, giving rise to global warming, which in turn has caused melting of the polar ice, substantial changes to climatic conditions, and depletion of the ozone layer.

Within a couple of centuries, the unchecked effects of global warming will not only change the makeup of the global land mass but will affect human's lifestyle on the planet.

Continued melting of the polar ice caps will increase seawater levels and will gradually cover some habitable areas of global shorelines. It will also result in unpredictable climatic changes, such as unusual precipitation, floods, hurricanes, and tornadoes.

In view of the rapid expansion of the world's economies, particularly those of developing countries with large populations, such as China and India, demand for fossil fuel and construction materials will become severe. Within the next few decades, if continued at the present projected pace, the excessive demand for fossil fuel energy resources, such as crude oil, natural gas, and coal, will result in the demise of the ecology of our planet and, if not mitigated, may be irreversible. Today China's enormous demand for energy and construction materials has resulted in considerable cost escalations of crude oil, construction steel, and lumber, all of which require the expenditure of fossil fuel energy.

Developing countries are the most efficient consumers of energy, since every scrap of material, paper, plastic, metal cans, rubber, and even common trash, is recycled and reused. However, when the 2.3 billion combined populations of China and India attain a higher margin of families with middle-class incomes, the new demand for electricity, manufacturing, and millions of automobiles will undoubtedly change the balance of ecological and social stability to a level beyond imagination.

The United States is the richest country in the world. With 5 percent of the world's population, the country consumes 25 percent of the global aggregate energy. As a result of its economic power, the United States enjoys one of the highest standards of living with the best medical care and human longevity. The relative affluence of the country as a whole has resulted in the cheapest cost of energy and its wastage.

Most consumption of fossil fuel energy is a result of inefficient and wasteful transportation and electric power generation technologies. Because of the lack of comprehensive energy control policies and lobbying efforts of special-interest groups, research and development funds to accelerate sustainable and renewable energy technologies have been neglected.

In order to curb the waste of fossil fuel energy, it is imperative that our nation, as a whole, from politicians and educators to the general public, be made aware of the dire consequences of our nation's energy policies and make every effort to promote the use of all available renewable energy technologies so that we can reduce the demand for nonrenewable energy and safeguard the environment for future generations.

The deterioration of our planet's ecosystem and atmosphere cannot be ignored or considered a matter that is not of immediate concern. Our planet's ozone layer according to scientists has been depleted by about 40 percent over the past century and greenhouse gases have altered meteorological conditions. Unfortunately, the collective social consciousness of the educated masses of our society has not concerned itself with the disaster awaiting our future generations and continues to ignore the seriousness of the situation.

About This Book

During years of practice as a research and design engineer, I have come to realize that the best way to promote the use of solar power as a sustainable energy design is to properly educate key professionals, such as architects, engineers, and program managers whose opinions direct project development.

I have found that even though solar power at present is a relatively mature technology, its use and application in the building industry is hampered due to lack of exposure and education. Regardless of present federal and state incentive programs, sustainable design by use of renewable energy will not be possible without a fundamental change in the way we educate our architects, engineers, and decision makers.

In two earlier books titled *Sustainable Energy Systems in Architectural Design* and *Sustainable Energy Systems Engineering,* I attempted to introduce architects, engineers, and scientists to a number of prevailing renewable energy technologies and their practical use, in the hopes that a measure of familiarity and understanding could perhaps encourage their deployment.

This book has been specifically written to serve as a pragmatic design resource for solar photovoltaic power systems engineering. When writing the manuscript, I attempted to minimize unnecessary mathematics and related theoretical photovoltaic physics, by only covering real-life, straightforward design techniques that are commonly practiced in the industry.

As scientists, engineers, and architects, we have throughout the last few centuries been responsible for the elevation of human living standards and contributed to advancements in technology. We have succeeded in putting a human on the moon, while ignoring the devastating side effects to the global ecology. In the process of creating the betterment and comforts of life, we have tapped into the most precious nonrenewable energy resources, miraculously created over the life span of our planet, and have been misusing them in a wasteful manner to satisfy our most rudimentary energy needs.

Before it is too late, as responsible citizens of our global village, it is high time that we assume individual and collective responsibility to resolve today's environmental issues and ensure that future life on Earth will continue to exist as nature intended.

Global Warming and Climate Change

Ever since the industrial revolution, human activities have constantly changed the natural composition of Earth's atmosphere. Concentrations of trace atmospheric gases, nowadays termed "greenhouse gases," are increasing at an alarming rate. There is conclusive evidence that the consumption of fossil fuels, conversion of forests to agricultural land, and the emission of industrial chemicals are principal contributing factors to air pollution.

According to the National Academy of Sciences, Earth's surface temperature has risen by about one degree Fahrenheit (°F)in the past century, with accelerated warming occurring in the past three decades. According to statistical review of the atmospheric and climatic records, there is substantial evidence that global warming over the past 50 years is directly attributable to human activities.

Under normal atmospheric conditions, energy from the sun controls Earth's weather and climate patterns. Heating of Earth's surface resulting from the sun radiates energy back into space. Atmospheric greenhouse gases, including carbon dioxide (CO_2), methane (CH_4), nitrous oxide (N_2O), tropospheric ozone (O_3), and water vapor (H_2O) trap some of this outgoing energy, retaining it in the form of heat, somewhat like a glass dome. This is referred to as the *greenhouse effect*.

Without the greenhouse effect, surface temperatures on Earth would be roughly 30°C [54 degrees Fahrenheit (°F)] colder than they are today—too cold to support life. Reducing greenhouse gas emissions depends on reducing the amount of fossil fuel–fired energy that we produce and consume.

Fossil fuels include coal, petroleum, and natural gas, all of which are used to fuel electric power generation and transportation. Substantial increases in the use of non-renewable fuels is a principal factor in the rapid increase in global greenhouse gas emissions. The use of renewable fuels can be extended to power industrial, commercial, residential, and transportation applications to substantially reduce air pollution.

Examples of zero-emission, renewable fuels include solar, wind, geothermal, and renewably powered fuel cells. These fuel types, in combination with advances in energy-efficient equipment design and sophisticated energy management techniques, can reduce the risk of climate change and the resulting harmful effects on the ecology. Keep in mind that natural greenhouse gases are a necessary part of sustaining life on Earth. It is the anthropogenic or human-caused increase in greenhouse gases that is of concern to the international scientific community and governments around the world.

Since the beginning of the modern industrial revolution, atmospheric concentrations of carbon dioxide have increased by nearly 30 percent, methane concentrations have more than doubled, and nitrous oxide concentrations have also risen by about 15 percent. These increases in greenhouse gas emissions have enhanced the heat-trapping capability of Earth's atmosphere.

Fossil fuels burned to operate electric power plants, run cars and trucks, and heat homes and businesses are responsible for about 98 percent of U.S. carbon dioxide emissions, 24 percent of U.S. methane emissions, and 18 percent of U.S. nitrous oxide emissions. Increased deforestation, landfills, large agricultural production, industrial production, and mining also contribute a significant share of emissions. In 2000, the United States produced about 25 percent of total global greenhouse gas emissions, the largest contributing country in the world.

Estimating future emissions depends on demographics, economics, technological policies, and institutional developments. Several emissions scenarios have been developed based on differing projections of these underlying factors. It is estimated that by the year 2100, in the absence of emission-control policies, carbon dioxide concentrations will be about 30 to 150 percent higher than today's levels.

Increasing concentrations of greenhouse gases are expected to accelerate global climate change. Scientists expect that the average global surface temperatures could rise an additional 1°F to 4.5°F within the next 50 years and 2.2°F to 10°F over the next century, with significant regional variation. Records show that the 10 warmest years of the twentieth century all occurred in the last 15 years of that century. The expected impacts of this weather warming trend include the following:

Water resources. A warming-induced decrease in mountain snowpack storage will increase winter stream flows (and flooding) and decrease summer flows. This along with an increased evapotranspiration rate is likely to cause a decrease in water deliveries.

Agriculture. The agricultural industry will be adversely affected by lower water supplies and increased weather variability, including extreme heat and drought.

Forestry. An increase in summer heat and dryness is likely to result in forest fires, an increase in insect populations, and disease.

Electric energy. Increased summer heat is likely to cause an increase in the demand for electricity due to an increased reliance on air conditioning. Reduced snowpack is likely to decrease the availability of hydroelectric supplies.

Regional air quality and human health. Higher temperatures may worsen existing air quality problems, particularly if there is a greater reliance on fossil fuel generated electricity. Higher heat would also increase health risks for some segments of the population.

Rising ocean levels. Thermal expansion of the ocean and glacial melting are likely to cause a 0.5 to 1.5 m (2 to 4 ft) rise in ocean levels by 2100.

Natural habitat. Rising ocean levels and reduced summer river flow are likely to reduce coastal and wetland habitats. These changes could also adversely affect spawning fish populations. A general increase in temperatures and accompanying increases in summer dryness could also adversely affect wildland plant and animal species.

Scientists calculate that without considering feedback mechanisms a doubling of carbon dioxide would lead to a global temperature increase of 1.2°C (2.2°F). But, the net effect of positive and negative feedback patterns would cause substantially more warming than would the change in greenhouse gases alone.

Pollution Abatement Consideration

According to a 1999 study report by the U.S. Department of Energy (DOE), one kilowatt of energy produced by a coal-fired electric power–generating plant requires about 5 pound (lb) of coal. Likewise, generation of 1.5 kilowatt-hours (kWh) of electric energy per year requires about 7400 lb of coal that in turn produces 10,000 lb of carbon dioxide (CO_2).

Roughly speaking, the calculated projection of the power demand for the project totals to about 2500 to 3000 kWh. This will require between 12 million and 15 million lb of coal, thereby producing about 16 million to 200 million lb of carbon dioxide. Solar power, if implemented as previously discussed, will substantially minimize the air pollution index. The Environmental Protection Agency (EPA) will soon be instituting an air pollution indexing system that will be factored into all future construction permits. All major industrial projects will be required to meet and adhere to the air pollution standards and offset excess energy consumption by means of solar or renewable energy resources.

Energy Escalation Cost Projection

According to the Energy Information Administration data source published in 1999, California consumes just as much energy as Brazil or the United Kingdom. The entire global crude oil reserves are estimated to last about 30 to 80 years, and over 50 percent of the nation's energy is imported from abroad. It is inevitable that energy costs will surpass historical cost escalations averaging projections. The growth of fossil fuel consumption is illustrated in Figure I.1. It is estimated that the cost of nonrenewable energy will, within the next decade, increase by approximately 4 to 5 percent by producers.

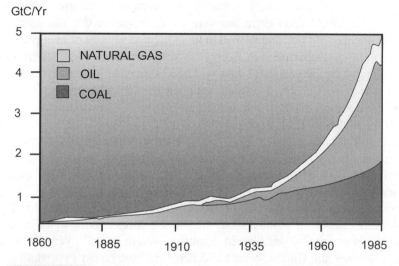

Figure I.1   **Growth in fossil fuel consumption.** *Courtesy of Geothermal Education Office.*

When compounded with a general inflation rate of 3 percent, the average energy cost increase, over the next decade, could be expected to rise at a rate of about 7 percent per year. This cost increase does not take into account other inflation factors, such as regional conflicts, embargoes, and natural catastrophes.

Solar power cogeneration systems require nearly zero maintenance and are more reliable than any human-made power generation devices. The systems have an actual life span of 35 to 40 years and are guaranteed by the manufacturers for a period of 25 years. It is my opinion that in a near-perfect geographic setting, the integration of the systems into the mainstream of architectural design will not only enhance the design aesthetics but also will generate considerable savings and mitigate adverse effects on the ecology and global warming.

Social and Environmental Concerns

Nowadays, we do not think twice about leaving lights on or turning off the television or computers, which run for hours. Most people believe that energy seems infinite, but in fact, that is not the case. World consumption of fossil fuels, which supply us with most of our energy, is steadily rising. In 1999, it was found that out of 97 quads of energy used (a quad is 3×1011 kWh) 80 quads came from coal, oil, and natural gas. As we know, sources of fossil fuels will undoubtedly run out within a few generations and the world has to be ready with alternative and new sources of energy. In reality, as early as 2020, we could be having some serious energy deficiencies. Therefore, interest in renewable fuels such as wind, solar, hydropower, and others is a hot topic among many people.

Renewable fuels are not a new phenomenon, although they may seem so. In fact, the industrial revolution was launched with renewable fuels. The United States and the world has, for a long time, been using energy without serious concern, until the 1973 and 1974 energy conferences, when the energy conservation issues were brought to the attention of the industrialized world. Ever since, we were forced to realize that the supply of fossil fuels would one day run out, and we had to find alternate sources of energy.

In 1999, the U.S. Department of Energy (DOE) published a large report in which it was disclosed that by the year 2020 there will be a 60 percent increase in carbon dioxide emissions which will create a serious strain on the environment, as it will further aggravate the dilemma with greenhouse gases. Figure I.2 shows the growth of carbon dioxide in the atmosphere.

A simple solution may seem to be to reduce energy consumption; however, it would not be feasible. It has been found that there is a correlation between high electricity consumption (4000 kWh per capita) and a high Human Development Index (HDI), which measures quality of life.

In other words there is a direct correlation between quality of life and the amount of energy used. This is one of the reasons that our standard of living in the industrialized countries is better than in third-world countries, where there is very little access to electricity. In 1999, the United States had 5 percent of the world's population and produced 30 percent of the gross world product. We also consumed 25 percent of the world's energy and emitted 25 percent of the carbon dioxide.

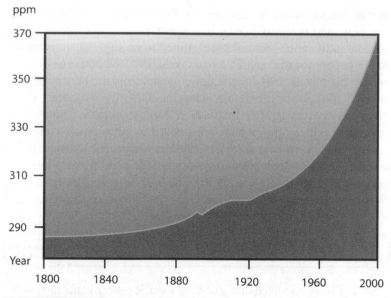

Figure I.2 **Growth of CO$_2$ in the atmosphere.** *Courtesy of Geothermal Education Office.*

It is not hard to imagine what countries, such as China and India, with increasing population and economic growth, can do to the state of the global ecology.

The most significant feature of solar energy is that it does not harm the environment. It is clean energy. Using solar power does not emit any of the extremely harmful greenhouse gases that contribute to global warming. There is a small amount of pollution when the solar panels are produced, but it is miniscule in comparison to fossil fuels. The sun is also a free source of energy. As technology advances, solar energy will become increasingly economically feasible because the price of the photovoltaic modules will go down. The only concern with solar power is that it is not energy on demand and that it only works during the day and when it is very sunny. The only way to overcome this problem is to build storage facilities to save up some of the energy in batteries; however, that adds more to the cost of solar energy.

A Few Facts about Coal-Based Electric Power Generation

At present the most abundant fossil fuel resource available in the United States is coal. Coal-based electric power generation represents about 50 percent of energy used and is the largest environmental pollution source. Coal burned in boilers generates an abundance of CO$_2$, SO$_x$ (sulfur dioxide), NO$_x$ (nitrogen dioxide), arsenic, cadmium, lead, mercury, soot particles, and tons of coal ash, all of which pollute the atmosphere and water. At present 40 percent of the world's CO$_2$ emissions comes from coal-burning power plants.

Under the advertising slogan of *Opportunity Returns,* the coal industry in the United States has recently attempted to convey an unsubstantiated message to the public that a new clean coal gasification technology, assumed to be superclean, is on the horizon to provide safe energy for the next 250 years. Whatever the outcome of the promised technology, at present coal-fired electric power generation plant construction is on the rise and currently 120 power generation plants are under construction.

So-called clean coal integrated gasification combined cycle (IGCC) technology converts coal into synthetic gas, which is supposed to be as clean as natural gas and 10 percent more efficient when used to generate electricity. The technology is expected to increase plant power efficiency by 10 percent, produce 50 percent less solid waste, and reduce water pollution by 40 percent. Even with all these coal power energy production improvements, the technology will remain a major source of pollution.

Coal Power Generation Industry Facts

- By the year 2030 it is estimated that coal-fired electric power generation will represent a very large portion of the world's power and provide 1,350,000 megawatts (MW) of electric energy, which in turn will inject 572 billion tons of CO_2 into the atmosphere, which will be equal to the pollution generated over the past 250 years.
- In the United States it takes 20 lb of coal by weight to generate sufficient energy requirements per person per day. When totaled, it represents approximately 1 billion tons of coal. The percent of coal-based energy production in the United States will be 50 percent, that for China 40 percent, and that for India 10 percent.
- By the year 2030 the world energy demand is projected to be doubled.
- The worldwide coal resource is estimated to be 1 trillion tons, and the United States holds 25 percent of the total resource or about 270 billion tons. China has 75 billion tons of coal, which is expected to provide 75 years of coal-based electric energy.
- The consequences of cheap electric power generated from coal will require large national expenditures to mitigate environmental pollution and related public health problems that will translate into medical bills for treating asthma, emphysema, heart attacks, and cancer. The effects of pollution will also be extended to global ecological demise and genetic changes in plant and animal life.

  According to a study conducted by Princeton University, the effects of U.S. coal-fired electric power generation plants on public health would add $130 billion per megawatt-hour of energy. At present the average cost of one megawatt of power is about $120.
- The Kyoto Protocol, ratified by 162 nations except the United States and Australia, calls for cutting greenhouse gas emissions by 5.2 percent by 2012. China and developing nations like India, which are exempt, will most likely generate twice the expected amount of atmospheric pollution.

ACKNOWLEDGMENTS

I would like to thank my colleagues and individuals who have encouraged and assisted me to write this book. I am especially grateful to all agencies and organizations that provided photographs and allowed the use of some textual material and my colleagues who read the manuscript and provided valuable insight.

Michael Lehrer FAIA, President Lehrer Architects; Vahan Garboushian, CEO of Amonix Solar Mega Concentrator; Dr. Danny Petrasek Ph.D., MD Professor of Biotechnology Sciences, Caltech; Mark Gangi, AIA President Gangi Architects; Mary Olson Khanian, Environmentalist; Armen Nahapetian, VP engineering, Teledyne Corp.; Gene Beck, President EnviroTech Financial, Inc.; Jim Zinner, President Environmental Consultant; Peter Carpenter, D.J. MWH Americas, Inc.; Lance A. Williams, Ph.D., executive director of U.S. Green Building Council (USGBC); Taylor & Company, Los Angeles, CA; Benny Chan Fotoworks Studio for book cover photographs; Dr. Farhat Iqbal, President Silica Solar; Vahe Peroomian, Ph.D., Professor of Physics, UCLA; Julie D. Taylor, Principal Communications for Creative Industries; Guadalupe Flores, President of Pasadena and Foothill AIA Chapter; Raju Yenamandra, Director of Marketing SolarWorld; Kevin Macakamul and Michael Doman, Regensis Power; Jovan Gayton, Gangi Architects; Sara Eva Winkle for the graphics; and to the following organizations for providing photographs and technical support materials.

Amonix, Inc.
3425 Fujita Street, Torrance, CA

Atlantis Energy Systems, Inc.
Sacramento, CA

California Energy Commission
1516 9th Street, MS-45, Sacramento, CA

EnviroTech Financial, Inc.
333 City Blvd., West 17th, Orange, CA 92868

Fotoworks Studio
Los Angeles, CA

Heliotronics, Inc.
1083 Main Street, Hingham, MA

Metropolitan Water District of Southern California
Museum of Water and Life, Hemet, CA
Center for Water Education

Nextek Power Systems, Inc.
89 Cabot Court, Suite L, Hauppauge, NY

Solargenix Energy
3501 Jamboree Road, Suite 606, Newport Beach, CA

SolarWorld California
4650 Adhor Lane, Camarillo, CA

Solar Integrated Technologies
1837 E. Martin Luther King Jr. Blvd., Los Angeles, CA

UMA/Heliocol
13620 49th Street, Clearwater, FL

U.S. Department of Energy
National Renewable Energy Laboratories
Sandia National Laboratories

U.S. Green Building Council, Los Angeles Chapter
315 W. 9th Street, Los Angeles, CA

DISCLAIMER NOTE

This book examines solar power generation and renewable energy sources, with the sole intent to familiarize the reader with the existing technologies and to encourage policy makers, architects, and engineers to use available energy conservation options in their designs.

The principal objective of the book is to emphasis solar power cogeneration design, application, and economics.

Neither the author, individuals, organizations or manufacturers referenced or credited in this book make any warranties, express or implied, or assume any legal liability or responsibility for the accuracy, completeness, or usefulness of any information, products, and processes disclosed or presented.

Reference to any specific commercial product, manufacturer, or organization does not constitute or imply endorsement or recommendation by the author.

SOLAR POWER SYSTEM PHYSICS

Introduction

Solar, or photovoltaic (PV), cells are electronic devices that essentially convert the solar energy of sunlight into electric energy or electricity. The physics of solar cells is based on the same semiconductor principles as diodes and transistors, which form the building blocks of the entire world of electronics.

Solar cells convert energy as long as there is sunlight. In the evenings and during cloudy conditions, the conversion process diminishes. It stops completely at dusk and resumes at dawn. Solar cells do not store electricity, but batteries can be used to store the energy.

One of the most fascinating aspects of solar cells is their ability to convert the most abundant and free form of energy into electricity, without moving parts or components and without producing any adverse forms of pollution that affect the ecology, as is associated with most known forms of nonrenewable energy production methods, such as fossil fuel, hydroelectric, or nuclear energy plants.

In this chapter we will review the overall solar energy conversion process, system configurations, and the economics associated with the technology. We will also briefly look into the mechanism of hydrogen fuel cells.

In Chapter 2 of this book we will review the fundamentals of solar power cogeneration design and explore a number of applications including an actual design of a 500-kilowatt (kW) solar power installation project, which also includes a detailed analysis of all system design parameters.

A Brief History of the Photoelectric Phenomenon

In the later part of the century, physicists discovered a phenomenon: when light is incident on liquids or metal cell surfaces, electrons are released. However, no one had an explanation for this bizarre occurrence. At the turn of the century, Albert Einstein

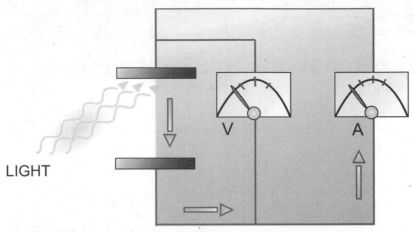

Figure 1.1   The photoelectric effect experiment.

provided a theory for this which won him the Nobel Prize in physics and laid the groundwork for the theory of the *photoelectric effect*. Figure 1.1 shows the photoelectric effect experiment. When light is shone on metal, electrons are released. These electrons are attracted toward a positively charged plate, thereby giving rise to a photoelectric current.

Einstein explained the observed phenomenon by a contemporary theory of *quantized energy levels,* which was previously developed by Max Planck. The theory described light as being made up of miniscule bundles of energy called *photons*. Photons impinging on metals or semiconductors knock electrons off atoms.

In the 1930s, these theorems led to a new discipline in physics called *quantum mechanics*, which consequently led to the discovery of transistors in the 1950s and to the development of semiconductor electronics.

APPLICATION OF DC SOLAR POWER

Historical AC/DC debate between Edison and Tesla   The application of direct current (dc) electric power is a century-old technology that took a backseat to alternating current (ac) in early 1900s when Edison and Tesla were having a feud over their energy transmission and distribution inventions. The following are some interesting historical notes that were communicated by two of the most brilliant inventors in the history of electrical engineering.

> Nicola Tesla: "Alternating Current will allow the transmission of electrical power to any point on the planet, either through wires or through the air, as I have demonstrated."

> Thomas Edison: "Transmission of ac over long distances requires lethally high voltages, and should be outlawed. To allow Tesla and Westinghouse to proceed with their proposals is to risk untold deaths by electricide."

Tesla: "How will the dc power a 1,000 horsepower electric motor as well as a single light bulb? With AC, the largest as well as the smallest load may be driven from the same line."

Edison: "The most efficient and proper electrical supply for every type of device from the light bulb to the phonograph is Direct Current at low voltage."

Tesla: "A few large AC generating plants, such as my hydroelectric station at Niagara Falls, are all you need: from these, power can be distributed easily wherever it is required."

Edison: "Small dc generating plants, as many as are required, should be built according to local needs, after the model of my power station in New York City."

EARLY AC DOMINANCE

After Edison introduced his dc power stations, the first of their kind in the world, the demand for electricity became overwhelming. Soon, the need to send power over long distances in rural and suburban America was paramount. How did the two power systems compare in meeting this need? Alternating current could be carried over long distances, via a relatively small line given an extremely high transmission voltage of 50,000 volts (V) or above. The high voltage could then be transformed down to lower levels for residential, office, and industrial use.

While higher in quality and more efficient than alternating current, dc power could not be transformed or transmitted over distances via small cables without suffering significant losses through resistance.

AC power became the standard of all public utilities, overshadowing issues of safety and efficiency and forcing manufacturers to produce appliances and motors compatible with the national grid.

THE 100-YEAR-OLD POWER SCHEME

With ac power the only option available from power utilities, the world came to rely almost exclusively on ac-based motors and other appliances, and the efficiencies and disadvantages of ac power became accepted as unavoidable. Nicola Tesla's development of the polyphase induction ac motor was a key step in the evolution of ac power applications. His discoveries contributed greatly to the development of dynamos, vacuum bulbs, and transformers, strengthening the existing ac power scheme 100 years ago. Compared to direct current and Edison's findings, ac power is inefficient because of the energy lost with the rapid reversals of the current's polarity. We often hear these reversals as the familiar 60 cycles per second [60 hertz (Hz)] hum of an appliance. AC power is also prone to harmonic distortion, which occurs when there is a disruption in the ideal ac sinusoidal power waveshape. Since most of today's technologically advanced on-site power devices use direct current, there is a need to use inverters to produce alternating current through the system and then convert it back to direct current into the end source of power. These inverters are inefficient; energy is lost (up to 50 percent) when these devices are used. This characteristic is evident in many of today's electronic devices that have internal converters, such as fluorescent lighting.

ALTERNATING AND DIRECT CURRENT: 1950 TO 2000

The discovery of semiconductors and the invention of the transistor, along with the growth of the American economy, triggered a quiet but profound revolution in how we use electricity. Changes over the last half-century have brought the world into the era of electronics with more and more machines and appliances operating internally on dc power and requiring more and more expensive solutions for the conversion and regulation of incoming ac supply. The following table reflects the use of ac and dc device applications of the mid-twentieth and twenty-first centuries.

AC DEVICES—1950	DC DEVICES—2000
Electric typewriters	Computers, printers, CRTs, scanners
Adding machines	CD-ROMs, photocopiers
Wired, rotary telephones	Wired, cordless, and touch-tone phones
Teletypes	Answering machines, modems, faxes
Early fluorescent lighting	Advanced fluorescent lighting with electronic ballasts
Radios, early TVs	Electronic ballast, gas discharge lighting
Record players	HDTVs, CD players, videocassettes
Electric ranges	Microwave ovens
Fans, furnaces	Electronically controlled HVAC systems

As seen from the preceding table, over the last 50 years we have moved steadily from an electromechanical to an electronic world—a world where most of our electric devices are driven by direct current and most of our non-fossil-fuel energy sources (such as photovoltaic cells and batteries) deliver their power as a dc supply.

Despite these changes, the vast majority of today's electricity is still generated, transported, and delivered as alternating current. Converting alternating current to direct current and integrating alternative dc sources with the mainstream ac supply are inefficient and expensive activities that add significantly to capital costs and lock us all into archaic and uncompetitive utility pricing structures. With the advent of progress in solar power technology, the world that Thomas Edison envisioned (one with clean, efficient, and less costly power) is now, after a century of dismissal, becoming a reality. The following exemplify the significance of dc energy applications from solar photovoltaic systems: first, on-site power using direct current to the end source is the most efficient use of power; second, there are no conversion losses resulting from the use of dc power which allows maximum harvest of solar irradiance energy potential.

Solar Cell Physics

Most solar cells are constructed from semiconductor material, such as silicon (the fourteenth element in the Mendeleyev table of elements). Silicon is a semiconductor that has the combined properties of a conductor and an insulator.

Metals such as gold, copper, and iron are conductors; they have loosely bound electrons in the outer shell or orbit of their atomic configuration. These electrons can be detached when subjected to an electric voltage or current. On the contrary, atoms of insulators, such as glass, have very strongly bonded electrons in the atomic configuration and do not allow the flow of electrons even under the severest application of voltage or current. Semiconductor materials, on the other hand, bind electrons midway between that of metals and insulators.

Semiconductor elements used in electronics are constructed by fusing two adjacently doped silicon wafer elements. Doping implies impregnation of silicon by positive and negative agents, such as phosphor and boron. Phosphor creates a free electron that produces so-called N-type material. Boron creates a "hole," or a shortage of an electron, which produces so-called P-type material. Impregnation is accomplished by depositing the previously referenced dopants on the surface of silicon using a certain heating or chemical process. The N-type material has a propensity to lose electrons and gain holes, so it acquires a positive charge. The P-type material has a propensity to lose holes and gain electrons, so it acquires a negative charge.

When N-type and P-type doped silicon wafers are fused together, they form a *PN junction*. The negative charge on P-type material prevents electrons from crossing the junction, and the positive charge on the N-type material prevents holes from crossing the junction. A space created by the P and N, or PN, wafers creates a potential barrier across the junction.

This PN junction, which forms the basic block of most electronic components, such as diodes and transistors, has the following specific operational uses when applied in electronics:

In *diodes*, a PN device allows for the flow of electrons and, therefore, current in one direction. For example, a battery, with direct current, connected across a diode allows the flow of current from positive to negative leads. When an alternating sinusoidal current is connected across the device, only the positive portion of the waveform is allowed to pass through. The negative portion of the waveform is blocked.

In *transistors*, a wire secured in a sandwich of a PNP-junction device (formed by three doped junctions), when properly polarized or biased, controls the amount of direct current from the positive to the negative lead, thus forming the basis for current control, switching, and amplification, as shown in Figure 1.2.

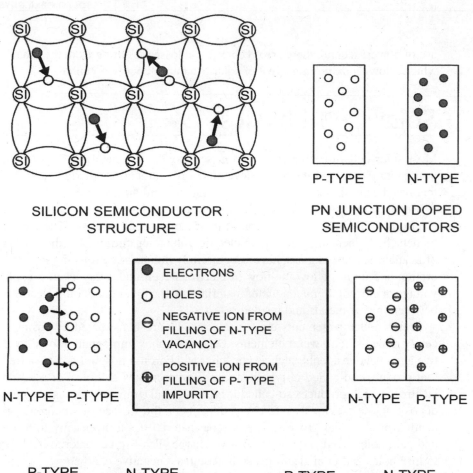

SILICON SEMICONDUCTOR
STRUCTURE

PN JUNCTION DOPED
SEMICONDUCTORS

P-TYPE   N-TYPE

● ELECTRONS
○ HOLES
⊖ NEGATIVE ION FROM
  FILLING OF N-TYPE
  VACANCY
⊕ POSITIVE ION FROM
  FILLING OF P- TYPE
  IMPURITY

N-TYPE   P-TYPE

N-TYPE   P-TYPE

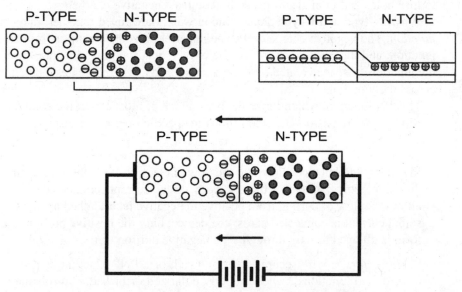

P-TYPE   N-TYPE

P-TYPE   N-TYPE

P-TYPE   N-TYPE

SEMICONDUCTOR JUNCTION CURRENT FLOW

Figure 1.2   NPN junction showing holes and electron flow in an NPN transistor.

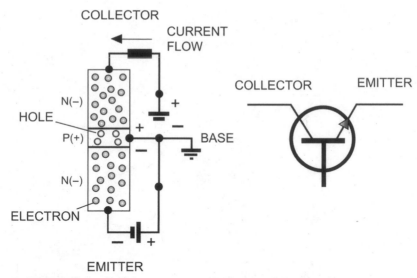

Figure 1.3   **Semiconductor depletion region formations.**

In *light-emitting diodes* (LEDs), a controlled amount and type of doping material in a PN-type device connected across a dc voltage source converts the electric energy to visible light with differing frequencies and colors, such as white, red, blue, amber, and green.

In *solar cells*, when a PN junction is exposed to sunshine, the device converts the stream of photons (packets of quanta) that form the visible light into electrons (the reverse of the LED function), making the device behave like a minute battery with a unique characteristic voltage and current, which is dependent on the material dopants and PN-junction physics. This is shown in Figure 1.3.

The bundles of photons that penetrate the PN junction randomly strike silicon atoms and give energy to the outer electrons. The acquired energy allows the outer electrons to break free from the atom. Thus, the photons in the process are converted to electron movement or electric energy as shown in Figure 1.4.

It should be noted that the photovoltaic energy conversion efficiency is dependent on the wavelength of the impinging light. Red light, which has a lower frequency, produces insufficient energy, whereas blue light, which has more energy than needed to break the electrons, is wasted and dissipates as heat.

Solar Cell Electronics

An electrostatic field is produced at a PN junction of a solar cell by impinging photons that create 0.5 V of potential energy, which is characteristic of most PN junctions and all solar cells. This miniscule potential resembles in function a small battery with

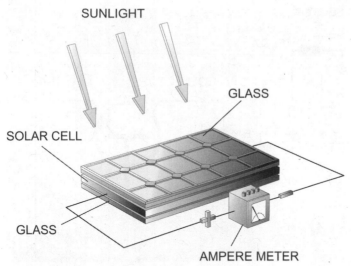

Figure 1.4   Photovoltaic module operational diagrams.

positive and negative leads. These are then connected front to back in series to achieve higher voltages.

For example, 48 solar cell modules connected in series will result in 24 V of output. An increase in the number of solar cells within the solar cell bank will result in a higher voltage. This voltage is employed to operate inverters which convert the dc power into a more suitable ac form of electricity.

In addition to the previously discussed PN-junction device, solar cells contain construction components, for mechanical assembly purposes, that are laid over a rigid or flexible holding platform or a substrate, such as a glass or a flexible film, and are interconnected by micron-thin, highly conductive metals. A typical solar panel used in photovoltaic power generation is constructed from a glass supportive plate that houses solar PV modules, each formed from several hundreds of interconnected PN devices. Depending on the requirements of a specific application, most solar panels manufactured today produce an output of 6, 12, 24, or 48 V dc. The amount of power produced by a solar panel, expressed in watts, represents an aggregate power output of all solar PN devices. For example, a manufacturer will express various panel characteristics by voltage, wattage, and surface area.

Types of Solar Cells Technologies

Solar cell technologies at present fall into three main categories: monocrystalline (single-crystal construction), polycrystalline (semicrystalline), and amorphous silicon thin film.

Solar cell manufacturing and packaging process   Currently, solar cells essentially are manufactured from mono crystalline, polycrystalline amorphous and thin-film-based materials. A more recent undisclosed solar technology, known as organic photovoltaics, is also currently under commercial development. Each of the technologies have unique physical, chemical, manufacturing, and performance characteristics and are best suited for specialized applications.

In this section we will discuss the basic manufacturing principles, and in subsequent chapters we will review the production and manufacturing process of several solar power cell technologies.

Introduction to monocrystalline and polycrystalline silicon cells   The heart of the most monocrystalline and polycrystalline photovoltaic solar cells is a crystalline silicon semiconductor. This semiconductor is manufactured by a silicon purification process, ingot fabrication, wafer slicing, etching, and doping which finally forms a PNP junction that traps photons, resulting in the release of electrons within the junction barrier, thereby creating a current flow.

The manufacturing of a solar photovoltaic cell in itself is only a part of the process of manufacturing a solar panel product. To manufacture a functionally viable product that will last over 25 years requires that the materials be specially assembled, sealed, and packaged to protect the cells from natural climatic conditions and to provide proper conductivity, electrical insulation, and mechanical strength.

One of the most important materials used in sealing solar cells is the fluoropolymer manufactured by DuPont called Elvax. This chemical compound is manufactured from ethylene vinyl acetate resin. It is then extruded into a film and used to encapsulate the silicon wafers that are sandwiched between tempered sheets of glass to form the solar panel. One special physical characteristic of the Elvax sealant is that it provides optical clarity while matching the refractive index of the glass and silicon, thereby reducing photon reflections. Figure 1.5 depicts various stages of monocrystalline solar power manufacturing process.

Another chemical material manufactured by DuPont, called Tedlar, is a polyvinyl fluoride film that is coextruded with polyester film and applied to the bottom of silicon-based photovoltaic cells as a backplane that provides electrical insulation and protection against climatic and weathering conditions. Other manufacturing companies such as Mitsui Chemical and Bridgestone also manufacture comparable products to Tedlar which are widely used in the manufacture and assembly of photovoltaic panels.

Another important product manufactured by DuPont chemical is Solamet, which is a silver metallization paste used to conduct electric currents generated by individual solar silicon cells within each module. Solamet appears as micronwide conductors that are so thin that they do not block the solar rays.

A dielectric silicon-nitride product used in photovoltaic manufacturing creates a sputtering effect that enhances silicon to trap sunlight more efficiently.

Major fabricators of polycrystalline silicon are Dow Corning and General Electric in the United States and Shin-Etsu Handotai and Mitsubishi Material in Japan.

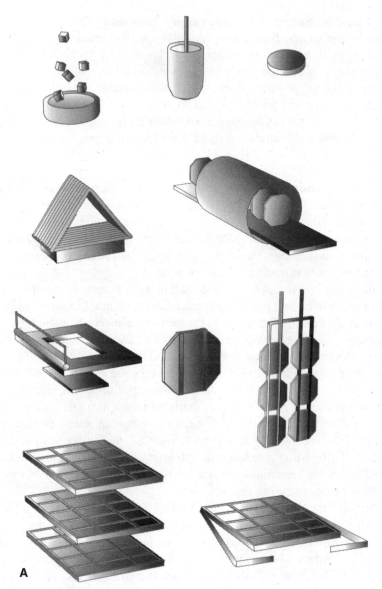

A

Figure 1.5   (a) Monocrystalline solar panel manufacturing process. (b) Solar power module frame assembly.

Because of the worldwide silicon shortage, the driving cost of solar cells has become a limiting factor for lowering the manufacturing cost. At present, silicon represents more than 50 percent of the manufactured solar panel. To reduce silicon costs, at present the industrial trend is to minimize the wafer thickness from 300 to 180 microns.

It should be noted that the process of ingot slicing results in 30 percent wasted material. To minimize this material waste General Electric is currently developing

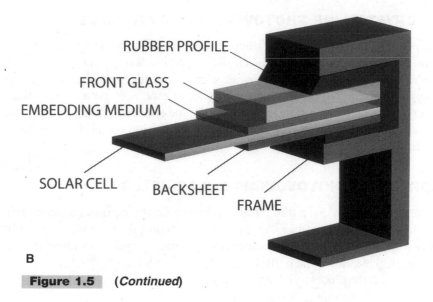

RUBBER PROFILE

FRONT GLASS

EMBEDDING MEDIUM

SOLAR CELL   BACKSHEET

FRAME

B

Figure 1.5 (*Continued*)

a technology to cast wafers from silicon powder. Cast wafers thus far have proven to be somewhat thicker and less efficient than the conventional sliced silicon wafers; however, they can be manufactured faster and avoid the 30 percent waste produced by wafer sawing.

Thin-film solar cell technology The core material of thin-film solar cell technology is amorphous silicon. This technology instead of using solid polycrystalline silicon wafers uses *silane gas*, which is a chemical compound that costs much less than crystalline silicon. Solar cell manufacturing involves a lithographic-like process where the silane film is printed on flexible substrates such as stainless steel or Plexiglas material on a roll-to-roll process.

Silane (SiH_4) is also called silicon tetrahydride, silicanel, or monosilane, which is a flammable gas with a repulsive odor. It does not occur in nature. Silane was first discovered in 1857 by F. Wohler and H. Buffy by reacting hydrochloric acid (HCL) with an Al-Si alloy.

Silane is principally used in the industrial manufacture of semiconductor devices for the electronic industry. It is used for polycrystalline deposition, interconnection or masking, growth of epitaxial silicon, chemical vapor deposition of silicon diodes, and production of amorphous silicon devices such as photosensitive films and solar cells.

Even though thin-film solar power cells have about 4 percent efficiency in converting sunlight to electricity compared to the 15 to 20 percent efficiency of polysilicon products, they have an advantage that they do not need direct sunlight to produce electricity, and as a result, they are capable of generating electric power over a longer period of time.

POLYCRYSTALLINE PHOTOVOLTAIC SOLAR CELLS

In the polycrystalline process, the silicon melt is cooled very slowly, under controlled conditions. The silicon ingot produced in this process has crystalline regions, which are separated by grain boundaries. After solar cell production, the gaps in the grain boundaries cause this type of cell to have a lower efficiency compared to that of the monocrystalline process just described. Despite the efficiency disadvantage, a number of manufacturers favor polycrystalline PV cell production because of the lower manufacturing cost.

AMORPHOUS PHOTOVOLTAIC SOLAR CELLS

In the amorphous process, a thin wafer of silicon is deposited on a carrier material and doped in several process steps. An amorphous silicon film is produced by a method similar to the monocrystalline manufacturing process and is sandwiched between glass plates, which form the basic PV solar panel module.

Even though the process yields relatively inexpensive solar panel technology, it has the following disadvantages:

- Larger installation surface
- Lower conversion efficiency
- Inherent degradation during the initial months of operation, which continues over the life span of the PV panels

The main advantages of this technology are

- Relatively simple manufacturing process
- Lower manufacturing cost
- Lower production energy consumption

Other Technologies

There are other prevalent production processes that are currently being researched and will be serious contenders in the future of solar power production technology. We discuss these here.

Thin-film cadmium telluride cell technology   In this process, thin crystalline layers of cadmium telluride (CdTe, of about 15 percent efficiency) or copper indium diselenide ($CuInSe_2$, of about 19 percent efficiency) are deposited on the surface of a carrier base. This process uses very little energy and is very economical. It has simple manufacturing processes and relatively high conversion efficiencies.

Gallium-arsenide cell technology   This manufacturing process yields a highly efficient PV cell. But as a result of the rarity of gallium deposits and the poisonous qualities of arsenic, the process is very expensive. The main feature of gallium-arsenide (GaAs) cells, in addition to their high efficiency, is that their output is relatively independent of the operating temperature and is primarily used in space programs.

Multijunction cell technology   This process employs two layers of solar cells, such as silicon (Si) and GaAs components, one on top of another, to convert solar power with higher efficiency. Staggering of two layer provides trapping of wider bandwidth of solar rays thus enhancing the solar cell solar energy conversion efficiency.

Concentrators

Concentrators are lenses or reflectors that focus sunlight onto the solar cell modules. Fresnel lenses, which have concentration ratios of 10 to 500 times, are mostly made of inexpensive plastic materials engineered with refracting features that direct the sunlight onto the small, narrow PN-junction area of the cells. Module efficiencies of most PV cells discussed above normally range from 10 to 18 percent, whereas the concentrator type solar cell technology efficiencies can exceed 30 percent.

In this technology reflectors are used to increase power output, by increasing the intensity of light on the module, or extend the time that sunlight falls on the modules. The main disadvantage of concentrators is their inability to focus scattered light, which limits their use to areas such as deserts.

Depending on the size of the mounting surface, solar panels are secured on tilted structures called *stanchions*. Solar panels installed in the northern hemisphere are mounted facing south with stanchions tilted to a specific degree angle. In the southern hemisphere solar panels are installed facing north.

Solar Panel Arrays

Serial or parallel interconnections in solar panels are called *solar panel arrays* (SPAs). Generally, a series of solar panel arrays are configured to produce a specific voltage potential and collective power production capacity to meet the demand requirements of a project.

Solar panel arrays feature a series of interconnected positive (+) and negative (−) outputs of solar panels in a serial or parallel arrangement that provides a required dc voltage to an inverter. Figure 1.6 Shows the internal wiring of a solar power cell.

The average daily output of solar power systems is entirely dependent on the amount of exposure to sunlight. This exposure is dependent on the following factors. An accurate north-south orientation of solar panels (facing the sun), as referenced earlier, has a significant effect on the efficiency of power output. Even slight shadowing will affect a module's daily output. Other natural phenomena that affect solar production include diurnal

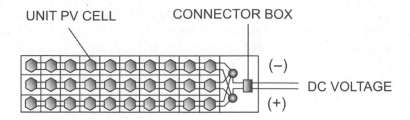

UNIT PV CELL   CONNECTOR BOX

PV CELL CONNECTION DETAIL

Figure 1.6   **Internal wiring of a solar power cell.**

variations (due to the rotation of Earth about its axis), seasonal variation (due to the tilt of Earth's axis), annual variation (due to the elliptical orbit of Earth around the sun), solar flares, solar sunspots, atmospheric pollution, dust, and haze. Figure 1.7 depicts a photo-voltaic panel module assembly mounted on a galvanized Unistrut channel.

Photovoltaic solar array installation in the vicinity of trees and elevated structures, which may cast a shadow on the panels, should be avoided. The geographic location of the project site and seasonal changes are also significant factors that must be taken into consideration.

Figure 1.7   **Photovoltaic panel module assembly mounted on a galvanized Unistrut channel.**

In order to account for the average daily solar exposure time, design engineers refer to world sunlight exposure maps. Each area is assigned an "area exposure time factor," which depending on the location may vary from 2 to 6 hours. A typical example for calculating daily watt-hours (Wh) for a solar panel array consisting of 10 modules with a power rating of 75 W in an area located with a multiplier of 5 will be (10 × 75 W) × 5 h = 3750 Wh of average daily power.

Solar Power System Components

Photovoltaic modules only represent the basic element of a solar power system. They work only in conjunction with complementary components, such as batteries, inverters, and transformers. Power distribution panels and metering complete the energy conversion process.

STORAGE BATTERIES

As mentioned previously, solar cells are devices that merely convert solar energy into a dc voltage. Solar cells do not store energy. To store energy beyond daylight, the dc voltage is used to charge an appropriate set of batteries.

The reserve capacity of batteries is referred to as the *system autonomy*. This varies according to the requirements of specific applications. Batteries in applications that require autonomy form a critical component of a solar power system. Battery banks in photovoltaic applications are designed to operate at deep-cycle discharge rates and are generally maintenance-free.

The amount of required autonomy time depends on the specific application. Circuit loads, such as telecommunication and remote telemetry stations, may require two weeks of autonomy, whereas a residential unit may require no more than 12 hours. Batteries must be properly selected to store sufficient energy for the daily demand. When calculating battery ampere-hours and storage capacity, additional derating factors, such as cloudy and sunless conditions, must be taken into consideration.

CHARGE REGULATORS

Charge regulators are electronic devices designed to protect batteries from overcharging. They are installed between the solar array termination boxes and batteries.

INVERTERS

As described earlier, photovoltaic panels generate direct current, which can only be used by a limited number of devices. Most residential, commercial, and industrial devices and appliances are designed to work with alternating current. Inverters are devices that convert direct current to alternating current. Although inverters are usually

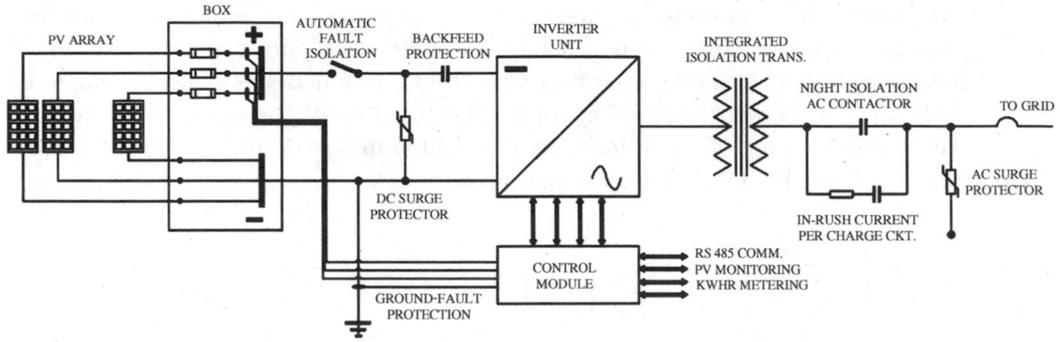

Figure 1.8   **An inverter single line diagram.** *Courtesy of SatCon, Canada.*

designed for specific application requirements, the basic conversion principles remain the same. Essentially, the inversion process consists of the following.

Wave formation process. Direct current, characterized by a continuous potential of positive and negative references (bias), is essentially chopped into equidistant segments, which are then processed through circuitry that alternately eliminates positive and negative portions of the chopped pattern, resulting in a waveform pattern called a *square wave.* Figure 1.8 shows an inverter single line diagram.

Waveshaping or filtration process. A square wave, when analyzed mathematically (by Fourier series analysis), consists of a combination of a very large number of sinusoidal (alternating) wave patterns called harmonics. Each wave harmonic has a distinct number of cycles (rise-and-fall pattern within a time period).

An electronic device referred to as a choke (magnetic coils) or filters discriminate passes through 60-cycle harmonics, which form the basis of sinusoidal current. Solid-state inverters use a highly efficient conversion technique known as envelope construction. Direct current is sliced into fine sections, which are then converted into a progressive rising (positive) and falling (negative) sinusoidal 60-cycle waveform pattern. This chopped sinusoidal wave is passed through a series of electronic filters that produce an output current, which has a smooth sinusoidal curvature.

Protective relaying systems. In general, most inverters used in photovoltaic applications are built from sensitive solid-state electronic devices that are very susceptible to external stray spikes, load short circuits, and overload voltage and currents. To protect the equipment from harm, inverters incorporate a number of electronic circuitry:

■ Synchronization relay
■ Undervoltage relay
■ Overcurrent relay
■ Ground trip or overcurrent relay

- Overvoltage relay
- Overfrequency relay
- Underfrequency relay

Most inverters designed for photovoltaic applications are designed to allow simultaneous paralleling of multiple units. For instance, to support a 60-kW load, outputs of three 20-kW inverters may be connected in parallel. Depending on the power system requirements, inverters can produce single- or three-phase power at any required voltage or current capacity. Standard outputs available are single-phase 120 V ac and three-phase 120/208 and 277/480 V ac. In some instances step-up transformers are used to convert the output of 120/208 V ac inverters to higher voltages.

Input and output power distribution   To protect inverters from stray spikes resulting from lightning or high-energy spikes, dc inputs from PV arrays are protected by fuses, housed at a junction box located in close proximity to the inverters. Additionally, inverter dc input ports are protected by various types of semiconductor devices that clip excessively high voltage spikes resulting from lightning activity.

To prevent damage resulting from voltage reversal, each positive (+) output lead within a PV cell is connected to a rectifier, a unidirectional (forward-biased) element. Alternating-current output power from inverters is connected to the loads by means of electronic or magnetic-type circuit breakers. These serve to protect the unit from external overcurrent and short circuits.

Grid-connected inverters   In the preceding we described the general function of inverters. Here we will review their interconnection to the grid, which requires a thorough understanding of safety regulations that are mandated by various state agencies. Essentially the goal of design safety standards for inverters used in grid-connected systems, whether they be deployed in photovoltaic, wind turbine, fuel cell, or any other type of power cogeneration system, is to have one unified set of guidelines and standards for the entire country. Standard regulations for manufacturing inverters address issues concerning performance characteristics and grid connectivity practices and are recommended by a number of national test laboratories and regulatory agencies.

Underwriters Laboratories   For product safety, the industry in the United States has worked with Underwriters Laboratories (UL) to develop *UL1741, Standard for Static Inverter and Charge Controller for Use in Independent Power Systems,* which has become the safety standard for inverters being used in the United States. Standard UL1741 covers many aspects of inverter design including enclosures, printed circuit board configuration, interconnectivity requirements such as the amount of direct current the inverters can inject into the grid, total harmonic distortion (THD) of the output current, inverter reaction to utility voltage spikes and variations, reset and recovery from abnormal conditions, and reaction to islanding conditions when the utility power is disconnected.

Islanding is a condition that occurs when the inverter continues to produce power during a utility outage. Under such conditions the power produced by a PV system becomes a safety hazard to utility workers who could be inadvertently exposed to hazardous electric currents; as such, inverters are required to include anti-islanding control circuitry to cut the power to the inverter and disconnect it from the grid network.

Anti-islanding also prevents the inverter output power from getting out of phase with the grid when the automatic safety interrupter reclosures reconnect the inverter to the grid, which could result in high voltage spikes that can cause damage to the conversion and utility equipment.

Institute of Electrical and Electronics Engineers   The Institute of Electrical and Electronics Engineers (IEEE) provides suggestions for customers and utilities alike regarding the control of harmonic power and voltage flicker, which frequently occur on utility buses, in its IEEE 929 guideline (not a standard), *Recommended Practice for Utility Interface of Photovoltaic (PV) Systems.* Excessive harmonic power flow and power fluctuation from utility buses can damage customers' equipment; therefore a number of states, including California, Delaware, New York, and Ohio, specifically require that inverters be designed to operate under abnormal utility power conditions.

Power limit conditions   The maximum size of a photovoltaic power cogeneration system is subject to limitations imposed by various states. Essentially most utilities are concerned about large sources of private grid-connected power generation, since most distribution systems are designed for unidirectional power flow. The addition of a large power cogeneration system on the other hand creates bidirectional current flow conditions on the grid, which in some instances can diminish utility network reliability. However, it is well known that, in practice, small amounts of cogenerated power do not usually create a grid disturbance significant enough to be a cause for concern. To regulate the maximum size of a cogeneration system, a number of states have set various limits and caps for systems that generate in excess of 100 kW of power.

Utility side disconnects and isolation transformers   In some states such as California, Delaware, Florida, New Hampshire, Ohio, and Virginia, utilities require that visible and accessible disconnect switches be installed outside for grid service isolation. It should be noted that several states such as California require that customers open the disconnect switches once every 4 years to check that the inverters are performing the required anti-islanding.

In other states such as New Mexico and New York, grid isolation transformers are required to reduce noise created by private customers that could be superimposed on the grid. This requirement is however not a regulation that is mandated by UL or the Federal Communication Commission (FCC).

PV Power cogeneration capacity   In order to protect utility companies' norm of operation, a number of states have imposed a cap on the maximum amount of power that can be generated by photovoltaic systems. For example, New Hampshire limits the maximum to 0.05 percent and Colorado to 1 percent of the monthly grid network peak demand.

INVERTER CAPABILITY TO WITHSTAND SURGES

In most instances power distribution is undertaken through a network of overhead lines that are constantly exposed to climatic disturbances, such as lightning, which result in power surges. Additional power surges could also result from switching capacitor banks used for power factor correction and from power conversion equipment or during load shedding and switching. The resulting power surges, if not clamped, could seriously damage inverter equipment by breaking down conductor insulation and electronic devices.

To prevent damage caused by utility spikes, IEEE has developed national recommended guidelines for inverter manufacturers to provide appropriate surge protection. A series of tests devised to verify IEEE recommendations for surge immunity are performed by UL as part of equipment approval.

PV system testing and maintenance log   Some states, such as California, Vermont, and Texas, require that comprehensive commissioning testing be performed on PV system integrators to certify that the system is operating in accordance with expected design and performance conditions. It is interesting to note that for PV systems installed in the state of Texas a log must be maintained of all maintenance performed.

EXAMPLE OF A UL1741 INVERTER

The following is an example of a UL1741-approved inverter manufactured by SatCon, Canada.

An optional combiner box, which includes a set of special ceramic overcurrent protection fuses, provides accumulated dc output to the inverter. At its dc input the inverter is equipped with an automatic current fault isolation circuit, a dc surge protector, and a dc backfeed protection interrupter. In addition to the preceding, the inverter has special electronic circuitry that constantly monitors ground faults and provides instant fault isolation. Upon conversion of direct current to alternating current, the internal electronics of the inverter provide precise voltage and frequency synchronization with the grid. Figure 1.9 depicts view of inverter electronics.

An integrated isolation transformer within the inverter provides complete noise isolation and filtering of the ac output power. A night isolation ac contactor disconnects the inverter at night or during heavy cloud conditions. The output of the inverter also includes an ac surge isolator and a manual circuit breaker that can disconnect the equipment from the grid.

A microprocessor-based control system within the inverter includes, in addition to waveform envelope construction and filtering algorithms, a number of program subsets that perform anti-islanding, voltage, and frequency control.

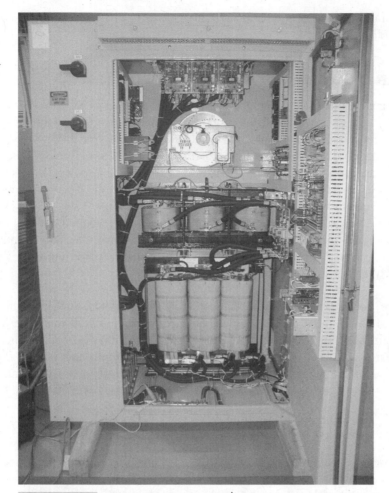

Figure 1.9   View of inverter electronics. *Courtesy of SatCon, Canada.*

As an optional feature, the inverter can also provide data communication by means of an RS 485 interface that can transmit equipment operational and PV measurement parameters such as PV output power, voltage, current, and totalized kilowatt-hour metering data for remote monitoring and display.

2

SOLAR POWER TECHNOLOGIES

Introduction

In Chapter 1 we briefly reviewed specifics of monocrystalline, polycrystalline, amorphous, and concentrator cell technologies. In this chapter the principal technologies reviewed are limited to four categories or classes of solar power photovoltaics, namely, monosilicon wafer, amorphous silicon, and thin-film technologies, and concentrator-type PV technologies and associated sun-tracking systems. We will review the basic physical and functional properties, manufacturing processes, and specific performance parameters of these technologies. In addition, we present some unique case studies that will provide a more profound understanding of the applications of these technologies.

Each of the technologies covered here have been developed and designed for a specific use and have unique application advantages and performance profiles. It should be noted that all the technologies presented here can be applied in a mixed-use fashion, each meeting special design criteria.

Crystalline Solar Photovoltaic Module Production

In this section we will review the production and manufacturing process cycle of a crystalline-type photovoltaic module. The product manufacturing process presented is specific to SolarWorld Industries; however, it is representative of the general fundamental manufacturing cycle for the monosilicon class of commercial solar power modules presently offered by a large majority of manufacturers.

The manufacture of monocrystalline photovoltaic cells starts with silicon crystals, which are found abundantly in nature in the form of flint stone. The word *silicon* is derived from the Latin *silex,* meaning flint stone, which is an amorphous substance found in nature consisting of one part silicon and two parts oxygen

(SiO_2). Silicon (Si) was first produced in 1823 by Berzelius when he separated the naturally occurring ferrous silica (SiF_4) by heat exposure with potassium metal. Commercial production of silicon commenced in 1902 and resulted in an iron-silicon alloy with an approximate weight of 25 percent iron which was used in steel production as an effective deoxidant. At present more than one million tons of metallurgical-grade 99 percent pure silicon is used by the steel industry. Approximately 60 percent of the referenced silicon is used in metallurgy, 35 percent in the production of silicones, and approximately 5 percent for the production of semiconductor-grade silicon.

In general, common impurities found in silicon are iron (Fe), aluminum (Al), magnesium (Mg), and calcium (Ca). The purest grade of silicon used in semiconductor applications contains about one part per billion (ppb) of contamination. Purification of silicon involves several different types of complex refining technologies such as chemical vapor deposition, isotopic enrichment, and a crystallization process. Figure 2.1 depicts silicon crystals prior to ingot manufacturing process.

Chemical vapor deposition   One of the earlier silicon refining processes, known as chemical vapor deposition, produced a higher grade of metallurgical silicon, which consisted of a chemical reaction of silicon tetrachloride ($SiCl_4$) and zinc (Zn) under

Figure 2.1 Silicon crystals.
Photo courtesy of SolarWorld.

high-temperature vaporization conditions which yielded pure silicon through the following chemical reaction:

$$SiCl_4 + 2(Zn) = Si + 2\ ZnCl_2$$

The main problem of this process was that $SiCl_4$ always contained boron chloride (BCl_3) when combined with zinc-produced boron, which is a serious contaminant. In 1943 a chemical vapor deposition was developed that involved replacement of the zinc by hydrogen (H), which gave rise to pure silicon since hydrogen, unlike zinc, does not reduce the boron chloride to boron. Further refinement involved replacement of silicon tetrachloride with trichlorosilane ($SiHCl_3$), which is readily reduced to silicon.

Czochralski crystal growth   In 1916 a Polish metallurgist, Jan Czochralski, developed a technique to produce silicon crystallization which bears his name. The crystallization process involved inserting a metal whisker into molten silicon and pulling it out with increasing velocity. This allowed for formation of pure crystal around the wire and was thus a successful method of growing single crystals. The process was further enhanced by attaching a small silicon crystal seed to the wire rod. Further production efficiency was developed by attaching the seed to a rotatable and vertically movable spindle. Incidentally the same crystallization processing apparatus is also equipped with special doping ports where P- or N-type dopants are introduced into the crystal for generation of PN- or NP-junction-type crystals (discussed in Chapter 1), used in the construction of NPN or PNP transistors, diodes, light-emitting diodes, solar cells, and virtually all high-density, large-scale integrated circuitry used in electronic technologies. Figure 2.2 depicts silicone crystallization melting and ingot manufacturing chamber.

Figure 2.2 Silicon ingot produced by the Czochralski crystallization process. *Photo courtesy of SolarWorld.*

The chemical vaporization and crystallization process described here is energy intensive and requires a considerable amount of electric power. To produce purified silicon ingots at a reasonable price, in general, silicon ingot production plants are located within the vicinity of major hydroelectric power plants, which produce an abundance of low-cost hydroelectric power. Ingots produced from this process are in either circular or square form and are cleaned, polished, and distributed to various semiconductor manufacturing organizations.

Solar photovoltaic cell production   The first manufacturing step in the production of photovoltaic modules involves incoming ingot inspection, wafer cleaning, and quality control. Upon completion of the incoming process, in a clean-room environment the ingots are sliced into millimeter-thick wafers and both surfaces are polished, etched, and diffused to form a PN junction. After being coated with antireflective film, the cells are printed with a metal-filled paste and fired at high temperature. Each individual cell is then tested for 100 percent functionality and is made ready for module assembly.

Photovoltaic module production   The photovoltaic module production process involves robotics and automatic controls where a series of robots assemble the solar cells step by step: laying the modules, soldering the cells in a predetermined pattern, and then laminating and framing the assembly as a finished product. Upon completion of framing, each PV module is tested under artificial insolation conditions and the results are permanently logged and serialized. The last step of production involves a secondary module test, cleaning, packaging, and crating. In general, the efficiency of the PV modules produced by this technique range from 15 to 18 percent. Figure 2.3 depicts silicon ingot inspection process.

Photovoltaic module life span and recycling   To extend the life span of solar power photovoltaic modules, PV cell assemblies are laminated between two layers of protective covering. In general the top protective cover is constructed from 1/4- to 5/8-inch (in) tempered glass and the lower protective cover either from a tempered glass or a hard plastic material. A polyurethane membrane is used as a gluing membrane, which holds the sandwiched PV assembly together. In addition to acting as the adhesive agent, the membrane hermetically seals the upper and lower covers preventing water penetration or oxidation. As a result of hermetical sealing, silicon-based PV modules are able to withstand exposure to harsh atmospheric and climatic conditions. Figure 2.4 depicts ingot production chamber.

Even though the life span of silicon-based PV modules is guaranteed for a period of at least 20 years, in practice it is expected that the natural life span of the modules will exceed 45 years without significant degradation. Figure 2.5 depicts fabricated monosilicon solar cell inspection.

In order to minimize environmental pollution, SolarWorld has adopted a material recovery process whereby obsolete, damaged, or old PV modules (including the aluminum framing, tempered glass, and silicon wafers) are fully recycled and reused to produce new solar photovoltaic modules.

Figure 2.3 **Formed silicon ingot cylinder in process chamber.** *Photo courtesy of SolarWorld.*

CONCENTRATOR TECHNOLOGIES

Concentrator-type solar technologies are a class of photovoltaic systems that deploys a variety of lenses to concentrate and focus solar energy on semiconductor material used in the manufacture of conventional PV cells. Figure 2.6 depicts solar panel lamination robotic machinery.

The advantage of these types of technologies is that for a comparable surface area of silicon wafer it becomes possible to harvest considerably more solar energy. Since silicon wafers used in the manufacture of photovoltaic systems represent a substantial portion of the product cost, by use of relatively inexpensive magnifying concentrator lenses it is possible to achieve a higher-efficiency product at a lower cost than conventional PV power systems.

One of the most efficient solar power technologies commercially available for large-scale power production is a product manufactured by Amonix. This concentrator technology has been specifically developed for ground installation only and is suitable only for solar farm-type power cogeneration. The product efficiency of this unique PV solar power concentrator technology under field test conditions in numerous applications in the United States (determined by over half a decade of testing by the Department of Energy; Arizona Public Service; Southern California Edison; and the University of Nevada, Las Vegas) has exceeded 26 percent, nearly twice that of comparable conventional solar power systems. At present Amonix is in the process of developing a multijunction concentrating cell that will augment the solar power energy production efficiency to 36 percent.

Figure 2.4 Silicon ingot production chamber.

Photo courtesy of SolarWorld.

WHY CONCENTRATION?

Before photovoltaic systems can provide a substantial part of the world's need for electric energy, there needs to be a large reduction in the cost. Studies conducted by the Department of Energy (DOE), Electrical Power Research Institute (EPRI), and others show that concentrating solar energy systems can eventually achieve lower costs than conventional PV power systems. The lower cost results from:

■ *Less expensive material.* Because the semiconductor material for solar cells is a major cost element of all photovoltaic systems, one approach to cost reduction is to reduce the required cell area by concentrating a relatively large area of solar insolation onto a relatively small solar cell. The solar power concentrator technology developed by Amonix deploys low-cost Fresnel lenses to focus the sun power onto the cells, which reduces the required cell area and material by nearly 250 times. A 6-in wafer used in a flat-plate PV system will produce about 2.5 W, but will produce 1000 W in the Amonix system as illustrated in Figure 2.7.

Figure 2.5 Fabricated monosilicon solar cell inspection.
Photo courtesy of SolarWorld.

Figure 2.6 Solar panel lamination robotic machinery. *Photo courtesy of SolarWorld.*

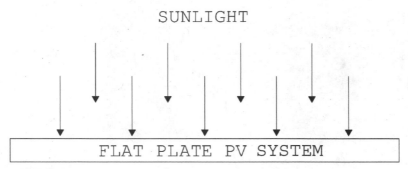

Figure 2.7 Flat-plate PV solar ray impact.

■ *Higher efficiency.* Concentrating PV cells achieve higher efficiencies than do non-concentrating PV cells. Flat-plate silicon cells have efficiencies in the range of 8 to 15 percent, while the Amonix concentrating silicon cell has an efficiency of 26 percent. Concentrating multijunction cells presently under development are expected to achieve an efficiency greater than 34 percent.

■ *More annual energy.* Further increased annual energy production is achieved by the incorporation of a two-axis sun-tracking system. All high-concentration system technologies require a sun-tracking control system. A computerized tracking system periodically adjusts the MegaConcentrator platform to optimize the insolation angle that results in additional annual energy generation production. The average annual energy for 19 different fixed flat-plate installations under test conditions in Phoenix, Arizona, ranged from 1000 to 1500 kWh per rated kilowatt. The same equivalent power-rated Mega Concentrator manufactured by Amonix generated in excess of 1900 to 2000 kWh per rated kilowatt.

Amonix Megaconcentrators

SYSTEM DESCRIPTION

The solar PV concentrator technology described here is a proprietary patented product of Amonix Corporation located in Torrance, California. Essentially the solar power system referred to as a Mega Concentrator consists of six specific components that have resulted in a reduction in product cost:

1 *MegaModule subsystem.* Concentrates the sun's energy on a solar cell that converts it into electric energy. It consists of Fresnel lenses, solar cells, and the support structure. Each system consists of five to seven MegaModules.

2 *Drive subsystem.* Rotates the MegaModules in azimuth and elevation to track the sun. The drive system consists of a foundation, pedestal, rotating bearing head, hydraulic actuators, and torque tube.

3 *Hydraulic subsystem.* Applies hydraulic pressure to one side of the hydraulic actuators to move the torque tube and MegaModules in elevation and azimuth in order
to keep the system pointing at the sun. The hydraulic system consists of hydraulic
valves, an accumulator, a pump, a reservoir, and pressure sensors.

4 *Tracking control subsystem.* Monitors sensors on the system, calculates the required
movement for the commanded operation, and applies signals to the hydraulic valves
to move the system to the commanded position. The commanded position could be
to track the sun, move to a night stow position, move to a wind stow position, or
move to a maintenance position.

5 *AC/DC control subsystem.* Combines the dc power, converts it to ac power, and
interfaces with the ac grid. It consists of dc fuses, circuit breakers, and an inverter.

6 *Control mechanism.* Sun-tracking platforms to be described later, results in additional annual energy generation per installed kilowatt. The average annual energy
test of Mega Concentrator installations in California and Arizona have augmented
50 percent more power output than comparable fixed flat-plate installations.

CONCENTRATOR OPTICS

Refractive optics is used to concentrate the sun's irradiance onto a solar cell, as illustrated in Figure 2.8. A square Fresnel lens, incorporating circular facets, is used to turn
sun rays to a central focal point. A solar cell is mounted at this focal point and converts the sun power into electric power. A number of Fresnel lenses are manufactured
as a single piece, or parquet.

The solar cells are mounted on a plate, at locations corresponding to the focus of
each Fresnel lens. A steel C-channel structure maintains the aligned positions of the

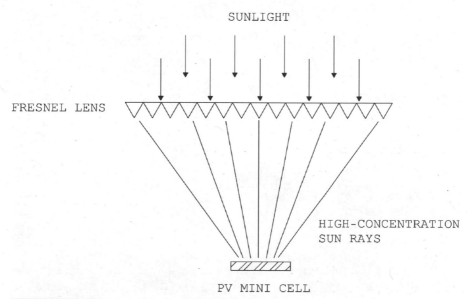

Figure 2.8 Fresnel lens concentrator.

lenses and cell plates. The lenses, cell plates, and steel structure are collectively referred to as an Amonix MegaModule. Each MegaModule is designed to produce 5 kW of dc power at 850 W/m^2 direct normal insolation and 20 degrees Celsius (°C) ambient temperature (IEEE standard). One to seven MegaModules are mounted on a sun-tracking structure to obtain a 5 kW of dc power.

SYSTEM OPERATION

The Mega Concentrator solar power systems are designed for unattended operation for either grid-connected or off-grid applications. As described previously, the system moves automatically from a night stow position to tracking the sun in the early morning. The system tracks the sun throughout the day, typically generating electric power whenever the direct normal irradiance (DNI) is above 400 W/m^2, until the sun sets in the evening. The controller monitors the sun position with respect to the centerline of the unit and adjusts the tracking position if required to maintain the required pointing accuracy. If clouds occur during the day, sun-position mathematical algorithms are used to keep the unit pointing at the expected sun position until the clouds dissipate. Figure 2.9 depicts Amonix MegaConcentrator module assembly and Figure 2.10 is a photograph of the two axis hydraulic tracking system.

A central unattended remote monitoring system provides diagnostic information that can be retrieved from a central or distant location. In addition to providing system operational diagnostics, the monitoring system also provides solar power output information such as currents, voltages, and power data that are stored in the central supervisory system memory. The cumulated data are then used to verify if there are any performance

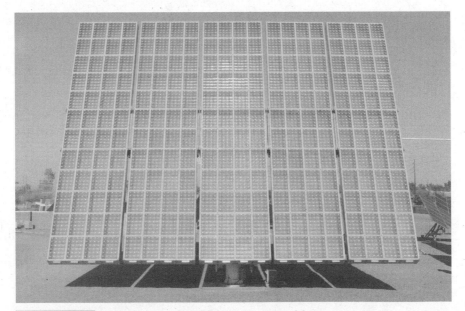

Figure 2.9 MegaConcentrator module assembly. *Photo courtesy of Amonix.*

Figure 2.10 MegaConcentrator module two axis hydraulic tracking system. *Photo courtesy of Amonix.*

malfunctions of the inverter, fuses, or PV strings; tracking anomalies; or poor environmental conditions. The monitoring system is also used to determine when the lenses have become soiled and need to be washed. The central supervisory system also provides diagnostic data acquisition from the hydraulic drive system operating parameters such as fluid level, pump cycling frequency, and deviations from normal operating range, which in turn are stored in the memory archives for monitoring and diagnostic purposes. These data are retrievable from a central operating facility and can be used to diagnose a current problem or to detect a potential future problem. The supervisory control program in addition to the preceding provides equipment parts replacement data for the site maintenance personnel. Figure 2.11 depicts a typical Amonix MegaConcentrator solar farm.

SPECIFIC CASE STUDIES

To date, over 570 kW of the fifth-generation MegaConcentrator system have been manufactured and installed over the last 6 years. The first three 20-kW units have been in operation since May of 2000. Over the same period of time several additional units have been installed at Arizona Public Service (APS) and the University of Nevada, Las Vegas, which have generated over 3.5 GWh of grid power.

APS STAR Center, west field, Tempe, Arizona   There are currently 145 kW in operation in the west field at the APS STAR facility in Tempe, Arizona. The field now consists of three 25-kW units and two 35-kW units. Initially there were three 20-kW

Figure 2.11   Amonix MegaConcentrator solar farm. *Photo courtesy of Amonix.*

units and three 25-kW units as shown in Figure 2.9 The MegaModules from the three 20-kW units were moved to a new 35-kW drive system that incorporates seven MegaModules. The units in the west field have produced over 1185 MWh of grid power since the start of operation.

A second field of Amonix units has been installed on the east side of the APS STAR facility. There are a total of five 25-kW units, which form a 125-kW solar power system. Ever since operation of the units since 2002 the solar power system has generated over 832 MWh of grid power.

SYSTEM PERFORMANCE

Part of the Mega Concentrator development plan has been to deploy one or more units at different possible solar sites in order to test the hardware under the various environmental conditions to determine the operating performance. Units have been deployed in Southern California, Nevada, Arizona, Texas, and Georgia, and there have been different lessons learned from each of the sites. Some of the units have been in field operation for 6 years, and the accumulated 3.5 GWh of electric grid energy.

A WORD ABOUT AMONIX

Amonix was formed in the 1990s to commercialize a high-concentration silicon solar cell developed at Stanford University. Several configurations of structure, drives, and controls were manufactured and field-tested in early 2000. To verify the manufacturability of the

Figure 2.12 Amonix MegaCaoncentrator platform assembly.
Photo courtesy of Amonix.

system and develop manufacturing procedures and processes, a commercial manufac-
turing center was established. To verify the field performance, units were installed at
various sites and operated to determine short-term and long-term operating problems.
Over 570 kW of the systems were manufactured, installed, and tested at various loca-
tions to verify the performance of the design. At present, units have been installed for
Arizona Public Service (APS), University of Nevada Las Vegas (UNLV), Nevada
Power Company, Southern California Edison (at California State Polytechnic
University, Pomona, California), and Southwest Solar Park in Texas. Some of the
installations have now been in operation for nearly 6 years. Figure 2.12 depicts Amonix
MegaConcentrator platform assembly.

Film Technologies

SOLAR INTEGRATED TECHNOLOGIES

Solar Integrated Technologies has developed a flexible solar power technology specifically for use in roofing applications. The product meets the unique requirements of
applications where the solar power cogeneration also serves as a roofing material. This
particular product combines solar film technology overlaid on a durable single-ply
polyvinyl chloride roofing material that offers an effective combined function as a roof

covering and solar power cogeneration that can be readily installed on a variety of flat and curved roof surfaces. Even though the output efficiency of this particular technology is considerably lower than that of conventional glass-laminated mono- or poly-crystalline silicon photovoltaic systems, its unique pliability and dual-function use as both a solar power cogenerator and a roof covering system make it indispensable in applications where roof material replacement and coincidental renewable energy generation become the only viable options.

Over the past years, Solar Integrated Technologies has evolved from a traditional industrial roofing company into the leading supplier of building-integrated photovoltaic (BIPV) roofing systems. The company's unique approach to the renewable energy market enables it to stand out from the competition by supplying a product that produces clean renewable energy, while offering a durable industrial-grade roof.

Innovative product   Solar Integrated Technologies has combined the world's first single-ply roofing membrane under the EPA's ENERGY STAR program with the most advanced, amorphous silicon photovoltaic cells. The result is an integrated, flexible solar roofing panel that rolls onto flat surfaces. Figure 2.13 depicts manufacturing process of film technology and single ply PVC lamination process.

Until the introduction of this product, the installation of solar panels on large-area flat or low-slope roofs was limited due to the heavy weight of traditional rigid crystalline solar panels. This lightweight solar product overcomes this challenge and eliminates any related roof penetrations.

The Solar Integrated Technologies BIPV roofing product is installed flat as an integral element of the roof and weighs only 12 ounces per square foot (oz/ft^2), allowing

Figure 2.13   Manufacturing process of film technology and single-ply PVC lamination process. *Photo courtesy of Solar Integrated Technologies, Los Angeles, California.*

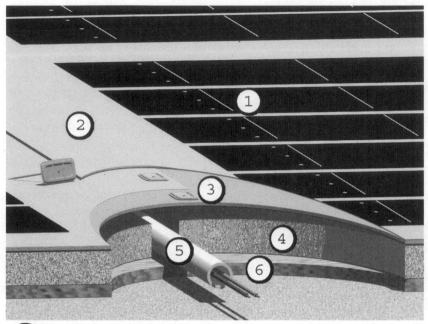

1 Flexible PhotoVoltaic Module

2 BIPV Roofing System

3 Gypsum Board

4 Rigid Foam Insulation

5 Electrical Wiring Conduit

6 Existing Roof Deck

Figure 2.14 Solar Integrated Technologies product configuration.
Graphics courtesy of Solar Integrated Technologies.

installation on existing and new facilities. Application of this technology offsets electric power requirements of buildings, and where permitted in net metering applications excess electricity can be sold to the grid. Figure 2.14 depicts Solar Integrated Technologies product configuration.

In addition to being lightweight, this product uses unique design features to increase the total amount of sunlight converted to electricity over each day including better performance in cloudy conditions.

Both the single-ply PVC roof material and BIPV solar power system are backed by an extensive maintenance 20-year package operations and maintenance service warranty. Similar to all solar power cogeneration systems, this technology also offers

Figure 2.15 **Coca-Cola Bottling Plant, Los Angeles, California.** *Photo courtesy of Solar Integrated Technologies.*

a comprehensive real-time data acquisition and monitoring system whereby customers are able to monitor exactly the amount of solar power being generated with real-time metering for effective energy management and utility bill reconciliation. Figure 2.15 depicts installation of Solar Integrated Technologies system in Coca-Cola Bottling Plant, Los Angeles, California

Custom-fabricated BIPV solar cells   Essentially BIPV is a term commonly used to designate a custom-made assembly of solar panels specifically designed and manufactured to be used as an integral part of building architecture. These panels are used as architectural ornaments such as window and building entrance canopies, solariums, curtain walls, and architectural monuments.

The basic fabrication of BIPV cells consists of lamination of mono- or polycrystalline silicon cells which are sandwiched between two specially manufactured tempered glass plates referred to as a glass-on-glass assembly. A variety of cells arranged in different patterns and spacing are sealed and packaged in the same process as described previously in this chapter. Prefabricated cell wafers used by BIPV fabricators are generally purchased from major solar power manufacturers. Figure 2.16 depicts Atlantis Energy system BIPV manufacturing robotics machine.

Fabrication of BIPV cells involves complete automation whereby the entire assembly is performed by special robotic equipment which can be programmed to implement solar cell configuration layout, lamination, sealing and framing, in a clean room environmental setting without any manual labor intervention. Some solar power fabricators

Figure 2.16 **BIPV manufacturing robotics machine.** *Photo courtesy of Atlantis Energy Systems.*

such as Sharp Solar of Japan, for aesthetic purposes, offer a limited variety of colored and transparent photovoltaic cells. Cell colors, which are produced in deep marine, sky blue, gold, and silverfish brown, usually are somewhat less efficient and are manufactured on an on-demand basis.

Because of their lower performance efficiency, BIPV panels are primarily used in applications where there is the presence of daylight, such as in solariums, rooms with skylights, or sunrooms. In these cases the panels becomes an essential architectural requirement. Figure 2.17 depicts a custom BIPV module manufactured by Atlantis energy Systems.

SUNSLATE SOLAR MODULES

SunSlate solar modules are photovoltaic products that serve two specific functions. They are constructed to be both a roof shingle and a solar power plant simultaneously. This class of products is specifically well-suited for residential and light commercial applications where a large portion of a tiled roof structure could be fitted with relative ease. Figure 2.18 depicts application of an Atlantis energy Systems BIPV system in CalTrans building, Los Angeles, California.

SunSlates are secured to roof rafters and structure by means of storm anchor hooks and anchor nails that rest on 2 ft × 2 ft sleepers. When the tiles are secured to the roof structure, they can withstand 120 mile per hour winds. Adjacent SunSlate tiles interconnect with specially designed gas-tight male-female connectors forming PV array

Figure 2.17   BIPV module. *Photo courtesy of Atlantis Energy Systems.*

strings that in turn are terminated in a splice box located under the roof. A grid-connected inverter, located within the vicinity of the main service meter, readily converts the direct current generated by the solar cells to alternating current. The entire wiring of such a system can be readily handled by one electrician in a matter of hours. Figure 2.19 depicts SunSlate solar power module manufactured by Atlantis Energy Systems.

Typically 100 ft^2 of SunSlate roofing weighs about 750 pounds (lb) or 7.5 lb/ft, which is comparatively much lighter than light concrete roof that can weigh nearly twice as much and is slightly heavier than an equivalent composition shingle roofing tile that can weigh about 300 lb. Figure 2.20 depicts application of SunSlate as roofing tile.

MEGASLATE CUSTOMIZED ROOF-MOUNT PV SYSTEM

MegaSlate is BIPV roofing system. It consists of frameless individual modules that combine into an overall system that is adaptable to the dimensions of a roof.

MegaSlate is the ideal, simple, and cost-effective solution for owners, architects, and planners who intend to implement functional yet also aesthetically appealing roofs. It provides a simple and fast system assembly that could be installed very rapidly, saving a significant amount of time for owners and contractors alike.

Power Production Requirements   To obtain maximum output production efficiency, MegaSlate system should conform to the following recommendations:

Figure 2.18 **Application of BIPV as window canopy in CalTrans building, los Angeles, California.** *Photo courtesy of Atlantis Energy Systems.*

- The orientation of the roof should preferably be facing south (east over south to west).
- The photovoltaic area exposure to sunlight should always remain unshaded.
- The PV-mounted area must be sufficiently large and unobstructed.
- For optimum efficiency the PV support rafters or platform must have a 20-degree tilt.

Potential Annual Power Production   Considering a system with optimal orientation and an installed power of 1 kWp (kilowatt peak), which corresponds to an area of 9 square meters (m^2) an average annual yield of around 980 to 1200 kWh plus can be anticipated in most parts of North America. Under adequate geoclimatic conditions a PV system conforming to this system configuration will have an annual power production capacity of about 120 to 128 kWh/m^2.

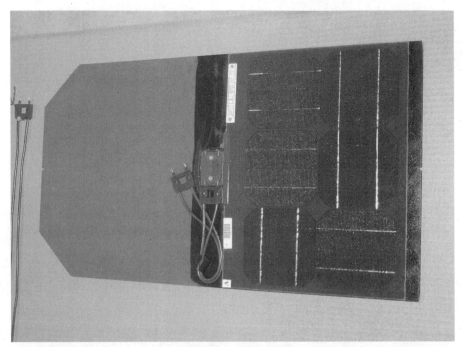

Figure 2.19   SunSlate solar power module. *Photo courtesy of Atlantis Energy Systems.*

Custom Solar Solution   MegaSlate roof elements are manufactured at an optimal size specifically designed for use on a wide variety of roof surfaces. PV units are manufactured in a fashion as to be optimally appealing when homogeneously integrated as a component of the building architecture. In most instances, MegaSlate PV units are

Figure 2.20   Application of SunSlate as roofing tile. *Photo courtesy of Atlantis Energy Systems.*

Figure 2.21 **Roof-integrated MegaSlate photovoltaic system.** *Photo courtesy of Atlantis Energy Systems.*

integrated in the construction of chimneys and skylights and for architectural effect whenever necessary. Dummy MegaSlate elements can be deployed to enhance the architectural aesthetic requirements. Figure 2.21 depicts a Roof-integrated MegaSlate photovoltaic installation by Atlantis Energy systems.

Photovoltaic Support Structure   The system-specific support structure consists of channel profiles that are mounted onto the roof substructure. These channels are fitted with specially designed rubber elements mounted on the sides that serve to support MegaSlate elements and also allow for rainwater drainage. Each MegaSlate element is secured within brackets specially coated to withstand long-term environmental exposures.

Field Cabling   The MegaSlate PV modules deploy male-female plug-in connectors that interconnect strings of arrays in a daisy-chain fashion which eventually terminate in a combiner box and finally connect to an inverter.

Installation   In contrast to standard roof tiling, MegaSlate elements are preferably laid from top to bottom. Before being secured into the brackets, they are connected with touch-safe electrical connectors. An appropriate functioning check is required before operating the system.

After the cables are connected to the terminal box on the dc side and to the inverter on the ac side, electricity produced by the building can be fed into the grid.

In the event of malfunction the MegaSlate PV elements are easy to replace or exchange. The MegaSlate roof-mount installations are walkable. It is recommended that system maintenance be undertaken by qualified personnel.

MegaSlate Technical Specifications

- Available in polycrystalline or monocrystalline cells.
- Horizontal length is 1.32 m (approximate).
- Vertical length is 1.0 m (approximate).
- Glass thickness is 6 to 10 mm.
- Overlap from upper to lower element is 15 cm.
- Electric power on each slate is 130 to 150 W.
- Has an electric touch-safe connector.

Solar Photovoltaic System Power Research and Development in the United States

DEPARTMENT OF ENERGY SOLAR ENERGY PROJECT FUNDING

The following is adapted from an excerpt from the U.S. Department of Energy (DOE) 2007 funding awards announcement for Solar America Initiative which is intended to make solar technology cost-competitive by 2015.

Thirteen development teams selected for negotiation have formed Technology Pathway Partnerships (TPP) to accelerate the drive toward commercialization of U.S.-produced solar photovoltaic systems. These partnerships are comprised of more than 50 companies, 14 universities, 3 nonprofit organizations, and 2 national laboratories. DOE funding is expected to begin in fiscal year 2007, with $51.6 million going to the TPPs.

In addition, the projects announced today will enable the projected expansion of the annual U.S. manufacturing capacity of PV systems from 240 MW in 2005 to as much as 2850 MW by 2010, representing more than a 10-fold increase. Such capacity would also put the U.S. industry on track to reduce the cost of electricity produced by photovoltaic cells from current levels of $0.18 to $0.23 per kilowatt-hours to $0.05 to $0.10 per kilowatt-hours by 2015—a price that is competitive in markets nationwide.

Teams selected for negotiations under the Solar America initiative

Amonix   At present Amonix is the manufacturer of the industry's most efficient solar power MegaConcentrator dual-axis tracking system. It manufactures a low-cost, high-concentration PV system for utility markets. This project will focus on manufacturing technology for high-concentrating PV and on low-cost production using multibandgap cells. Partners for the project include CYRO Industries, Xantrex, the Imperial Irrigation District, Hernandez Electric, the National Renewable Energy Laboratory (NREL), Spectrolab, Micrel, Northstar, JOL Enterprises, the University of Nevada at Las Vegas, and Arizona State University. Subject to negotiations, DOE funding for the first year of the project is expected to be roughly $3,200,000, with approximately $14,800,000 available over 3 years if the team meets its goals.

Boeing   Boeing is currently developing a high-efficiency concentrating photovoltaic power system. This project will focus on cell fabrication research that is expected to yield very high efficiency systems. The partners for the project will be Light Prescription Innovators, PV Powered, Array Technologies, James Gregory Associates, Sylarus, Southern California Edison, NREL, the California Institute of Technology, and the University of California at Merced. Subject to negotiations, DOE funding for the first year of the project is expected to be approximately $5,900,000, with approximately $13,300,000 available over 3 years if the team meets its goals.

BP Solar   British Petroleum will be developing a low-cost approach to grid parity using crystalline silicon. This project's research will focus on reducing wafer thickness while improving yield of multicrystalline silicon PV for commercial and residential markets. Project partners include Dow Corning, Ceradyne, Bekaert, Ferro, Specialized Technology Resources, Komax, Palo Alto Research Center, AFG Industries, Automation Tooling Systems Ohio, Xantrex, Fat Spaniel, the Sacramento Municipal Utility District, Recticel, the Georgia Institute of Technology, the University of Central Florida, and Arizona State University. Subject to negotiations, DOE funding for the first year of the project is expected to be approximately $7,500,000, with approximately $19,100,000 available over 3 years if the team meets its goals.

Dow Chemical   Dow chemical is currently developing PV-integrated residential and commercial building solutions. This project will employ Dow's expertise in encapsulates, adhesives, and high-volume production to develop integrated PV-powered technologies for roofing products. Partners include Miasole, SolFocus, Fronius, IBIS Associates, and the University of Delaware. Subject to negotiations, funding for the first year of the project is expected to be roughly $3,300,000, with approximately $9,400,000 available over 3 years if the team meets its goals.

General Electric   General Electric will be assuming a value chain partnership responsibility to accelerate U.S. PV growth. This project will develop various cell technologies—including a new bifacial, high-efficiency silicon cell that could be incorporated into systems solutions that can be demonstrated across the industry. Partners include REC Silicon, Xantrex, Solaicx, the Georgia Institute of Technology, North Carolina State University, and the University of Delaware. Subject to negotiations, DOE funding for the first year of the project is expected to be roughly $8,100,000, with approximately $18,600,000 available over 3 years if the team meets its goals.

Greenray   Greenray is a manufacturer and developer of a dc-to-ac module conversion system. This team will design and develop a high-powered, ultra-high-efficiency solar module that contains an inverter, eliminating the need to install a separate inverter and facilitating installation by homeowners. Research will focus on increasing the lifetime of the inverter. Partners include Sanyo, Tyco Electronics, Coal Creek Design, BluePoint Associates, National Grid, and Sempra Utilities. Subject to negotiations, DOE funding for the first year of the project is expected to be roughly $400,000, with approximately $2,300,000 available over 3 years if the team meets its goals.

Konarka   Konarka is currently developing integrated organic photovoltaics. This project will focus on manufacturing research and product reliability assurance for extremely low-cost photovoltaic cells using organic dyes that convert sunlight to electricity. Partners for this project include NREL and the University of Delaware. Subject to negotiations, DOE funding for the first year of the project is expected to be $1,200,000, with approximately $3,600,000 available over 3 years if the team meets its goals.

Miasole   Miasole is a manufacturer of low-cost, scalable, flexible PV systems with integrated electronics. This project will develop high-volume manufacturing technologies and PV component technologies. Research will focus on new types of flexible thin-film modules with integrated electronics and advances in technologies used for installation and maintenance. Project partners include Exeltech, Carlisle SynTec, Sandia National Laboratories, NREL, the University of Colorado, and the University of Delaware. Subject to negotiations, DOE funding for the first year of the project is expected to be $5,800,000, with approximately $20,000,000 available over 3 years if the team meets its goals.

Nanosolar   Nanosolar is conducting research on low-cost, scalable PV systems for commercial rooftops. This project will work on improved low-cost systems and components using back-contacted thin-film PV cells for commercial buildings. Research will focus on large-area module deposition, inverters, and mounting. Partners include SunLink, SunTechnics, and Conergy. Subject to negotiations, DOE funding for the first year of the project is expected to be roughly $1,100,000, with approximately $20,000,000 available over 3 years if the team meets its goals.

PowerLight   PowerLight, a subsidiary of SunPower, will be undertaking product development of a PV cell-independent effort to improve automated manufacturing systems. This project will focus on reducing noncell costs by making innovations with automated design tools and with modules that include mounting hardware. Partners include Specialized Technology Resources and Autodesk. Subject to negotiations, first-budget period funding for this project is expected to be approximately $2,800,000, with approximately $6,000,000 available over 3 years if the team meets its goals.

Practical Instruments   Practical Instruments Corporation will be developing a low-concentration CPV system for rooftop applications. This project will explore a novel concept for low-concentration optics to increase the output of rooftop PV systems. The project will also explore designs using multijunction cells to allow for very high efficiency modules. Project partners include Spectrolab, Sandia National Laboratories, SunEdison, and the Massachusetts Institute of Technology. Subject to negotiations, funding for the first year of the project is expected to be roughly $2,200,000, with approximately $4,000,000 available over 3 years if the team meets its goals.

SunPower   SunPower will be developing a grid-competitive residential solar power generating system. This project will research lower-cost ingot and wafer fabrication technologies, automated manufacture of back-contact cells, and new module designs to lower costs. Project partners include Solaicx, the Massachusetts Institute of Technology, NREL, and Xantrex. Subject to negotiations, first-budget period funding for this project is expected to be approximately $7,700,000, with approximately $17,900,000 available over 3 years if the team meets its goals.

United Solar Ovonic   United Solar Ovonic is currently in the process of developing low-cost thin-film building-integrated PV systems. This project will focus on increasing the efficiency and deposition rate of multibandgap, flexible, thin-film photovoltaic cells and reducing the cost of inverters and balance-of-system components. Partners include SMA America, Sat Con Technology Corporation, PV Powered, the ABB Group, Solectria Renewables, Developing Energy Efficient Roof Systems, Turtle Energy, Sun Edison, the University of Oregon, Syracuse University, the Colorado School of Mines, and NREL. Subject to negotiations, funding for the first year of the project is expected to be roughly $2,400,000, with approximately $19,300,000 available over 3 years if the team meets its goals.

For more information on DOE PV development support, the reader should visit the Solar America Initiative Web page: www.eere.energy.gov/solar/solar_america/.

The Energy Policy Act of 2005 (EPAct), signed by President Bush in August of 2005, provides incentives for purchasing and using solar equipment. At present the act extends through 2008, which provides incentives credit equal to 30 percent of qualifying expenditures for the purchase of commercial solar installations, with no cap on the total credit allowed. EPAct also provides a 30 percent tax credit for qualified PV property and solar water heating property used exclusively for purposes other than heating swimming pools and hot tubs. Under this act owners of qualified private property could be eligible for a credit up to $2000 for either property, with a maximum of $4000 allowed, if both photovoltaic and solar hot water qualified properties are installed. More information on available incentives for solar installations in the United States is available at the U.S. government Web page at: http://energystar.gov/index.cfm?c= products.pr_tax_credits.

3

SOLAR POWER SYSTEM
DESIGN CONSIDERATIONS

Introduction

This section is intended to acquaint the reader with the basic design concepts of solar power applications. The typical solar power applications that will be reviewed include stand-alone systems with battery backup, commonly used in remote telemetry; vehicle charging stations; communication repeater stations; and numerous installations where the installation cost of regular electrical service becomes prohibitive. An extended design application of stand-alone systems also includes the integration of an emergency power generator system.

Grid-connected solar power systems, which form a large majority of residential and industrial applications, are reviewed in detail. To familiarize the reader with the prevailing state and federal assistance rebate programs, a special section is devoted to reviewing the salient aspects of existing rebates.

Solar power design essentially consists of electronics and power systems engineering, which requires a thorough understanding of the electrical engineering disciplines and the prevailing standards outlined in Article 690 of the National Electrical Code (NEC).

The solar power design presented, in addition to reviewing the various electrical design methodologies, provides detailed insight into photovoltaic modules, inverters, charge controllers, lightning protection, power storage, battery sizing, and critical wiring requirements. To assist the reader with the economic issues of solar power cogeneration, a detailed analysis of a typical project, including system planning, photovoltaic power system cogeneration estimates, economic cost projection, and payback analysis, is covered later in Chapter 8.

Solar Power System Components and Materials

As described later in this chapter (see the section entitled "Ground-Mount Photovoltaic Module Installation and Support Hardware"), solar power photovoltaic (PV) modules are constructed from a series of cross-welded solar cells, each typically producing a specific wattage with an output of 0.5 V.

Effectively, each solar cell could be considered as a 0.5-V battery that produces current under adequate solar ray conditions. To obtain a desired voltage output from a PV panel assembly, the cells, similar to batteries, are connected in series to obtain a required output.

For instance, to obtain a 12-V output, 24 cell modules in an assembly are connected in tandem. Likewise, for a 24-V output, 48 modules in an assembly are connected in series. To obtain a desired wattage, a group of several series-connected solar cells are connected in parallel.

The output power of a unit solar cell or its efficiency is dependent on a number of factors such as crystalline silicon, polycrystalline silicon, and amorphous silicon materials, which have specific physical and chemical properties, details of which were discussed in Chapter 2.

Commercially available solar panel assemblies mostly employ proprietary cell manufacturing technologies and lamination techniques, which include cell soldering. Soldered groups of solar cells are in general sandwiched between two tempered-glass panels, which are offered in framed or frameless assemblies.

Solar Power System Configuration and Classifications

There are four types of solar power systems:

1 Directly connected dc solar power system
2 Stand-alone dc solar power system with battery backup
3 Stand-alone hybrid solar power system with generator and battery backup
4 Grid-connected solar power cogeneration system

DIRECTLY CONNECTED DC SOLAR POWER SYSTEM

As shown in Figure 3.1, the solar system configuration consists of a required number of solar photovoltaic cells, commonly referred to as PV modules, connected in series or in parallel to attain the required voltage output. Figure 3.2 shows four PV modules that have been connected in parallel.

The positive output of each module is protected by an appropriate overcurrent device, such as a fuse. Paralleled output of the solar array is in turn connected to a dc motor via a two-pole single throw switch. In some instances, each individual PV module is also

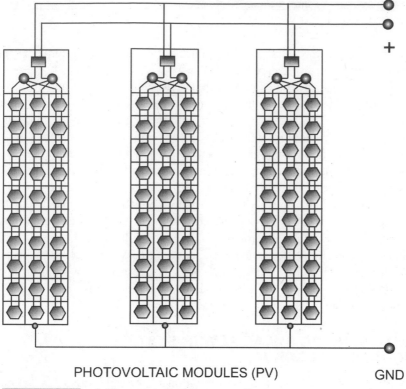

PHOTOVOLTAIC MODULES (PV)   GND

Figure 3.1   A three-panel solar array diagram.

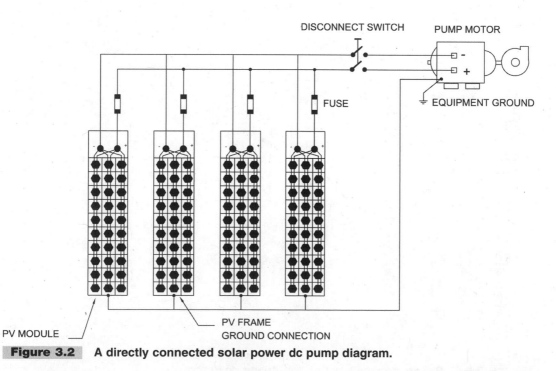

Figure 3.2   A directly connected solar power dc pump diagram.

protected with a forward-biased diode connected to the positive output of individual solar panels (not shown in Figure 3.2).

An appropriate surge protector connected between the positive and negative supply provides protection against lightning surges, which could damage the solar array system components. In order to provide equipment-grounding bias, the chassis or enclosures of all PV modules and the dc motor pump are tied together by means of grounding clamps. The system ground is in turn connected to an appropriate grounding rod. All PV interconnecting wires are sized and the proper type selected to prevent power losses caused by a number of factors, such as exposure to the sun, excessive wire resistance, and additional requirements that are mandated by the NEC.

The photovoltaic solar system described is typically used as an agricultural application, where either regular electrical service is unavailable or the cost is prohibitive. A floating or submersible dc pump connected to a dc PV array can provide a constant stream of well water that can be accumulated in a reservoir for farm or agricultural use. In subsequent sections we will discuss the specifications and use of all system components used in solar power cogeneration applications.

STAND-ALONE DC SOLAR POWER SYSTEM WITH BATTERY BACKUP

The solar power photovoltaic array configuration shown in Figure 3.3, a dc system with battery backup, is essentially the same as the one without the battery except that there are a few additional components that are required to provide battery charge stability.

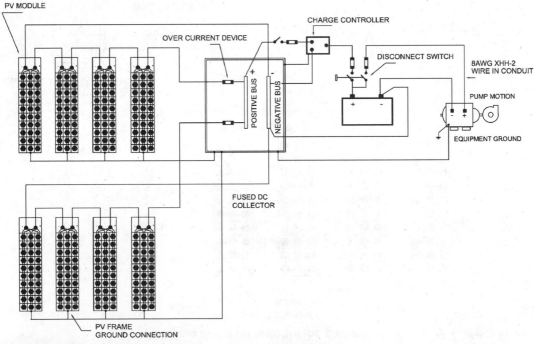

Figure 3.3　Battery-backed solar power–driven dc pump.

Stand-alone PV system arrays are connected in series to obtain the desired dc voltage, such as 12, 24, or 48 V; outputs of that are in turn connected to a dc collector panel equipped with specially rated overcurrent devices, such as ceramic-type fuses. The positive lead of each PV array conductor is connected to a dedicated fuse, and the negative lead is connected to a common neutral bus. All fuses as well are connected to a common positive bus. The output of the dc collector bus, which represents the collective amperes and voltages of the overall array group, is connected to a dc charge controller, which regulates the current output and prevents the voltage level from exceeding the maximum needed for charging the batteries.

The output of the charge controller is connected to the battery bank by means of a dual dc cutoff disconnect. As depicted in Figure 3.3, the cutoff switch, when turned off for safety measures, disconnects the load and the PV arrays simultaneously.

Under normal operation, during the daytime when there is adequate solar insolation, the load is supplied with dc power while simultaneously charging the battery. When sizing the solar power system, take into account that the dc power output from the PV arrays should be adequate to sustain the connected load and the battery trickle charge requirements.

Battery storage sizing depends on a number of factors, such as the duration of an uninterrupted power supply to the load when the solar power system is inoperative, which occurs at nighttime or during cloudy days. Note that battery banks inherently, when in operation, produce a 20 to 30 percent power loss due to heat, which also must be taken into consideration.

When designing a solar power system with a battery backup, the designer must determine the appropriate location for the battery racks and room ventilation, to allow for dissipation of the hydrogen gas generated during the charging process. Sealed-type batteries do not require special ventilation.

All dc wiring calculations discussed take into consideration losses resulting from solar exposure, battery cable current derating, and equipment current resistance requirements, as stipulated in NEC 690 articles.

STAND-ALONE HYBRID AC SOLAR POWER SYSTEM WITH GENERATOR AND BATTERY BACKUP

A stand-alone hybrid solar power configuration is essentially identical to the dc solar power system just discussed, except that it incorporates two additional components, as shown in Figure 3.4. The first component is an inverter. Inverters are electronic power equipment designed to convert direct current into alternating current. The second component is a standby emergency dc generator, which will be discussed later.

Alternating-current inverters The principal mechanism of dc-to-ac conversion consists of chopping or segmenting the dc current into specific portions, referred to as square waves, which are filtered and shaped into sinusoidal ac waveforms.

Any power waveform, when analyzed from a mathematical point of view, essentially consists of the superimposition of many sinusoidal waveforms, referred to as harmonics. The first harmonic represents a pure sinusoidal waveform, which has a unit base wavelength, amplitude, and frequency of repetition over a unit of time called

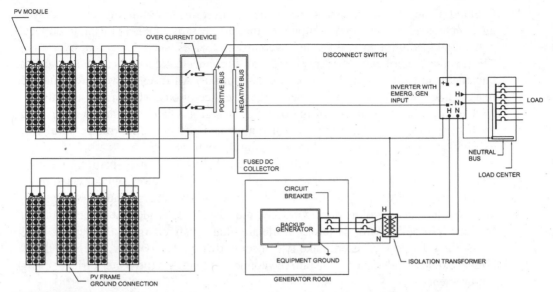

Figure 3.4   Stand-alone hybrid solar power system with standby generator.

a cycle. Additional waveforms with higher cycles, when superimposed on the base waveform, add or subtract from the amplitude of the base sinusoidal waveform.

The resulting combined base waveform and higher harmonics produce a distorted waveshape that resembles a distorted sinusoidal wave. The higher the harmonic content, the squarer the waveshape becomes.

Chopped dc output, derived from the solar power, is considered to be a numerous superimposition of odd and even numbers of harmonics. To obtain a relatively clean sinusoidal output, most inverters employ electronic circuitry to filter a large number of harmonics. Filter circuits consist of specially designed inductive and capacitor circuits that trap or block certain unwanted harmonics, the energy of which is dissipated as heat. Some types of inverters, mainly of earlier design technology, make use of inductor coils to produce sinusoidal waveshapes.

In general, dc-to-ac inverters are intricate electronic power conversion equipment designed to convert direct current to a single- or three-phase current that replicates the regular electrical services provided by utilities. Special electronics within inverters, in addition to converting direct current to alternating current, are designed to regulate the output voltage, frequency, and current under specified load conditions. As discussed in the following sections, inverters also incorporate special electronics that allow them to automatically synchronize with other inverters when connected in parallel. Most inverters, in addition to PV module input power, accept auxiliary input power to form a standby generator, used to provide power when battery voltage is dropped to a minimum level.

A special type of inverter, referred to as the *grid-connected* type, incorporates synchronization circuitry that allows the production of sinusoidal waveforms in unison with the electrical service grid. When the inverter is connected to the electrical service grid, it can effectively act as an ac power generation source. Grid-type inverters used in grid-connected solar power systems are strictly regulated by utility agencies that provide net metering.

Some inverters incorporate an internal ac transfer switch that is capable of accepting an output from an ac-type standby generator. In such designs, the inverters include special electronics that transfer power from the generator to the load.

Standby generators   A standby generator consists of an engine-driven generator that is used to provide auxiliary power during solar blackouts or when the battery power discharge reaches a minimum level. The output of the generator is connected to the auxiliary input of the inverter.

Engines that drive the motors operate with gasoline, diesel, natural gas, propane, or any type of fuel. Fuel tank sizes vary with the operational requirements. Most emergency generators incorporate underchassis fuel tanks with sufficient fuel storage capacity to operate the generator up to 48 hours. Detached tanks could also be designed to hold much larger fuel reserves, which are usually located outside the engine room. In general, large fuel tanks include special fuel-level monitoring and filtration systems. As an option, the generators can be equipped with remote monitoring and annunciation panels that indicate power generation data and log and monitor the functional and dynamic parameters of the engine, such as coolant temperature, oil pressure, and malfunctions.

Engines also incorporate special electronic circuitry to regulate the generator output frequency, voltage, and power under specified load conditions.

Hybrid system operation   As previously discussed, the dc output generated from the PV arrays and the output of the generator can be simultaneously connected to an inverter. The ac output of the inverter is in turn connected to an ac load distribution panel, which provides power to various loads by means of ac-type overcurrent protection devices.

In all instances, solar power design engineers must ensure that all chassis of equipment and PV arrays, including stanchions and pedestals, are connected together via appropriate grounding conductors that are connected to a single-point service ground bus bar, usually located within the vicinity of the main electrical service switchgear.

In grid-connected systems, switching of ac power from the standby generator and the inverter to the service bus or the connected load is accomplished by internal or external automatic transfer switches.

Standby power generators must always comply with the National Electrical Code requirements outlined in the following articles:

- Electrical Service Requirement, NEC 230
- General Grounding Requirements, NEC 250
- Generator Installation Requirements, NEC 445
- Emergency Power System Safety Installation and Maintenance Requirements, NEC 700

GRID-CONNECTED SOLAR POWER COGENERATION SYSTEM

With reference to Figure 3.5, a connected solar power system diagram, the power cogeneration system configuration is similar to the hybrid system just described. The essence of a grid-connected system is *net metering*. Standard service meters are odometer-type counting wheels that record power consumption at a service point by means of a rotating disc, which is connected to the counting mechanism. The rotating discs operate by an electrophysical principle called *eddy current*, which consists of voltage and current measurement sensing coils that generate a proportional power measurement.

New electric meters make use of digital electronic technology that registers power measurement by solid-state current- and voltage-sensing devices that convert analog measured values into binary values that are displayed on the meter bezels by liquid-crystal display (LCD) readouts.

In general, conventional meters only display power consumption; that is, the meter counting mechanism is unidirectional.

Net metering   The essential difference between a grid-connected system and a stand-alone system is that inverters, which are connected to the main electrical service, must have an inherent line frequency synchronization capability to deliver the excess power to the grid.

Net meters, unlike conventional meters, have a capability to record consumed or generated power in an exclusive summation format; that is, the recorded power registration is the net amount of power consumed—the total power used minus the amount of power that is produced by the solar power cogeneration system. Net meters are supplied

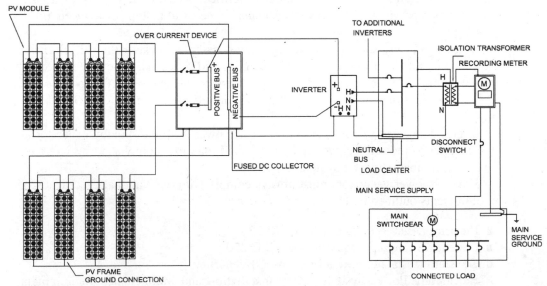

Figure 3.5   **Grid-connected hybrid solar power system with standby generator.**

and installed by utility companies that provide grid-connection service systems. Net-metered solar power cogenerators are subject to specific contractual agreements and are subsidized by state and municipal governmental agencies. The major agencies that undertake distribution of the state of California's renewable energy rebate funds for various projects are the California Energy Commission (CEC), Southern California Edison, Southern California Gas (Sempra Power), and San Diego Gas and Electric (SG&E), as well as principal municipalities, such as the Los Angeles Department of Water and Power. When designing net metering solar power cogeneration systems, the solar power designers and their clients must familiarize themselves with the CEC rebate fund requirements. Essential to any solar power implementation is the preliminary design and economic feasibility study needed for project cost justification and return-on-investment analysis. The first step of the study usually entails close coordination with the architect in charge and the electrical engineering consultant. A preliminary PV array layout and a computer-aided shading study are essential for providing the required foundation for the design. Based on the preceding study, the solar power engineer must undertake an econometrics study to verify the validity and justification of the investment. Upon completion of the study, the solar engineer must assist the client to complete the required CEC rebate application forms and submit it to the service agency responsible for the energy cogeneration program.

Grid-connection isolation transformer   In order to prevent spurious noise transfer from the grid to the solar power system electronics, a delta-y isolation transformer is placed between the main service switchgear disconnects and the inverters. The delta winding of the isolation transformer, which is connected to the service bus, circulates noise harmonics in the winding and dissipates the energy as heat.

Isolation transformers are also used to convert or match the inverter output voltages to the grid. Most often, in commercial installations, inverter output voltages range from 208 to 230 V (three phase), which must be connected to an electric service grid that supplies 277/480 V power.

Some inverter manufacturers incorporate output isolation transformers as an integral part of the inverter system, which eliminates the use of external transformation and ensures noise isolation.

Storage Battery Technologies

One of the most significant components of solar power systems consist of battery backup systems that are frequently used to store electric energy harvested from solar photovoltaic systems for use during the absence of sunlight, such as at night and during cloudy conditions. Because of the significance of storage battery systems it is important for design engineers to have a full understanding of the technology since this system component represents a notable portion of the overall installation cost. More importantly, the designer must be mindful of the hazards associated with handling, installation, and maintenance. To provide an in-depth knowledge about the battery technology, this section covers the physical and chemical principles, manufacturing, design application, and maintenance procedures of the storage battery. In this section we will also attempt to analyze and discuss the advantages and disadvantages of different types of commercially available solar power batteries and their specific performance characteristics.

HISTORY

In 1936, while excavating the ruins of a 2000-year-old village near Baghdad, called Khujut Rabu, workers discovered a mysterious small jar identified as a Sumerian artifact dated to 250 BC. This jar, which was identified as the earliest battery, was a 6-in-high pot of bright yellow clay that included a copper-enveloped iron rod capped with an asphalt-like stopper. The edge of the copper cylinder was soldered with a lead-tin alloy comparable to today's solder. The bottom of the cylinder was capped with a crimped-in copper disk and sealed with bitumen or asphalt. Another insulating layer of asphalt sealed the top and also held in place the iron rod that was suspended into the center of the copper cylinder. The rod showed evidence of having been corroded with an agent. The jar when filled with vinegar produces about 1.1 V of electric potential.

A German archaeologist, Wilhelm Konig, who examined the object (see Figures 3.6 and 3.7), came to the surprising conclusion that the clay pot was nothing less than

Figure 3.6   The Baghdad battery.

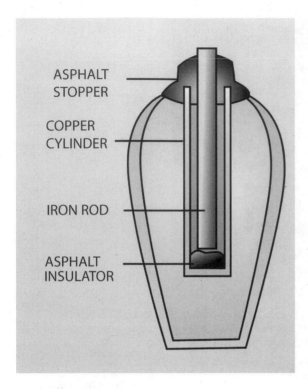

ASPHALT
STOPPER

COPPER
CYLINDER

IRON ROD

ASPHALT
INSULATOR

Figure 3.7   The Baghdad battery elements.

an ancient electric battery. It is stipulated that the Sumerians made use of the battery for electroplating inexpensive metals such as copper with silver or gold.

Subsequent to the discovery of this first battery, several other batteries were unearthed in Iraq, all of which dated from the Parthian occupation between 248 BCE and 226 CE.

In the 1970s, German Egyptologist Arne Eggebrecht built a replica of the Baghdad battery and filled it with grape juice, which he deduced ancient Sumerians might have used as an electrolyte. The replica generated 0.87 V of electric potential. Current generated from the battery was then used to electroplate a silver statuette with gold.

However, the invention of batteries is associated with the Italian scientist Luigi Galvani, an anatomist who, in 1791, published works on animal electricity. In his experiments, Galvani noticed that the leg of a dead frog began to twitch when it came in contact with two different metals. From this phenomenon he concluded that there is a connection between electricity and muscle activity. Alessandro Conte Volta, an Italian physicist, in 1800, reported the invention of his electric battery or "pile." The battery was made by piling up layers of silver, paper or cloth soaked in salt, and zinc (see Figure 3.8). Many triple layers were assembled into a tall pile, without paper or cloth between the zinc and silver, until the desired voltage was reached. Even today the French word for battery is *pile,* pronounced "peel" in English. Volta also developed the concept of the electrochemical series, which ranks the potential produced when various metals come in contact with an electrolyte.

Figure 3.8 Alessandro Volta's pile.

The battery is an electric energy storage device that in physics terminology can be described as a device or mechanism that can hold kinetic or static energy for future use. For example, a rotating flywheel can store dynamic rotational energy in its wheel, which releases the energy when the primary mover such as a motor no longer engages the connecting rod. Similarly, a weight held at a high elevation stores static energy embodied in the object mass, which can release its static energy when dropped. Both of these are examples of energy storage devices or batteries.

Energy storage devices can take a wide variety of forms, such as chemical reactors and kinetic and thermal energy storage devices. Note that each energy storage device is referred to by a specific name; the word *battery,* however, is solely used for electrochemical devices that convert chemical energy into electricity by a process referred to as galvanic interaction. A galvanic cell is a device that consists of two electrodes, referred to as the anode and the cathode, and an electrolyte solution. Batteries consist of one or more galvanic cells.

Note that a battery is an electrical storage reservoir and not an electricity-generating device. Electric charge generation in a battery is a result of chemical interaction, a process that promotes electric charge flow between the anode and the cathode in the presence of an electrolyte. The electrogalvanic process that eventually results in depletion of the anode and cathode plates is resurrected by a recharging process that can be repeated numerous times. In general, batteries when delivering stored energy incur energy losses as heat when discharging or during chemical reactions when charging.

The Daniell cell   The Voltaic pile was not good for delivering currents over long periods of time. This restriction was overcome in 1820 with the Daniell cell. British researcher John Frederick Daniell developed an arrangement where a copper plate was located at the bottom of a wide-mouthed jar. A cast-zinc piece commonly referred to as a crowfoot, because of its shape, was located at the top of the plate, hanging on the rim of the jar. Two electrolytes, or conducting liquids, were employed. A saturated copper-sulfate solution covered the copper plate and extended halfway up the remaining distance toward the zinc piece. Then, a zinc-sulfate solution, which is a less dense liquid, was carefully poured over a structure that floated above the copper sulfate and immersed zinc.

In a similar experiment, instead of zinc sulfate, magnesium sulfate or dilute sulfuric acid was used. The Daniell cell was also one of the first batteries that incorporated mercury, which was amalgamated with the zinc anode to reduce corrosion when the batteries were not in use. The Daniell battery, which produced about 1.1 V, was extensively used to power telegraphs, telephones, and even to ring doorbells in homes for over a century.

Plante's battery   In 1859 Raymond Plante invented a battery that used a cell by rolling up two strips of lead sheet separated by pieces of flannel material. The entire assembly when immersed in diluted sulfuric acid produced an increased current that was subsequently improved upon by insertion of separators between the sheets.

The carbon-zinc battery   In 1866, Georges Leclanché, in France developed the first cell battery. The battery, instead of using liquid electrolyte, was constructed from moist ammonium chloride paste and a carbon and zinc anode and cathode. It was sealed and sold as the first dry battery. The battery was rugged and easy to manufacture and had a good shelf life. Carbon-zinc batteries were in use over the next century until they were replaced by alkaline-manganese batteries. Figure 3.9 depicts graphics of a lead acid battery current flow process.

Lead-acid battery suitable for autos   In 1881 Camille Faure produced the first modern lead-acid battery, which he constructed from cast-lead grids that were packed with lead oxide paste instead of lead sheets. The battery had a larger current-producing capacity. Its performance was further improved by the insertion of separators between the positive and negative plates to prevent particles falling from these plates, which could short out the positive and negative plates from the conductive sediment.

The Edison battery   Between the years 1898 and 1908, Thomas Edison developed an alkaline cell with iron as the anode material (–) and nickel oxide as the cathode material (+). The electrolyte used was potassium hydroxide, the same as in modern nickel-cadmium and alkaline batteries. The cells were extensively used in industrial

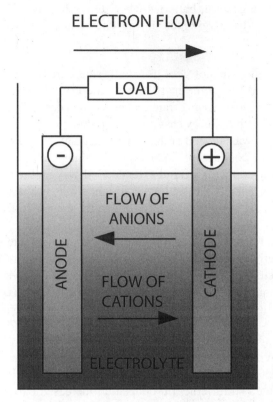

ELECTRON FLOW

LOAD

FLOW OF ANIONS

FLOW OF CATIONS

ANODE

CATHODE

ELECTROLYTE

Figure 3.9   **Lead-acid battery current.**

and railroad applications. Nickel-cadmium batteries are still being used and have remained unchanged ever since.

In parallel with Edison's work, Jungner and Berg in Sweden were working on the development of the nickel-cadmium cell. In place of the iron used in the Edison cell, they used cadmium, with the result that the cell operated better at low temperatures and was capable of self-discharge to a lesser degree than the Edison cell, and in addition the cell could be trickle-charged at a reduced rate. In 1949 the alkaline-manganese battery, also referred to as the alkaline battery, was developed by Lew Urry at the Eveready Battery Company laboratory in Parma, Ohio. Alkaline batteries are capable of storing higher energy within the same package size than comparable conventional dry batteries.

Zinc-mercuric oxide alkaline batteries   In 1950 Samuel Ruben invented the zinc-mercuric oxide alkaline battery (see Figure 3.10), which was licensed to the P.R. Mallory Co. The company later became Duracell, International. Mercury compounds have since been eliminated from batteries to protect the environment.

Deep-discharge batteries used in solar power backup applications in general have lower charging and discharging rate characteristics and are more efficient. A battery rated at 4 ampere-hours (Ah) over 6 hours might be rated at 220 Ah at the 20-hour rate

Figure 3.10 Alkaline batteries.

and 260 Ah at the 48-hour rate. The typical efficiency of a lead-acid battery is 85 to 95 percent, and that of alkaline and nickel-cadmium (NiCd) batteries is about 65 percent.

Practically all batteries used in PV systems and in all but the smallest backup systems are lead-acid type batteries. Even after over a century of use, they still offer the best price-to-power ratio. It is not recommended, however, to use NiCd batteries in systems that operate in extremely cold temperatures such as at –50°F or below.

NiCd batteries are expensive to buy and very expensive to dispose off due to the hazardous nature of cadmium. We have had almost no direct experience with these (alkaline) batteries, but from what we have learned from others we do not recommend them—one major disadvantage is that there is a large voltage difference between the fully charged and discharged states. Another problem is that they are very inefficient— there is a 30 to 40 percent heat loss just during charging and discharging.

It is important to note that all batteries commonly used in deep-cycle applications are lead-acid batteries. This includes the standard flooded (wet), gelled, and absorbed glass mat (AGM) batteries. They all use the same chemistry, although the actual construction of the plates and so forth can vary considerably.

Nickel-cadmium, nickel-iron, and other types of batteries are found in some systems but are not common due to their expense and/or poor efficiency.

MAJOR BATTERY TYPES

Solar power backup batteries are divided into two categories based on what they are used for and how they are constructed. The major applications where batteries are used as solar backup include automotive systems, marine systems, and deep-cycle discharge systems.

The major manufactured processes include flooded or wet construction, gelled, and AGM types. AGM batteries are also referred to as "starved electrolyte" or "dry" type, because instead of containing wet sulfuric acid solution, the batteries contain a fiber-glass mat saturated with sulfuric acid, which has no excess liquid.

Common flooded-type batteries are usually equipped with removable caps for maintenance-free operation. Gelled-type batteries are sealed and equipped with a small vent valve that maintains a minimal positive pressure. AGM batteries are also equipped with a sealed regulation-type valve that controls the chamber pressure within 4 pounds per square inch (lb/in^2).

As described earlier, common automobile batteries are built with electrodes that are grids of metallic lead containing lead oxides that change in composition during charging and discharging. The electrolyte is dilute sulfuric acid. Lead-acid batteries, even though invented nearly a century ago, are still the battery of choice for solar and backup power systems. With improvements in manufacturing, batteries can last as long as 20 years.

Nickel-cadmium, or alkaline, storage batteries, in which the positive active material is nickel oxide and the negative material contains cadmium, are generally considered very hazardous due to the cadmium. The efficiency of alkaline batteries ranges from 65 to 80 percent compared to 85 to 90 percent for lead-acid batteries. Their nonstandard voltage and charging current also make them very difficult to use.

Deep-discharge batteries used in solar power backup applications in general have lower charging and discharging rate characteristics and are more efficient.

In general, all batteries used in PV systems are lead-acid type batteries. Alkaline-type batteries are used only in exceptionally low temperature conditions of below –50°F. Alkaline batteries are expensive to buy and due to the hazardous contents are very expensive to dispose off.

BATTERY LIFE SPAN

The life span of a battery will vary considerably with how it is used, how it is maintained and charged, the temperature, and other factors. In extreme cases, it can be damaged within 10 to 12 months of use when overcharged. On the other hand if the battery is maintained properly, the life span could be extended over 25 years. Another factor that can shorten the life expectancy by a significant amount is when the batteries are stored uncharged in a hot storage area. Even dry charged batteries when sitting on a shelf have a maximum life span of about 18 months; as a result most are shipped from the factory with damp plates. As a rule, deep-cycle batteries can be used to start and run marine engines. In general, when starting, engines require a very large inrush of current for a very short time. Regular automotive starting batteries have a large number of thin plates for maximum surface area. The plates, as described earlier, are constructed from impregnated lead paste grids similar in appearance to a very fine foam sponge. This gives a very large surface area, and when deep-cycled, the grid plates quickly become consumed and fall to the bottom of the cells in the form of sediment. If automotive batteries are deep-cycled, they will generally fail after 30 to 150 deep cycles, whereas they may last for thousands of cycles in normal starting use discharge conditions. Deep-cycle batteries are designed to be discharged down time after time and are designed with thicker plates.

The major difference between a true deep-cycle battery and regular batteries is that the plates in a deep-cycle battery are made from solid lead plates and are not impregnated with lead oxide paste. Figure 3.11 shows a typical solar battery bank system.

Figure 3.11   **Deep-cycle battery bank system.** *Courtesy of Solar Integrated Technologies.*

Stored energy in batteries in general is discharged rapidly. For example, short bursts of power are needed when starting an automobile on a cold morning, which results in high amounts of current being rushed from the battery to the starter. The standard unit for energy or work is the joule (J), which is defined as 1 watt-second of mechanical work performed by a force of 1 newton (N) or 0.227 pound (lb) pushing or moving a distance of 1 meter (m). Since 1 hour has 3600 seconds, 1 watt-hour (Wh) is equal to 3600 J. The stored energy in batteries is either measured in milliampere-hours if small or ampere-hours if large. Battery ratings are converted to energy if their average voltages are known during discharge. In other words, the average voltage of the battery is maintained relatively unchanged during the discharge cycle. The value in joules can also be converted into various other energy values as follows:

Joules divided by 3,600,000 yields kilowatt-hours.

Joules divided by 1.356 yields English units of energy foot-pounds.

Joules divided by 1055 yields British thermal units.

Joules divided by 4184 yields calories.

BATTERY POWER OUTPUT

In each instance when power is discharged from a battery, the battery's energy is drained. The total quantity of energy drained equals the amount of power multiplied by the time the power flows. Energy has units of power and time, such as kilowatt-hours or watt-seconds. The stored battery energy is consumed until the available voltage and current levels of the battery are exhausted. Upon depletion of stored energy, batteries are recharged over and over again until they deteriorate to a level where they must be replaced by new units. High-performance batteries in general have the following notable characteristics. First, they must be capable of meeting the power demand requirements of the connected loads by supplying the required current while maintaining a constant voltage, and they must have sufficient energy storage capacity to maintain the load power demand as long as required. In addition, they must be as inexpensive and economical as possible and be readily replaced and recharged.

BATTERY INSTALLATION AND MAINTENANCE

Unlike many electrical apparatuses, standby batteries have specific characteristics that require special installation and maintenance procedures, which if not followed can impact the quality of the battery performance.

Battery types   As mentioned earlier, the majority of today's emergency power systems make use of two types of batteries, namely, lead-acid and nickel-cadmium (NiCd). Within the lead-acid family, there are two distinct categories, namely, flooded or vented (filled with liquid acid) and valve-regulated lead acid (VRLA, immobilized acid).

Lead-acid and NiCd batteries must be kept dry at all times and in cool locations, preferably below 70°F, and must not be stored for long in warm locations. Materials such as conduit, cable reels, and tools must be kept away from the battery cells.

Battery installation safety   What separates battery installers from the layperson is the level of awareness and respect for dc power. Energy stored in the battery cell is quite high, and sulfuric acid (lead-acid batteries) or potassium hydroxide (a base used in NiCd batteries) electrolytes could be very harmful if not handled professionally. Care should always be exercised when handling these cells. Use of chemical-resistant gloves, goggles, and a face shield, as well as protective sleeves, is highly recommended. The battery room must be equipped with an adequate shower or water sink to provide for rinsing of the hands and eyes in case of accidental contact with the electrolytes. Stored energy in a single NiCd cell of 100-Ah capacity can produce about 3000 A if short circuited between the terminal posts. Also, a fault across a lead-acid battery can send shrapnel and terminal post material flying in any direction, which can damage the cell and endanger workers.

Rack cabinet installation   Stationary batteries must be mounted on open racks or on steel or fiberglass racks or enclosures. The racks should be constructed and maintained in a level position and secured to the floor and must have a minimum of 3 feet of walking space for egress and maintenance.

Open racks are preferable to enclosures since they provide a better viewing of electrolyte levels and plate coloration, as well as easier access for maintenance. For multistep or bleacher-type racks, batteries should always be placed at the top or rear of the cabinet to avoid anyone having to reach over the cells. Always use the manufacturer-supplied connection diagram to ensure the open positive and negative terminals when charging the cells. In the event of installation schedule delays, if possible, delay delivery.

BATTERY SYSTEM CABLES

Appendix A provides code-rated dc cable tables for a variety of battery voltages and feed capacities. The tables provide American Wire Gauge (AWG) conductor gauges and voltage drops calculated for a maximum of a 2 percent drop. Whenever larger drops are permitted, the engineer must refer to NEC tables and perform specific voltage drop calculations.

CHARGE CONTROLLERS

A charge controller is essentially a current-regulating device that is placed between the solar panel array output and the batteries. These devices are designed to keep batteries charged at peak power without overcharging. Most charge controllers incorporate special electronics that automatically equalize the charging process.

DC FUSES

All fuses used as overcurrent devices, which provide a point of connection between PV arrays and collector boxes, must be dc rated. Fuse ratings for dc branch circuits, depending on wire ampacities, are generally rated from 15 to 100 A. The dc-rated fuses familiar to solar power contractors are manufactured by a number of companies such as Bussman, Littlefuse, and Gould and can be purchased from electrical suppliers.

Various manufacturers identify the fuse voltage by special capital letter designations. The following are a sample of time-delay type fuse designations used by various manufacturers.

Bussman

Voltage rating up to 125 V dc and current ampacity range of 1 to 600 A– Special fuse designation for this class of fuse is FRN-R.

Voltage rating up to 300 V dc and current ampacity range of 1 to 600 A– Special fuse designation for this class of fuse is FRS-R.

Littlefuse

Voltage rating up to 600 V dc and current ampacity range of 1 to 600 A– IDSR.

Photovoltaic output as a rule must be protected with extremely fast-acting fuses. The same fuses can also be utilized within solar power control equipment and collector boxes. Some of the fast-acting fuses used are manufactured by the same companies listed before:

Bussman

Midget fast-acting fuse, ampacity rating 0.1 to 30 A– Special fuse designation for this class of fuse is ATM.

JUNCTION BOXES AND EQUIPMENT ENCLOSURES

All junction boxes utilized for interconnecting raceways and conduits must be of waterproof construction and be designed for outdoor installation. All equipment boxes, such as dc collectors must either be classified as MENA 3R or NEMA 4X.

Solar Power System Wiring

This section covers solar power wiring design and is intended to familiarize engineers and system integrators with some of the most important aspects related to personnel safety and hazards associated with solar power projects.

Residential and commercial solar power systems, up until a decade ago, because of a lack of technology maturity and higher production costs, were excessively expensive and did not have sufficient power output efficiency to justify a meaningful return on investment. Significant advances in solar cell research and manufacturing technology have recently rendered solar power installation a viable means of electric power cogeneration in residential and commercial projects.

As a result of solar power rebate programs available throughout the United States, Europe, and most industrialized countries, solar power industries have flourished and expanded their production capacities in the past 10 years and are currently offering reasonably cost effective products with augmented efficiencies.

In view of constant and inevitable fossil fuel-based energy cost escalation and availability of worldwide sustainable energy rebate programs, solar power because of its inherent reliability and longevity, has become an important contender as one of the most viable power cogeneration investments afforded in commercial and industrial installations.

In view of the newness of the technology and constant emergence of new products, installation and application guidelines controlled by national building and safety organizations such as the National Fire Protection Association, which establishes the guidelines for the National Electrical Code (NEC), have not been able to follow up with a number of significant matters related to hazards and safety prevention issues.

In general, small-size solar power system wiring projects, such as residential installations commonly undertaken by licensed electricians and contractors who are trained in life safety installation procedures, do not represent a major concern. However, large installations where solar power produced by photovoltaic arrays generates several hundred volts of dc power require exceptional design and installation measures.

An improperly designed solar power system in addition to being a fire hazard can cause very serious burns and in some instances result in fatal injury. Additionally, an improperly designed solar power system can result in a significant degradation of power production efficiency and minimize the return on investment.

Some significant issues related to inadequate design and installation include improperly sized and selected conductors, unsafe wiring methods, inadequate overcurrent protection, unrated or underrated choice of circuit breakers, disconnect switches, system grounding, and numerous other issues that relate to safety and maintenance.

At present the NEC in general covers various aspects of photovoltaic power generation systems; however, it does not cover special application and safety issues. For example, in a solar power system a deep-cycle battery backup with a nominal 24 V and 500 Ah can discharge thousands of amperes of current if short circuited. The enormous energy generated in such a situation can readily cause serious burns and fatal injuries.

Unfortunately most installers, contractors, electricians, and even inspectors who are familiar with the NEC most often do not have sufficient experience and expertise with dc power system installation; as such requirements of the NEC are seldom met. Another significant point that creates safety issues is related to material and components used, which are seldom rated for dc applications.

Electrical engineers and solar power designers who undertake solar power system installations of 10 kWh or more (nonpackaged systems) are recommended to review 2005 NEC Section 690 and the suggested solar power design and installation practices report issued by Sandia National Laboratories.

To prevent the design and installation issues discussed, system engineers must ensure that all material and equipment used are approved by Underwriters Laboratories. All components such as overcurrent devices, fuses, and disconnect switches are dc rated. Upon completion of installation, the design engineer should verify, independently of the inspector, whether the appropriate safety tags are permanently installed and attached to all disconnect devices, collector boxes, and junction boxes and verify if system wiring and conduit installation comply with NEC requirements. The recognized materials and equipment testing organizations that are generally accredited in the United States and Canada are Underwriters Laboratories (UL), Canadian Standards Association (CSA), and Testing Laboratories (ETL), all of which are registered trademarks that commonly provide equipment certification throughout the North American continent.

Note that the NEC, with the exception of marine and railroad installation, covers all solar power installations, including stand-alone, grid-connected, and utility-interactive cogeneration systems. As a rule, the NEC covers all electrical system wiring and installations and in some instances has overlapping and conflicting directives that may not be suitable for solar power systems, in which case Article 690 of the code always takes precedence.

In general, solar power wiring is perhaps considered one of the most important aspects of the overall systems engineering effort; as such it should be understood and applied with due diligence. As mentioned earlier, undersized wiring or a poor choice of material application cannot only diminish system performance efficiency but can also create a serious safety hazard for maintenance personnel.

WIRING DESIGN

Essentially solar power installations include a hybrid of technologies consisting of basic ac and dc electric power and electronics—a mix of technologies, each requiring specific technical expertise. Systems engineering of a solar power system requires an intimate knowledge of all hardware and equipment performance and application requirements. In general, major system components such as inverters, batteries, and emergency power generators, which are available from a wide number of manufacturers, each have a unique performance specification specially designed for specific applications.

The location of a project, installation space considerations, environmental settings, choice of specific solar power module and application requirements, and numerous other parameters usually dictate specific system design criteria that eventually form the basis for the system design and material and equipment selection.

Issues specific to solar power relate to the fact that all installations are of the outdoor type, and as a result all system components, including the PV panel, support structures, wiring, raceways, junction boxes, collector boxes, and inverters must be selected and designed to withstand harsh atmospheric conditions and must operate under extreme temperatures, humidity, and wind turbulence and gust conditions. Specifically, the electrical wiring must withstand, in addition to the preceding environmental adversities, degradation under constant exposure to ultraviolet radiation and heat. Factors to be taken into consideration when designing solar power wiring include the PV module's short-circuit current (Isc) value, which represents the maximum module output when output leads are shorted. The short-circuit current is significantly higher than the normal or nominal operating current. Because of the reflection of solar rays from snow, a nearby body of water or sandy terrain can produce unpredicted currents much in excess of the specified nominal or Isc current. To compensate for this factor, interconnecting PV module wires are assigned a multiplier of 1.25 (25 percent) above the rated Isc.

PV module wires as per the NEC requirements are allowed to carry a maximum load or an ampacity of no more than 80 percent; therefore, the value of current-carrying capacity resulting from the previous calculation is multiplied by 1.25, which results in a combined multiplier of 1.56.

The resulting current-carrying capacity of the wires if placed in a raceway must be further derated for specific temperature conditions as specified in NEC wiring tables (Article 310, Tables 310.16 to 310.18).

All overcurrent devices must also be derated by 80 percent and have an appropriate temperature rating. Note that the feeder cable temperature rating must be the same as that for overcurrent devices. In other words, the current rating of the devices should be 25 percent larger than the total sum of the amount of current generated from a solar array. For overcurrent device sizing NEC Table 240-6 outlines the standard ampere ratings. If the calculated value of a PV array somewhat exceeds one of the standard ratings of this table, the next highest rating should be chosen.

All feeder cables rated for a specific temperature should be derated by 80 percent or the ampacity multiplied by 1.25. Cable ratings for 60, 75, and 90°C are listed in NEC Tables 310.16 and 310.17. For derating purposes it is recommended that cables rated for 75°C ampacity should use 90°C column values. Various device terminals, such as terminal block overcurrent devices must also have the same insulation rating as the cables. In other words, if the device is in a location that is exposed to a higher temperature than the rating of the feeder cable, the cable must be further derated to match the terminal connection device. The following example is used to illustrate these design parameter considerations.

A wiring design example   Assuming that the short-circuit current Isc from a PV array is determined to be 40 A, the calculation should be as follows:

1 PV array current derating = $40 \times 1.25 = 50$ A.
2 Overcurrent device fuse rating at 75°C = $50 \times 1.25 = 62.5$ A.
3 Cable derating at 75°C = $50 \times 1.25 = 62.5$. Using NEC Table 310-16, under the 75°C columns we find a cable AWG #6 conductor that is rated for 65-A capacity. Because of ultraviolet (UV) exposure, XHHW-2 or USE-2 type cable, which has a 75-A capacity, should be chosen. Incidentally, the "–2" is used to designate UV exposure protection. If the conduit carrying the cable is populated or filled with four to six conductors, it is suggested, as previously, by referring to NEC Table 310-15(B)(2)(a), that the conductors be further derated by 80 percent. At an ambient temperature of 40 to 45°C a derating multiplier of 0.87 is also to be applied: $75 \text{ A} \times 0.87 = 52.2$ A. Since the AWG #6 conductor chosen with an ampacity of 60 is capable of meeting the demand, it is found to be an appropriate choice.
4 By the same criteria the closest overcurrent device, as shown in NEC Table 240.6, is 60 A; however, since in step 2 the overcurrent device required is 62.5 A, the AWG #6 cable cannot meet the rating requirement. As such, an AWG #4 conductor must be used. The chosen AWG #4 conductor under the 75°C column of Table 310-16 shows an ampacity of 95.

If we choose an AWG #4 conductor and apply conduit fill and temperature derating, then the resulting ampacity is $95 \times 0.8 \times 0.87 = 66$ A; therefore, the required fuse per NEC Table 240-6 will be 70 A.

Conductors that are suitable for solar exposure are listed as THW-2, USE-2, and THWN-2 or XHHW-2. All outdoor installed conduits and wireways are considered to be operating in wet, damp, and UV-exposed conditions. As such, conduits should be capable of withstanding these environmental conditions and are required to be of a thick wall type such as rigid galvanized steel (RGS), intermediate metal conduit (IMC), thin wall electrical metallic (EMT), or schedule 40 or 80 polyvinyl chloride (PVC) nonmetallic conduits.

For interior wiring, where the cables are not subjected to physical abuse, special NEC code approved wires must be used. Care must be taken to avoid installation of underrated cables within interior locations such as attics where the ambient temperature can exceed the cable rating.

Conductors carrying dc current are required to use color coding recommendations as stipulated in Article 690 of the NEC. Red wire or any color other than green and white is used for positive conductors, white for negative, green for equipment grounding, and bare copper wire for grounding. The NEC allows nonwhite grounded wires, such as USE-2 and UF-2, that are sized #6 or above to be identified with a white tape or marker.

As mentioned earlier, all PV array frames, collector panels, disconnect switches, inverters, and metallic enclosures should be connected together and grounded at a single service grounding point.

PHOTOVOLTAIC SYSTEM GROUND-FAULT PROTECTION

When a photovoltaic system is mounted on the roof of a residential dwelling, NEC requirements dictate the installation of ground-fault protection (detection and interrupting) devices (GFPD). However, ground-mounted systems are not required to have the same protection since most grid-connected system inverters incorporate the required GFPDs.

Ground-fault detection and interruption circuitry perform ground-fault current detection, fault current isolation, and solar power load isolation by shutting down the inverter. This technology is currently going through a developmental process, and it is expected to become a mandatory requirement in future installations.

PV SYSTEM GROUNDING

Photovoltaic power systems that have an output of 50 V dc under open-circuit conditions are required to have one of the current-carrying conductors grounded. In electrical engineering, the terminologies used for grounding are somewhat convoluted and confusing. In order to differentiate various grounding appellations it would be helpful to review the following terminologies as defined in NEC Articles 100 and 250.

Grounded. Means that a conductor connects to the metallic enclosure of an electrical device housing that serves as earth.

Grounded conductor. A conductor that is intentionally grounded. In PV systems it is usually the negative of the dc output for a two-wire system or the center-tapped conductor of an earlier bipolar solar power array technology.

Equipment grounding conductor. A conductor that normally does not carry current and is generally a bare copper wire that may also have a green insulator cover. The conductor is usually connected to an equipment chassis or a metallic enclosure that provides a dc conduction path to a ground electrode when metal parts are accidentally energized.

Grounding electrode conductor. A conductor that connects the grounded conductors to a system grounding electrode, which is usually located only in a single location within the project site, and does not carry current. In the event of the accidental shorting of equipment the current is directed to the ground, which facilitates actuation of ground-fault devices.

Grounding electrode. A grounding rod, a concrete-encased ultrafiltration rate (UFR) conductor, a grounding plate, or simply a structural steel member to which a grounding electrode conductor is connected. As per the NEC all PV systems—whether grid-connected or stand-alone, in order to reduce the effects of lightning and provide a measure of personnel safety—are required to be equipped with an adequate grounding system. Incidentally, grounding of PV systems substantially reduces radio-frequency noise generated by inverter equipment.

In general, grounding conductors that connect the PV module and enclosure frames to the ground electrode are required to carry full short-circuited current to the ground; as such, they should be sized adequately for this purpose. As a rule, grounding conductors larger than AWG #4 are permitted to be installed or attached without special protection measures against physical damage. However, smaller conductors are required to be installed within a protective conduit or raceway. As mentioned earlier, all ground electrode conductors are required to be connected to a single grounding electrode or a grounding bus.

EQUIPMENT GROUNDING

Metallic enclosures, junction boxes, disconnect switches, and equipment used in the entire solar power system, which could be accidentally energized are required to be grounded. NEC Articles 690, 250, and 720 describe specific grounding requirements. NEC Table 25.11 provides equipment grounding conductor sizes. Equipment grounding conductors similar to regular wires are required to provide 25 percent extra ground current-carrying capacity and are sized by multiplying the calculated ground current value by 125 percent. The conductors must also be oversized for voltage drops as defined in NEC Article 250.122(B).

In some installations bare copper grounding conductors are attached along the railings that support the PV modules. In installations where PV current-carrying conductors are routed through metallic conduits, separate grounding conductors could be eliminated since the metallic conduits are considered to provide proper grounding when adequately coupled. It is, however, important to test conduit conductivity to ensure that there are no conduction path abnormalities or unacceptable resistance values.

Entrance Service Considerations for Grid-Connected Solar Power Systems

When integrating a solar power cogeneration within an existing or new switchgear, it is of the utmost importance to review NEC 690 articles related to switchgear bus capacity.

As a rule, when calculating switchgear or any other power distribution system bus ampacity, the total current-bearing capacity of the bus bars is not allowed to be loaded more than 80 percent of the manufacturer's equipment nameplate rating. In other words, a bus rated at 600 A cannot be allowed to carry a current burden of more than 480 A.

When integrating a solar power system with the main service distribution switchgear, the total bus current-bearing capacity must be augmented by the same amount as the current output capacity of the solar system. For example, if we were to add a 200-A solar power cogeneration to the switchgear, the bus rating of the switchgear must in fact be augmented by an extra 250 A. The additional 50 A represents an 80 percent safety margin for the solar power output current. Therefore, the service entrance switchgear bus must be changed from 600 to 1000 A or at a minimum to 800 A.

As suggested earlier, the design engineer must be fully familiar with the NEC 690 articles related to solar power design and ensure that solar power cogeneration system electrical design documents become an integral part of the electrical plan check submittal documents.

The integrated solar power cogeneration electrical documents must incorporate the solar power system components such as the PV array systems, solar collector distribution panels, overcurrent protection devices, inverters, isolation transformers, fused service disconnect switches, and net metering within the plans and must be considered as part of the basic electrical system design.

Electrical plans should incorporate the solar power system configuration in the electrical single-line diagrams, panel schedule, and demand load calculations. All exposed, concealed, and underground conduits must also be reflected on the plans with distinct design symbols and identification that segregate the regular and solar power system from the electrical systems.

Note that the solar power cogeneration and electrical grounding should be in a single location, preferably connected to a specially designed grounding bus, which must be located within the vicinity of the main service switchgear.

Lightning Protection

In geographic locations, such as Florida, where lightning is a common occurrence, the entire PV system and outdoor-mounted equipment must be protected with appropriate lightning arrestor devices and special grounding that could provide a practical mitigation and a measure of protection from equipment damage and burnout.

LIGHTNING'S EFFECT ON OUTDOOR EQUIPMENT

Lightning surges are comprised of two elements, namely voltage and the quantity of charge delivered by lightning. The high voltage delivered by lightning surges can cause serious damage to equipment since it can break down the insulation that isolates circuit elements and the equipment chassis. The nature and the amount of damage are directly proportional to the amount of current resulting from the charge.

In order to protect equipment damage from lightning, devices known as surge protectors or arrestors are deployed. The main function of a surge arrestor is to provide a direct conduction path for lightning charges to divert them from the exposed equipment chassis to the ground. A good surge protector must be able to conduct a sufficient current charge from the stricken location and lower the surge voltage to a safe level quickly enough to prevent insulation breakdown or damage.

In most instances all circuits have a capacity to withstand certain levels of high voltages for a short time; however, the thresholds are so narrow that if charges are not removed or isolated in time, the circuits will sustain an irreparable insulation breakdown.

The main purpose of a surge arrestor device is, therefore, to conduct the maximum amount of charge and reduce the voltage in the shortest possible time. Reduction of a voltage surge is referred to as clamping, shown in Figures 3.12 and 3.13. Voltage clamping in

(a)

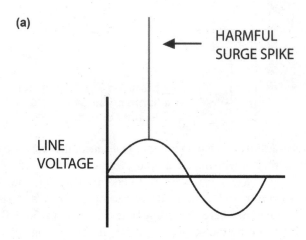

HARMFUL
SURGE SPIKE

LINE
VOLTAGE

(b) SURGE SPIKE WITH DELTA SURGE
ARRESTOR/SUPRESSOR

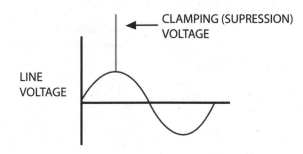

CLAMPING (SUPRESSION)
VOLTAGE

LINE
VOLTAGE

(c) SURGE SPIKE WITH
DELTA SURGE CAPACITOR

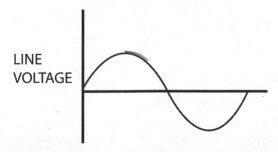

LINE
VOLTAGE

Figure 3.12 Effect of
(a) lightning surge spike
(b) lightning surge spike
clamping, and (c) lightning
surge spike suppression.
Courtesy of Delta Surge Arrestor.

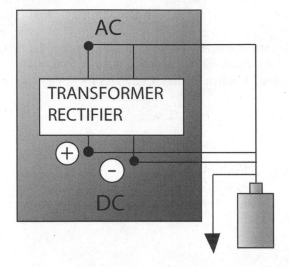

Figure 3.13 Deployment of a lightning surge arrestor in a rectifier circuit.

general depends on device characteristics such as internal resistance, the response speed of the arrestor, and the point in time at which the clamping voltage is measured.

When specifying a lightning arrestor, it is necessary to take into account the clamping voltage and the amount of current to be clamped, for example, 500 V and 1000 A. Let us consider a real-life situation where the surge rises from 0 to 50,000 V in 5 nanoseconds (ns). At any time during the surge, say at 100 ns, the voltage clamping would be different from say the lapsed time, at 20 ns, where the voltage could have been 25,000 V; nevertheless, the voltage will be arrested, since high current rating will cause adequate conductivity which will remove the surge current from the circuit rapidly and will therefore provide better protection.

The following is a specification for a Delta lightning arrestor rated for 2300 V and designed for secondary service power equipment such as motors, electrical panels, transformers, and solar power cogeneration systems.

Model 2301–2300 series specification

Type of design: silicone oxide varistor

Maximum current capacity: 100,000 A

Maximum energy dissipations: 3000 J per pole

Maximum time of 1-mA test: 5 ns

Maximum number of surges: unlimited

Response time to clamp 10,000 A: 10 ns

Response time to clamp 25,000 A: 25 ns

Leak current at double the rated voltage: none

Case material: PVC

Central Monitoring and Logging System Requirements

In large commercial solar power cogeneration systems, power production from the PV arrays is monitored by a central monitoring system that provides a log of operation performance parameters. The central monitoring station consists of a PC-type computer that retrieves operational parameters from a group of solar power inverters by means of an RS-232 interface, a power line carrier, or wireless communication system. Upon receipt of performance parameters, a supervisory software program processes the information and provides data in display or print format. Supervisory data obtained from the file can also be accessed from distant locations through Web networking.

Some examples of monitored data are:

- Weather-monitoring data
- Temperature
- Wind velocity and direction
- Solar power output
- Inverter output
- Total system performance and malfunction
- Direct-current power production
- Alternating-current power production
- Accumulated, daily, monthly, and yearly power production

The following is an example of a data acquisition system by Heliotronics referred to as Sunlogger, which has been specifically developed to monitor and display solar power cogeneration parameters.

SUN VIEWER DATA ACQUISITION SYSTEM

The following solar power monitoring system by Heliotronics, called the Sun Viewer, is an example of an integrated data acquisition system that has been designed to acquire and display real-time performance parameters by filed installed electric power and atmospheric measurement sensors. The system, in addition to providing vital system performance data monitoring and measurement, provides the means to view instantaneous real-time and historical statistical energy measurement data essential for system performance evaluation, research, and education. Figure 3.14 depicts a typical presentation of field measurement on a display monitor.

The system hardware configuration of the Sun Viewer consists of a desk to computer based data logging software that processes and displays measured solar power photovoltaic array and atmospheric output parameters from the following sensors and equipment:

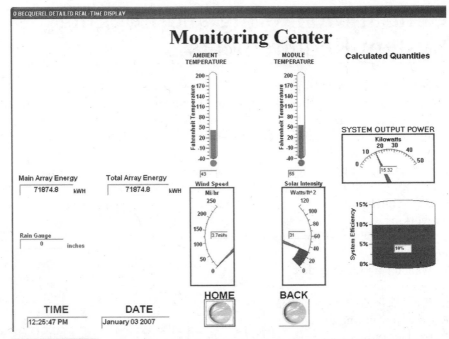

Figure 3.14   **Depiction of a typical presentation of field measurement on a display monitor.** *Graphics courtesy of Heliotronics.*

Meteorological data measurements An anemometer is a meteorological instrument that provides the following meteorological measurements:

■ Ambient air temperature sensor
■ Wind speed
■ Outdoor air temperature sensor
■ Pyrometer for measuring solar insolation

Photovoltaic power output performance measurement sensors

■ AC current and voltage transducer
■ DC current and voltage transducer
■ Kilowatt-hour meter transducer
■ Optically isolated RS-422 or RS-232C modem

Sun Viewer display and Sun Server monitoring software The Sun Viewer display and Sun Server monitoring software provide acquisition and display of real-time data every second and display the following on a variety of display monitors:

■ DC current
■ DC voltage

- AC current
- AC voltage
- AC kilowatt-hours
- Solar plane of array irradiance
- Ambient temperature
- Wind speed

The calculated parameters displayed include:

- AC power output
- Sunlight conversion efficiency to ac power
- Sunlight conversion efficiency to dc power
- Inverter dc-to-ac power conversion efficiency
- Avoided pollutant emissions of CO_2, SO_x, NO_x gases

The preceding information and calculated parameter are displayed on monitors and updated once every second. The data are also averaged every 15 minutes and stored in a locally accessible database. The software includes a "Virtual Array Tour" that allows observers to analyze the solar photovoltaic component of the photovoltaic array and monitoring system. It also provides an optional portal Web capability whereby the displayed data could be monitored from a remote distance over the Internet.

The monitoring and display software can also be customized to incorporate descriptive text, photographs, schematic diagrams, and user-specific data. Some of the graphing capabilities of the system include the following:

- Average plots of irradiance, ambient temperature, and module temperature that are updated every 15 minutes and averaged over one day.
- Daily values or totals of daily energy production, peak daily power, peak daily module temperature, and peak daily irradiance plotted over a specified month.
- Monthly values of energy production; incident solar irradiance; and avoided emission of CO_2, NO_x, and SO_x plotted over a specified year.

GENERAL DESCRIPTION OF A MONITORING SYSTEM

The central monitoring system discussed here reflects the actual configuration of the Water and Life Museum project, located in Hemet, California, and was designed by the author. This state-of-the-art monitoring system provides a real-time interactive display for education and understanding of photovoltaics and the solar electric installation as well as monitoring of the solar electric system for maintenance and troubleshooting purposes.

The system is made up of wireless inverter data transmitters, a weather station, a data storage computer, and a data display computer with a 26-in LCD screen. In the Water and Life Museum project configuration, the inverters, which are connected in parallel, output data to wireless transmitters located in their close proximity. Wireless transmitters throughout the site transmit data to a single central receiver located in a central data gathering and monitoring center. The received data are stored and analyzed using the sophisticated software in computer-based supervisory systems that also serve as a data-maintenance interface for the solar power system. A weather station also transmits weather-related information to the central computer.

The stored data are analyzed and forwarded to a display computer that is used for data presentation and storing information, such as video, sound, pictures, and text file data.

DISPLAYED INFORMATION

A standard display will usually incorporate a looping background of pictures from the site, graphical overlays of the power generation in watts and watt-hours for each building, and the environmental impact from the solar system. The display also shows current meteorological conditions.

Displayed data in general should include the following combination of items:

Project location (on globe coordinates—zoom in and out)

Current and historic weather conditions

Current positions of the sun and moon, with the date and time

Power generation from the total system and/or the individual solar power arrays

Historic power generation

Solar system environmental impact

Looping background solar system photos and videos

Educational PowerPoint presentations

Installed solar electric power overview

Display of renewable energy system environmental impact statistics

The display should also be programmed to periodically show additional information related to the building's energy management or the schedule of maintenance relevant to the project:

Weather station transmitted data. Transmitted data from the weather monitoring station should include air temperature, solar cell temperature, wind speed, wind direction, and sun intensity measured using a pyrometer.

Inverter monitoring transmitted data. Each inverter must incorporate a watt-hour transducer that will measure dc and ac voltage, current, and power; ac frequency; watt-hour accumulation; and inverter error codes and operation.

Typical central monitoring computer. The central supervising system must be configured with a CPU with a minimum of 3 GHz of processing power, 512 kilobytes of random-access memory (RAM), and a 60-gigabyte (Gbyte) hard drive. The operating system should preferably be based on Windows XP or an equivalent system operating software platform.

Wireless transmission system specification. Data communication system hardware must be based upon a switch selectable RS-232/422/485 communication transmission protocol, have a software selectable data transmission speed of 1200 to 57,600 bits/s, and be designed to have several hop sequences share multiple frequencies. The system must also be capable of frequency hopping from 902 to 928 MHz on the FM bandwidth and be capable of providing transparent multipoint drops.

ANIMATED VIDEO AND INTERACTIVE PROGRAMMING REQUIREMENTS

A graphical program builder must be capable of animated video and interactive programming and have an interactive animation display feature for customizing the measurements listed earlier. The system must also be capable of displaying various customizable chart attributes, such as labels, trace color and thickness, axis scale, limits, and ticks. The interactive display monitor should preferably have a 30- to 42-in LCD or light-emitting diode (LED) flat monitor and a 17- to 24-in touch screen display system.

Ground-Mount Photovoltaic Module Installation and Support Hardware

Ground-mount outdoor photovoltaic array installations can be configured in a wide variety of ways. The most important factor when installing solar power modules is the PV module orientation and panel incline. A ground-mount solar power installation is shown in Figure 3.15.

In general, the maximum power from a PV module is obtained when the angle of solar rays impinge directly perpendicular (at a 90-degree angle) to the surface of the panels. Since solar ray angles vary seasonally throughout the year, the optimum average tilt angle for obtaining the maximum output power is approximately the local latitude minus 9 or 10 degrees (see Appendix B for typical PV support platforms and hardware and Appendix A for tilt angle installations for the following cities in California: Los Angeles, Daggett, Santa Monica, Fresno, and San Diego).

In the northern hemisphere, PV modules are mounted in a north-south tilt (high end north) and in the southern hemisphere, in a south-north tilt. Appendix A also includes U.S. and world geographic location longitudes and latitudes.

To attain the required angle, solar panels are generally secured on tilted prefabricated or field-constructed frames that use rustproof railings, such as galvanized Unistrut or commercially available aluminum or stainless-steel angle channels, and

Figure 3.15   A typical ground mount solar power installation used in solar farms. *Courtesy of UniRac.*

fastening hardware, such as nuts, bolts, and washers. Prefabricated solar power support systems are also available from UniRac and several other manufacturers.

When installing solar support pedestals, also known as stanchions, attention must be paid to structural design requirements. Solar power stanchions and pedestals must be designed by a qualified, registered professional engineer. Solar support structures must take into consideration prevailing geographic and atmospheric conditions, such as maximum wind gusts, flood conditions, and soil erosion.

A typical ground-mount solar power installation includes agricultural grounds; parks and outdoor recreational facilities; carports; sanitariums; and large commercial solar power-generating facilities, also known as solar farms (see Figure 3.15). Most solar farms are owned and operated by electric energy-generating entities such as Edison. Prior to the installation of a solar power system, structural and electrical plans must be reviewed by local electrical service authorities, such as building and safety departments. Solar power installation must be undertaken by a qualified licensed electrical contractor with special expertise in solar power installations.

A solar mounting support system profile, shown in Figure 3.16, consists of a galvanized Unistrut railing frame that is field-assembled with standard commercially available manufactured components used in the construction industry. Basic frame components in general include a 2-in galvanized Unistrut channel, 90-degree and T-type connectors, spring-type channel nuts and bolts, and panel hold-down T-type or fender washers.

Figure 3.16   **A single PV frame ground-mount solar power support.**
Courtesy of UniRac.

The main frame that supports the PV modules is welded or bolted to a set of galvanized rigid metal round pipes or square channels. The foundation support is built from 12- to 18-in-diameter reinforced concrete cast in a sauna tube. Then, the metal support structure is secured to the concrete footing by means of expansion bolts. The depth of the footing and dimensions of channel hardware and method of PV module frame attachment are designed by a qualified structural engineer.

A typical solar power support structural design should withstand wind gusts from 80 to 120 miles per hour (mi/h). Prefabricated structures that are specifically designed for solar power applications are available from a number of manufacturers. Prefabricated solar power support structures, although somewhat more expensive, are usually designed to withstand 120-mi/h wind gusts and are manufactured from stainless steel, aluminum, or galvanized steel materials.

Roof-Mount Installations

Roof-mount solar power installations are made of either tilted or flat-type roof support structures or a combination of both. Installation hardware and methodologies also differ depending on whether the building already exists or is a new construction. Roof attachment hardware material also varies for wood-based and concrete constructions.

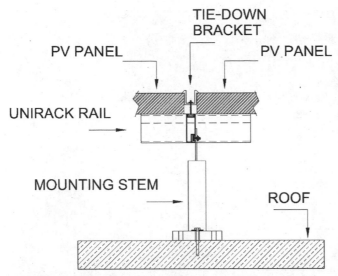

Figure 3.17   **Typical roof-mount solar power installation detail.** *Courtesy of Vector Delta Design Group.*

Figure 3.17 depicts a prefabricated PV module support railing system used for roof-mount installations.

WOOD-CONSTRUCTED ROOFING

In new constructions, the PV module support system installation is relatively simple since locations of solar array frame pods, which are usually secured on roof rafters, can be readily identified. Prefabricated roof-mount stands that support railings and associated hardware, such as fasteners, are commercially available from a number of manufacturers. Solar power support platforms are specifically designed to meet physical configuration requirements for various types of PV module manufacturers.

Some types of PV module installation, such as in Figure 3.18 and 3.19, have been designed for direct mounting on roof framing rafters without the use of specialty railing or support hardware. As mentioned earlier, when installing roof-mount solar panels, care must be taken to meet the proper directional tilt requirement. Another important factor to be considered is that solar power installations, whether ground or roof mounted, should be located in areas free of shade caused by adjacent buildings, trees, or air-conditioning equipment. In the event of unavoidable shading situations, the solar power PV module location, tilt angle, and stanchion separations should be analyzed to prevent cross shading. Figure 3.20 depicts a prefabricated PV module support railing for roof-mount by UniRac.

LIGHTWEIGHT CONCRETE-TYPE ROOFING

Solar power installation PV module support systems for concrete roofs are configured from prefabricated support stands and railing systems similar to the ones used in wooden

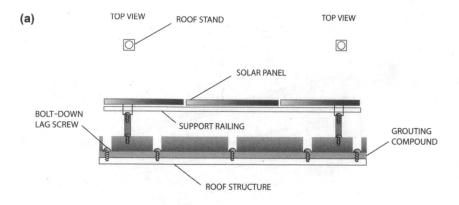

(a)

TOP VIEW   ROOF STAND

TOP VIEW

SOLAR PANEL

BOLT-DOWN
LAG SCREW

SUPPORT RAILING

GROUTING
COMPOUND

ROOF STRUCTURE

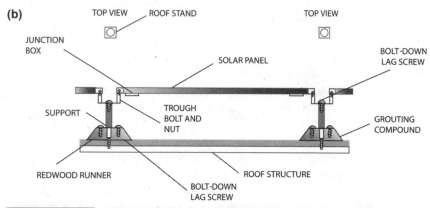

(b)

TOP VIEW   ROOF STAND

TOP VIEW

JUNCTION
BOX

SOLAR PANEL

BOLT-DOWN
LAG SCREW

SUPPORT

TROUGH
BOLT AND
NUT

GROUTING
COMPOUND

REDWOOD RUNNER

ROOF STRUCTURE

BOLT-DOWN
LAG SCREW

Figure 3.18   Typical roof-mount solar power railing installation
detail. (a) Side view. (b) Front view. *Courtesy of Vector Delta Design Group.*

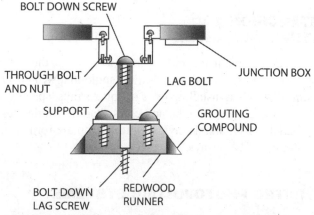

BOLT DOWN SCREW

THROUGH BOLT
AND NUT

JUNCTION BOX

LAG BOLT

SUPPORT

GROUTING
COMPOUND

BOLT DOWN
LAG SCREW

REDWOOD
RUNNER

Figure 3.19   Typical roof mount solar power
railing installation penetration detail. *Courtesy of Vector
Delta design Group.*

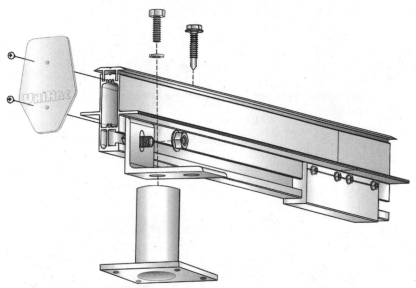

Figure 3.20   Prefabricated PV module support railing for roof-mount system. *Courtesy of UniRac.*

roof structures. Stanchions are anchored to the roof by means of rust-resistant expansion anchors and fasteners.

In order to prevent water leakage resulting from roof penetration, both wood and concrete standoff support pipe anchors are thoroughly sealed with waterproofing compounds. Each standoff support is fitted with thermoplastic boots that are in turn thermally welded to roof cover material, such as single-ply PVC. Figure 3.21 depicts a wood roof-mount standoff support railing system assembly detail.

PHOTOVOLTAIC STANCHION AND SUPPORT STRUCTURE TILT ANGLE

As discussed earlier, in order to obtain the maximum output from the solar power systems, PV modules or arrays must have an optimum tilt angle that will ensure a perpendicular exposure to sun rays. When installing rows of solar arrays, spacing between stanchions must be such that there should not be any cross shading. In the design of a solar power system, the available roof area is divided into a template format that compartmentalizes rows or columns of PV arrays.

BUILDING—INTEGRATED PHOTOVOLTAIC SYSTEMS

A custom-designed and manufactured photovoltaic module is called a building-integrated photovoltaic module (BIPV) shown in Figure 2.17. This type of solar panel is constructed by laminating individual solar cells in a desired configuration, specifically

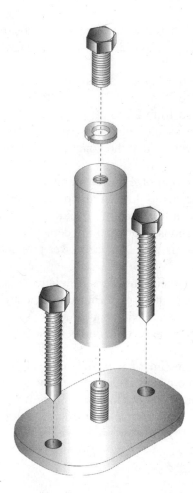

Figure 3.21   **Wood roof-mount standoff support railing system assembly detail.** *Courtesy of UniRac.*

designed to achieve a special visual effect, and are typically deployed in a solarium or trellis type of structure. Because of the separation gap between the adjacent cells, BIPV modules when compared to standard PV modules produce less energy per square foot of area.

Under operational conditions, when solar power systems actively generate power, a line carrying current at several hundred volts could pose serious burns or bodily injury and electric shock if exposed during the roof demolition process. To prevent injury under fire hazard conditions, all roof-mount equipment that can be accessed must be clearly identified with large red-on-white labels. Additionally, the input to the inverter from the PV collector boxes must be equipped with a crowbar disconnect switch that will short the output of all solar arrays simultaneously.

Solar Tracking Systems   Tracking systems are support platforms that orient solar photovoltaic module assemblies by keeping track of the sun's movement from dusk to dawn, thus maximizing solar energy power generation efficiency. Trackers are classified as passive or active and may be constructed to track in single or dual axis.

Single-axis trackers usually have a single-axis tilt movement, whereas dual-axis trackers also move in regular intervals adjusting for an angular position as shown in Figure 3.23 through 3.27.

In general single-axis trackers compared to fixed stationary tilted PV support systems increase solar power capture by about 20 to 25 percent. Dual-axis trackers on the other hand can increase solar power production from 30 to 40 percent. Solar power concentrators which use Fresnel lenses to focus the sun's energy on a solar cell, require a high degree of tracking accuracy to ensure that the concentrated sunlight is focused precisely on the PV cell.

Fixed-axis systems orient the PV modules to optimize power production for a limited time performance and generally have a relatively low annual power production. On the other hand, single-axis trackers, although less accurate than dual-axis tracker applications, produce strong power in the afternoon hours and are deployed in applications such as grid-connected solar power farms that enhance power production in the morning and afternoon hours.

Compared to the overall cost of photovoltaic systems, trackers are relatively inexpensive devices that significantly increase the power output performance efficiency of the PV panels. Even though some tracker systems operate with some degree of reliability, they usually require seasonal position adjustments, inspection, and periodic lubrication.

Basic physics of solar intensity   The amount of solar intensity of light that impinges upon the surface of solar photovoltaic panels is determined by an equation referred to as Lambert's cosine law, which states that the intensity of light (I) falling on a plane is directly proportional to the cosine of the angle (A) made by the direction of the light source to the normal of the plane:

$$I = k \times \cos A$$

where k is Lambert's constant. This equation is depicted in Figure 3.22. In other words, during the summer when the angle of the sun is directly overhead, the magnitude of intensity is at its highest, since the cosine of the angle is zero; therefore, $\cos 0 = 1$, which implies $I = k$.

The main objective of all solar trackers is to minimize the value of the cosine angle and maximize the solar intensity on the PV planes.

Polar trackers   Polar trackers are designed to have one axis rotate in the same pattern as Earth, hence the name. Essentially polar trackers are in general aligned perpendicular to an imaginary ecliptic disc that represents the apparent mathematical path of the sun. To maintain relative accuracy, these types of tracker are manually adjusted to compensate for the seasonal ecliptic shifts that occur with the seasons. Polar trackers are usually used in astronomical telescope mounts where high-accuracy solar tracking is an absolute requirement.

Horizontal—axle trackers   Horizontal trackers are designed to orient a horizontal axle by either passive or active mechanisms. Essentially a long tubular axle is supported

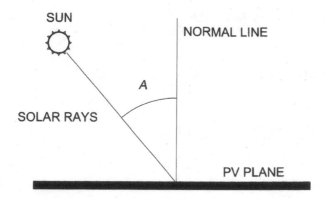

SUN

NORMAL LINE

A

SOLAR RAYS

PV PLANE

$I = k \cos A$

I = SOLAR INTENSITY

k = LAMBERT'S CONSTANT

A = SOLAR ANGLE

Figure 3.22   Solar intensity equation diagram.

by several bearings that are secured to some type of wooden, metallic, or concrete pylon structure frame. The tubular axles are installed in a north-south orientation, whereas PV panels are mounted on the tubular axle that rotates on an east-west axis and tracks the apparent motion of the sun throughout daylight hours. Note that single-axis trackers do not tilt toward the equator. As a result their power tracking efficiency is significantly reduced in midwinter; however, their productivity increases substantially during the spring and summer seasons when the sun path is directly overhead in the sky. Because of the simplicity of their mechanism, horizontal-axle single-axis trackers are considered to be very reliable, easy to clean and maintain, and not subject to self-shading.

Passive trackers   The rotational mechanism of a passive tracker is based on the use of low-boiling-point compressed gas fluids that are moved or displaced from the east to the west side by solar heat that converts the liquid to gas causing the tracker to tilt from one side to another. The imbalance created by the movement of the liquid-gas material creates the fundamental principle of bidirectional movement. Note that various climatic conditions such as temperature fluctuations, wind gusts, and solar clouding adversely affect the performance of passive solar trackers. As such they are considered to have unreliable tracking efficiency; however, they do provide better solar output performance capability than fixed-angle solar support platforms. Figure 3.23 depicts Zomeworks passive solar tracker component diagram.

One of the major passive solar tracker manufacturers is Zomeworks, which manufactures a series of tracking devices called Track Rack. Tracking devices begin tracking the sun by facing the racks westward. As the sun rises in the east, it heats an unshielded west side liquid-gas filled canister, forcing the liquid into the shaded east-side canister. As the liquid moves through a copper tube to the east-side canister, the tracker rotates so that it faces east.

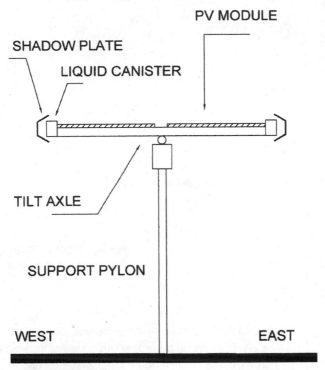

Figure 3.23   Zomeworks passive solar tracker.

The heating of the liquid is controlled by aluminum shadow plates. When one of the canisters is exposed to the sun more than the other, its vapor pressure is increased, hence forcing the liquid to the cooler, shaded side. The shifting weight of the liquid causes the rack to rotate until the canisters are equally shaded. Figure 3.24 depicts Solar tracker eastern sunrise position of Zomeworks passive solar tracker.

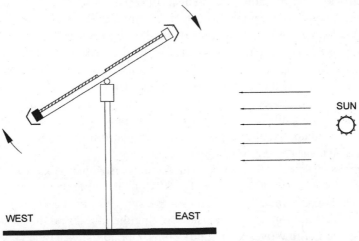

Figure 3.24   Solar tracker eastern sunrise position.

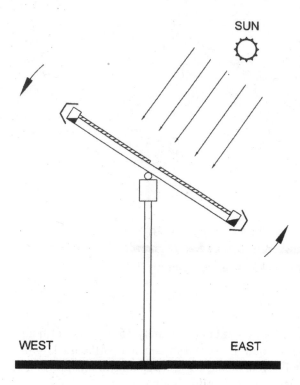

Figure 3.25   Sunrise shifting position of tracker.

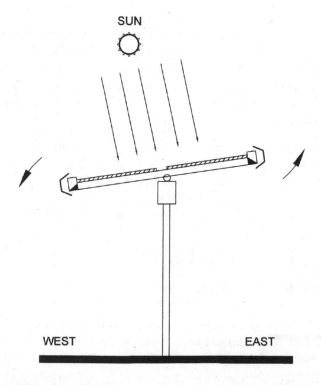

Figure 3.26   Liquid movement shifting position of tracker.

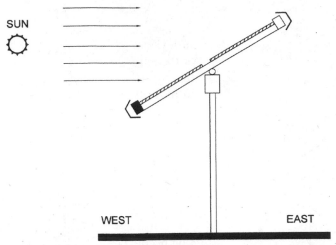

Figure 3.27   Position of tracker after completing daily cycle.

As the sun moves, the rack follows at approximately 15 degrees per hour continually seeking equilibrium as the liquid moves from one side of the track to the other.

The rack completes its daily cycle facing west. It remains in this position overnight until it is awakened by the rising sun the following morning.

Active trackers   Active trackers use motors and gear trains to control axle movements by means of programmable controlled timers, programmable logic controllers, and microprocessor-based controllers or global-positioning-based control devices that provide precise power drive data to a variety of electromechanical movement mechanisms. Programs within the control computational systems use a combination of solar movement algorithms that adjust rotational axis movement in orientations that constantly maintain a minimal cosine angle throughout all seasons.

Vertical-axle trackers   Vertical-axle trackers are constructed in such a manner as to allow pivotal movement of PV panels mounted about a vertical axis. These types of trackers have a limited use and are usually deployed in high latitudes, where the solar path travels in a long arc. PV panels mounted on a vertical-axle system are suitable for operation during long summer days in northern territories, which have extended solar days.

Electric Shock Hazard and Safety Considerations

Power arrays, when exposed to the sun, can produce several hundred volts of dc power. Any contact with an exposed or uninsulated component of the PV array can produce serious burns and fatal electric shock. The electrical wiring design and installation methodology are subject to rigorous guidelines, which are outlined in the National Electrical Code (NEC) Article 690.

System components, such as overcurrent devices, breakers, disconnect switches, and enclosures, are specifically rated for the application. All equipment that is subject to maintenance and repair is marked with special caution and safety warning tags to prevent inadvertent exposure to hazards (see Appendix B for typical sign details).

SHOCK HAZARD TO FIREFIGHTERS

An important safety provision, which has been overlooked in the past, is collaborating with local fire departments when designing roof-mount solar power systems on wood structures. In the event of a fire, the possibility of a serious shock hazard to firefighters will exist in instances when roof penetration becomes necessary.

SAFETY INSTRUCTIONS

- Do *not* attempt to service any portion of the PV system unless you understand the electrical operation and are fully qualified to do so.
- Use modules for their intended purpose only. Follow all the module manufacturer's instructions. Do *not* disassemble modules or remove any part installed by the manufacturer.
- Do *not* attempt to open the diode housing or junction box located on the back side of any factory-wired modules.
- Do *not* use modules in systems that can exceed 600 V open circuit.
- Do *not* connect or disconnect a module unless the array string is open or all the modules in the series string are covered with nontransparent material.
- Do *not* install during rainy or windy days.
- Do *not* drop or allow objects to fall on the PV module.
- Do *not* stand or step on modules.
- Do *not* work on PV modules when they are wet. Keep in mind that wet modules when cracked or broken can expose maintenance personnel to very high voltages.
- Do *not* attempt to remove snow or ice from modules.
- Do *not* direct artificially concentrated sunlight on modules.
- Do *not* wear jewelry when working on modules.
- Avoid working alone while performing field inspection or repair.
- Wear suitable eye protection goggles and insulating gloves rated at 1000 V.
- Do *not* touch terminals while modules are exposed to light without wearing electrically insulated gloves.
- Always have a fire extinguisher, a first-aid kit, and a hook or cane available when performing work around energized equipment.
- Do *not* install modules where flammable gases or vapors are present.

Maintenance

In general, solar power system maintenance is minimal, and PV modules often only require a rinse and mopping with mild detergent once or twice a year. They should be visually inspected for cracks, glass damage, and wire or cable damage. A periodic check of the array voltage by a voltmeter may reveal malfunctioning solar modules.

TROUBLESHOOTING

All photovoltaic modules become active and produce electricity when illuminated in the presence of natural solar or high ambient lighting. Solar power equipment should be treated with the same caution and care as regular electric power service. Unlicensed electricians or inexperienced maintenance personnel should not be allowed to work with solar power systems.

In order to determine the functional integrity of a PV module, the output of one module must be compared with that of another under the same field operating conditions. One of the best methods to check module output functionality is to compare the voltage of one module to that of another. A difference of greater than 20 percent or more will indicate a malfunctioning module. Note that the output of a PV module is a function of sunlight and prevailing temperature conditions, and as such, electrical output can fluctuate from one extreme to another.

When electric current and voltage output values of a solar power module are measured, short-circuit current (Isc) and open-circuit voltage (Voc) values must be compared with the manufacturer's product specifications.

To obtain the Isc value a multimeter ampere meter must be placed between the positive and negative output leads shorting the module circuit. To obtain the Voc reading a multimeter voltmeter should simply be placed across the positive and negative leads of the PV module.

For larger current-carrying cables and wires, current measurements must be carried out with a clamping meter. Since current clamping meters do not require circuit opening or line disconnection, different points of the solar arrays can be measured at the same time. An excessive differential reading will be an indication of a malfunctioning array.

Note that problems resulting from module malfunction or failure seldom occur when a PV system is put into operation; rather most malfunctions result from improper connections or loose or corroded terminals. In the event of a damaged connector or wiring, a trained or certified technician should be called upon to perform the repairs. PV modules, which are usually guaranteed for an extended period of time, that are malfunctioning should be sent back to the manufacturer or installer for replacement. *Caution:* Do not to disconnect dc feed cables from the inverters unless the entire solar module is deactivated or covered with a canvas or a nontransparent material.

The following safety warning signs must be permanently secured to solar power system components.

WARNING SIGNS

For a solar installation system:

Electric shock hazard—Do not touch terminals—Terminals on both line and load sides may be energized in open position.

For a switchgear and metering system:

Warning—Electric shock hazard—Do not touch terminals—Terminals on both the line and load side may be energized in the open position.

For pieces of solar power equipment:

Warning—Electric shock hazard—Dangerous voltages and currents—No user-serviceable parts inside—Contact qualified service personnel for assistance.

For battery rooms and containers:

Warning—Electric shock hazard—Dangerous voltages and currents—Explosive gas—No sparks or flames—No smoking—Acid burns—Wear protective clothing when servicing typical solar power system safety warning tags.

Photovoltaic Design Guidelines

When designing solar power generation systems, the designer must pay specific attention to the selection of PV modules, inverters, and installation material and labor expenses, and specifically be mindful of the financial costs of the overall project. The designer must also assume responsibility to assist the end user with rebate procurement documentation. The following are major highlights that must be taken into consideration.

Photovoltaic module design parameters

1 Panel rated power (185, 175, 750 W, etc.)
2 Unit voltage (6, 12, 24, 48 V, etc.)
3 Rated amps
4 Rated voltage
5 Short-circuit amperes
6 Short-circuit current
7 Open-circuit volts
8 Panel width, length, and thickness
9 Panel weight
10 Ease of cell interconnection and wiring

11 Unit protection for polarity reversal
12 Years of warranty by the manufacturer
13 Reliability of technology
14 Efficiency of the cell per unit surface
15 Degradation rate during the expected life span (warranty period) of operation
16 Longevity of the product
17 Number of installations
18 Project references and contacts
19 Product manufacturer's financial viability

Inverter and automatic transfer system

1 Unit conversion efficiency
2 Waveform harmonic distortion
3 Protective relaying features (as referenced earlier)
4 Input and output protection features
5 Service and maintenance availability and cost
6 Output waveform and percent harmonic content
7 Unit synchronization feature with utility power
8 Longevity of the product
9 Number of installations in similar types of application
10 Project references and contacts
11 Product manufacturer's financial viability

Note that solar power installation PV cells and inverters that are subject to the California Energy Commission's rebate must be listed in the commission's eligible list of equipment.

Installation contractor qualification

1 Experience and technical qualifications
2 Years of experience in solar panel installation and maintenance
3 Familiarity with system components
4 Amount of experience with the particular system product
5 Labor pool and number of full-time employees
6 Troubleshooting experience
7 Financial viability
8 Shop location
9 Union affiliation
10 Performance bond and liability insurance amount
11 Previous litigation history
12 Material, labor, overhead, and profit markups
13 Payment schedule
14 Installation warranty for labor and material

4

INTRODUCTION TO SOLAR POWER SYSTEM DESIGN

Essential steps required for solar power systems engineering design include site evaluation, feasibility study, site shading analysis, photovoltaic mapping or configuration analysis, dc-to-ac power conversion calculations, PV module and inverter system selection, and total solar power array electric power calculations.

In previous chapters we reviewed the physics, manufacturing technologies, and design considerations applied to photovoltaic solar power cogeneration systems. This chapter is intended to provide a pragmatic approach for designing solar power systems.

In order to have a holistic understanding of solar power cogeneration systems, designers must have a basic appreciation of insolation concepts, shading analysis, and various design parameters that affect the output performance and efficiency of the overall system. In view of the California Solar Initiative (CSI) and other state rebate programs, which will be discussed in future chapters, the importance of system performance and efficiency form the foundation that determines whether a project becomes financially viable.

Insolation

The amount of energy that is received from the sun rays that strike the surface of our planet is referred to as *insolation (I)*. The amount of energy that reaches the surface of Earth is by and large subject to climatic conditions such as seasonal temperature changes, cloudy conditions, and the angle at which solar rays strike the ground.

Because our planet revolves around the sun in an oval-shaped orbit with its axis tilted at approximately 23.5 degrees, the *solar declination angle (i)* (shown in Figure 4.1) constantly varies throughout the revolution, gradually changing from +23.5 degrees on June 21–22, when Earth's axis is tilted toward the sun, to –23.5 degrees by December 21–22, when Earth's axis is tilted away from the sun. Earth's

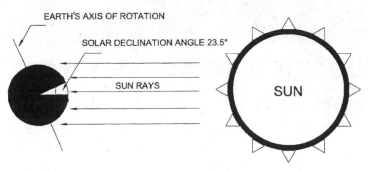

EARTH'S AXIS OF ROTATION

SOLAR DECLINATION ANGLE 23.5°

SUN RAYS

SUN

Figure 4.1   Solar declination angle.

axis at these two seasonal changes, referred to as the summer and winter equinoxes, is 0 degrees.

The solar declinations described in below result from seasonal cyclic variations and solar variations in insolation. For the sake of discussion if we consider Earth as being a sphere of 360 degrees, within a 24-hour period Earth rotates 15 degrees around its axis every hour, commonly referred to as the hour angle. It is the daily rotation of Earth around its axis that gives the notion of sunrise and sunset.

The *hour angle (H)* (shown in Figure 4.2) is the angle through which Earth has rotated since midday or the solar noon. At the noon hour when the sun is exactly above our heads and does not cast any shadow of vertical objects, the hour angle equals 0 degrees.

By knowing the solar declination angle and the hour angle we can apply geometry and find the angle from the observer's zenith point looking at the sun, which is referred to as the *zenith angle (Z)* (shown in Figure 4.3).

The amount of average solar energy striking the surface of Earth is established by measuring the sun's energy rays that impact perpendicular to a square meter area, which is referred to as the *solar constant (S)*. The amount of energy on top of Earth's atmosphere measured by satellite instrumentation is 1366 watts per square meter (W/m^2). Because of the scattering and reflection of solar rays when they enter the atmosphere,

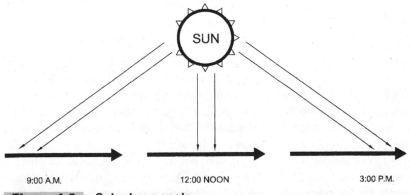

SUN

9:00 A.M.   12:00 NOON   3:00 P.M.

Figure 4.2   Solar hour angle.

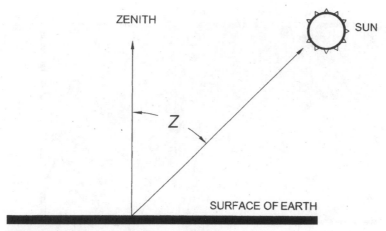

Figure 4.3   Solar zenith angle (Z).

solar energy loses 30 percent of its power; as a result on a clear, sunny day the energy received on Earth's surface is reduced to about 1000 W/m². The net solar energy received on the surface of Earth is also reduced due to cloudy conditions and is also subject to the incoming angle of radiation. Figure 4.3 solar zenith angle (Z) and Figure 4.4 depicts solar declination angle.

Solar insolation is calculated as follows:

$$I = S \times \cos Z$$

where $S = 1000$ W/m²
 $Z = (1/\cos) \times (\sin L \times \sin i + \cos L \times \cos I \times \cos H)$
 $L = $ latitude
 $H = 15° \times (\text{time} - 12)$

Time in the preceding formula is the hour of the day from midnight.

Magnetic Declination   As is commonly known, ordinary magnetic compasses do not normally point to true north. As a matter of fact, over the entire surface of our globe compasses do point at some angles either east or west of true geographic north. The

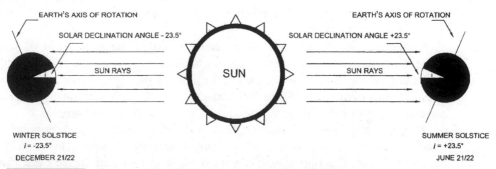

Figure 4.4   Solar declination angle in northern hemisphere.

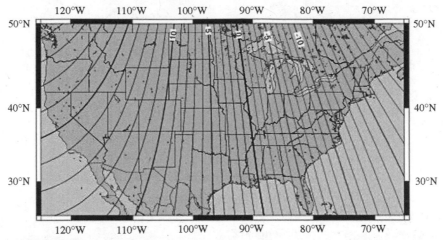

Figure 4.5   **U.S. magnetic declination map.** *Photo courtesy of Solar Pathfinder.*

direction that a compass needle points is referred to as *magnetic north* and the angle between magnetic north and true north is called the *magnetic declination.* They are also sometimes referred to as the variation, magnetic variation, or simply compass variation.

The magnetic declination is not a constant figure and varies in an unpredictable fashion over time. Earth's magnetic field is a result of complex movements of fluid magnetic elements such as iron, nickel, and cobalt that flow in the outer core of Earth, which lies about 2800 to 5000 km below Earth's crust. The poles resulting from the rotation of the molten matter generate a magnetic field that surrounds Earth. The poles of Earth's magnetic field do not coincide with true north and the south axis of rotation of Earth. The resulting changes to the magnetic field are referred to as *secular variation.* Figure 4.5 depicts the U.S. magnetic declination map.

Local magnetic material inconstancies in the upper crust of Earth's mantle or surface also cause unpredictable variations and distortions of the magnetic field. Some of the anomalies of Earth's magnetic field can be attributed to ferromagnetic ore deposits, volcanic lava beds with high concentrations of iron, and human-made structures (such as power lines, pipes, railroad railings, and building metallic structures; small ferrous manufactured items such as iron stoves; and even very small metallic objects like kettles, knifes, and forks). These items can induce a small percentage of error in magnetic field measurements.

Movements of the tectonic plates causes Earth's magnetic field to constantly undergo a gradual displacement, and over millennia cause a reversal in Earth's north and south poles, which is evidenced in the planet's paleomagnetic records.

So when orienting the Solar Pathfinder, keep the instrument away from metallic objects such as air-conditioning equipment, metallic ducts, and pipes.

After the magnetic declination angle from the given Web site is determined, turn the platform that holds the solar shading graph clockwise or anticlockwise accordingly, which is in turn lined up against a marked white dot located on the outer rim of the device. As an optional item Solar Pathfinder also provides software that allows insertion of a topographical site diagram of the solar platform, geographic latitude and longitude, or simply the area zip code. After you have entered the required parameters, the software automatically calculates the mean yearly shading efficiency multiplier.

Shading Analysis and Solar Energy Performance Multiplier

One of the significant steps prior to designing a solar power is to investigate the location of the solar installation platform where the solar PV arrays will be located. In order to harvest the maximum amount of solar energy, theoretically all panels in addition to being mounted at the optimum tilt angle must be totally exposed to sun rays without any shading that may be cast by surrounding building, object, trees, or vegetation.

To achieve the preceding objective, the solar power mounting terrain or the platform must be analyzed for year-round shading. Note that the casting of shadows due to the seasonal rise and fall of the solar angle must be taken into account as it has a significant impact on the shadow direction and surface area. For instance, a shadow cast by a building or tree will vary from month to month changing in length, width, and shape.

In order to analyze yearly shading of a solar platform, solar power designers and integrators make use of a commercial shading analysis instrument called the Solar Pathfinder shown in Figure 4.6. The Solar Pathfinder is used for shade analysis in areas

Figure 4.6   **Solar Pathfinder and shading graphs.**

Courtesy of Solar Pathfinder.

Figure 4.7 **Spherical dome showing the reflection of surrounding buildings and vegetation.** *Photo courtesy of Solar Pathfinder.*

that are surrounded by trees, buildings, and other objects that could cast shadows on
a designated solar platform. The device is essentially comprised of a semispherical
plastic dome shown in Figure 4.7, and latitude-specific yearly solar intensity time
interval semicircular removable or disposable plates shown in Figure 4.8. Figure 4.9
represents latitude map of the United states.

The disposable semicircular plates shown in Figure 4.8 have 12-month imprinted
curvatures that show percentage daily solar energy intensity from sunrise (around 5 a.m.)
to sunset (around 7 p.m.). Each of the solar energy intensity curves from January to
December is demarcated with vertical latitude lines denoting the separation of daily
hours. A percentage number ranging from 1 to 8 percent is placed between adjacent
hourly latitude lines.

Percentage values progress upward at sunrise from a value of 1 percent to a maximum value of 8 percent during midday at 12:00 p.m. and then drop down to 1 percent
at sunset. Depending on the inclination angle of the sun, the percentage solar energy
values depicted on the monthly curvatures vary for each month. For instance, the maximum percentage value for the months of November, December, and January is 8 percent
at solar noon 12:00 p.m. and for the other months from February through October the
percentage value is 7 percent.

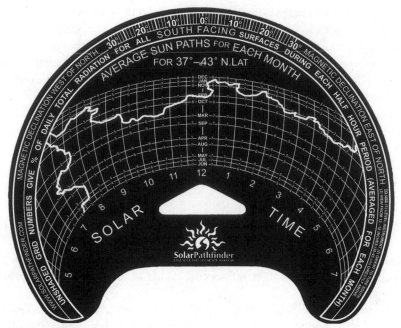

Figure 4.8   **Solar shading graphic insert.** *Photo courtesy of Solar Pathfinder.*

Figure 4.9   **Latitude map of the United States.** *Courtesy of Solar Pathfinder.*

The total sum of percentage points shown on the monthly solar energy curves represent the maximum percent of solar insolation (100 percent) on the platform. For instance, the total energy percent shading multiplier for the month of December or January, or any other month when summed up, totals to a 100 percent multiplier. For example, according to the 31 to 37 north latitude chart, the percent multiplier for the month of December when summed up equals.

$$\text{Efficiency multiplier }\% = 2 + 2 + 3 + 4 + 6 + 7 + 7 + 8 + 8 + 8 + 8 + 7 + 7 + 6 + 5 \\ + 4 + 3 + 2 + 2 + 1 = 100\%$$

The same summation for the month of June equals.

$$\text{Efficiency multiplier }\% = 1 + 1 + 1 + 2 + 2 + 3 + 4 + 5 + 5 + 6 + 6 + 7 + 7 + 7 + 7 \\ + 7 + 6 + 6 + 5 + 5 + 4 + 3 + 2 + 2 + 1 + 1 + 1 = 100\%$$

Note that the insolation angle of the sum increases and decreases for each latitude; hence, each plate is designed to cover specific bands of latitudes for the northern and southern hemispheres.

When the plastic dome is placed on top of the platform that holds the curved solar energy pattern, the shadows cast by the surrounding trees, buildings, and objects are reflected in the plastic dome, clearly showing the shading patterns at the site which are in turn cast on the pattern. The reflected shade on the solar pattern distinctly defines the jagged pattern of shading that covers the plate throughout the 12 months of the year.

A 180-degree opening on the lower side of the dome allows the viewer to mark the shading on the pattern by means of an erasable pen. To determine the total yearly percent shading multiplier, each portion of the 12 monthly curves not exposed to shading are totaled. When taking the mean average percent from all 12 months, a representative solar shading multiplier is derived that is applied in dc-to-ac conversion calculations.

Note that a number of rebate disbursement agencies such as the Los Angeles Department of Water and Power require inclusion of the solar performance multiplier in ac output power calculations and some do not.

USING THE SOLAR PATHFINDER INSTRUMENT SETTING AND MAGNETIC DECLINATION

The Solar Pathfinder dome and shading pattern assembly are mounted on a tripod as shown in Figure 4.10. For leveling purposes, the base of the assembly at its center has a fixed leveling bubble that serves to position the platform assembly on a horizontal level. At the lower part of the platform, which holds the pattern plate, is a fixed compass that indicates the geographic orientation of the unit. The pattern plate is in turn secured to the platform by a raised triangular notch.

To record the shading, place the platform on level ground and adjust the pathfinder for the proper magnetic declination (the deviation angle of the compass needle from

Figure 4.10 The Solar Pathfinder showing the removable shading graph, the leveling bubble at the center and the compass. *Photo courtesy of Solar Pathfinder.*

the true magnetic pole) to orient the device toward the true magnetic pole. Pulling down on the small brass lever allows you to use the center triangle to pivot or rotate the shading pattern toward the proper magnetic declination angle.

Global magnetic declination angle charts are available through magnetic declination Web sites for all countries. U.S. magnetic declination maps can be accessed at www. ga.gov.us/oracle/geomag/agrfform.jsp. For Australia, go to www.ga.gov.au/oracle/ geomag/agrfform.jsp. For Canada, go to www.ga.gov.ca/oracle/geomag/agrfform.jsp

Site Evaluation

Perhaps the most important step when undertaking a solar power design is to thoroughly investigate the site conditions. This section discusses the items for which the solar platform where PV units are to be installed must be evaluated.

Roof-mount solar power systems   For roof-mount systems evaluate the roof-mount obstacles such as heating, ventilating, and air-conditioning (HVAC) or air-handling units, vents, roof-mount chiller, ducts, surrounding trees and buildings, parapets, or any object that may cast a shadow. Solar design engineers must familiarize themselves with the use of the Pathfinder shading evaluation and calculation methods described earlier in this chapter.

Roofs, whether made of wood or concrete, require a structural integrity evaluation. The decision as to the choice of PV support structures must be made to meet specific requirements of the roof structure. In some instances, spacing of wooden roof rafters would necessitate special footing support reinforcement and structural engineering intervention. In other instances, existing old roof-covering material such as asphalt fiber or shingles must be completely replaced since the life expectancy of the PV system installation will significantly outlast the expected life of the roofing material. In order to prevent water penetration, PV support platforms and stanchion anchors must be covered by specially designed waterproof boots.

Regardless of the type of roof structure, a registered professional engineer must evaluate the solar power support structure for roof loading, penetration, and wind shear calculations.

Ground-mount solar power systems   For ground-mount solar power systems (more specifically for solar farm type installations), in addition to the site evaluation measures described previously, the designer must evaluate the site conditions for soil erosion, earthquake fault lines, and periodic floods.

Shading analysis   As discussed earlier in this chapter, sites that are susceptible to shading must be evaluated for the seasonal performance multiplier as per the procedure described in the shading calculation example.

Photovoltaic mapping or configuration analysis   After completing the field evaluation and shading analysis, the solar power designer must construct the topological configuration of the solar power arrays and subarrays in a fashion that would allow maximum harvest of solar energy. Upon choosing the most appropriate or suitable type of PV product, the solar platform footprint must be populated or mapped with the specific dimensional mosaic of the PV modules. Note that the tilting angle of solar arrays must be weighed against the available solar platform footprint. In some instances, the performance efficiency resulting from tilting a PV support structure that casts a shadow on adjacent arrays should be sacrificed for a flat-mounting configuration to increase the total output power generation capacity of the overall solar system.

In some other instances, climatic conditions may dictate the specific PV array tilt angle requirement. For example, in northern territories, to avoid accumulation of snow or ice and to allow natural self-cleaning, PV units must be mounted at the maximum latitude angle. However, in southern states, when summer electric energy tariff charges are high, it may be perhaps advisable to install the PV arrays in a flat configuration since in such a configuration seasonal solar insolation will allow harvesting of the maximum amount of solar energy. In the winter season, when electric energy tariffs are low, lower solar power harvesting may be justified, since there is much less air-conditioning system use, which in some instances represents 50 to 60 percent of the electric energy use. Even though optimum tilting of PV arrays results in superior average yearly energy production for the same number of PV modules, lower efficiency resulting from flat array installation may constitute a reasonable alternative.

DC-to-AC power conversion calculations After completing the preceding steps, the designer must evaluate the PV module's electrical performance parameter and configure PV strings in a manner most appropriate for use with a dc-to-ac inverter system. Note that upon preliminary configuration of the PV arrays and subarrays, the design engineer must coordinate solar power dc and ac wiring details with the inverter manufacturer. In view of specific electrical design performance characteristics of PV modules, inverter manufacturers provide dc input boundary limitations for various types of array configurations.

In general, the maximum allowable dc power voltages produced by a string of PV modules for a specific type of inverter may be limited within a 300- to 600-V dc bandwidth at which the inverter may perform power conversion within safe margins and yield the highest conversion efficiency. The excursion of output voltage produced by PV strings beyond the safe boundaries is determined by Vmp, or the combined series PV string maximum peak voltage when measured in an open-circuit condition. For instance, 11 SolarWorld PV modules (AG SW 175 mono) when connected in series at an average ambient temperature of 90°F (Vmp = 35.7 V) produce a swing voltage of 387 V, which will be within the input voltage boundaries of the inverter. Upon determining the allowed number of series PV string, the designer will be in a position to configure the topology of the PV array and subarrays that would conform to inverter power input requirements.

In most instances inverter manufacturers provide a Web-based solar array power calculator where the designer is allowed to choose the type of inverter power rating, PV module manufacturer, and model number. The preceding data along with the ambient operational temperature, tilt angle, and array derating coefficient figures (as outlined earlier in this chapter) can be inserted into the calculator. The calculations provide accurate inverter string connectivity and ac power output performance results.

The following is an example of a SatCon PV calculator used to determine the allowable string connectivity and power output performance for a 75-kW dc solar power system.

PV Module Specification

PV module	SolarWorld AG SW 175 mono
STC W (standard test conditions)	175 W
CEC W (California Energy Commission test)	162.7 W
Voc	44.4 V
Vmp	35.7 V

Isc (short-circuit current)	5.30 A
Imp (maximum peak current)	4.90 A
Maximum system voltage (V dc)	600 V

Input Assumptions

Inverter model	75 kW, 480 V ac
PV module	SolarWorld AG SW 175 mono
Temperature scale	Fahrenheit
Minimum ambient temperature	25
Maximum ambient temperature	90
Mounting method	Ground mount/tilt
CEC or STC module power	STC (standard test condition)
Optimum array derating coefficient	0.8
Voltage drop in array wiring	1.5

Resulting Design Parameters

Ideal number of series modules (strings)	11
Nominal Vmp w/11 series modules (V dc)	387
Minimum number of modules	11
Maximum number of modules	11
Maximum allowed number of modules	560
Maximum number of series of module strings	48

Inverter Output

Continuous ac power rating	75 kW
AC voltage (V ac line—line)	480 V, 3 phase
Nominal ac output current	91 A
Maximum fault ac output current/phase	115 A
Minimum dc input voltage	330 V dc
Maximum dc input voltage	600 V dc

Peak efficiency	97%
CEC efficiency	96%
Number of subarrays	6

Photovoltaic System Power Output Rating In general when designing a solar power cogeneration system, the designer must have a thorough understanding of PV system characteristics and associated losses when integrated in the array configuration.

Essentially the power output rating of a PV module is the dc rating that appears on the manufacturer's nameplate. For example, SolarWorld SW175 mono is rated at 175 W dc. The dc power output of the PV module is usually listed on the back of the units in watts per square meter or kilowatts per square meter (watts divided by 1000). The rating of the module is established according to international testing criteria referred to as the standard test condition (STC) of 1000 W/m^2 of solar irradiance at 25°C temperature discussed previously.

Another testing standard that is used in the United States is based on the dc rating of the nameplate which is defined as a 1000 W/m^2 square plane of array irradiance at 20°C ambient temperature at a wind speed of 1 m/s, which is referred to as the PVUSA test condition (PTC).

Note that the difference between the PTC and STC is that in the former the ambient temperature and wind speed can result in PV module temperatures of about 50°C as opposed to 25°C for the STC. As a result, under these PTC test conditions a crystalline-based PV module will result in −0.5 percent degradation per each degree Celsius; hence the power rating of silicon-type PV modules is reduced to 88 percent of the nameplate rating.

Note also that energy calculations of photovoltaic systems evaluated by the California Energy Commission and State of Nevada for rebate consideration standards take into account the PTC rating of PV modules and not the dc power output. However, manufacturers always rate their photovoltaic product based on the dc output power.

Photovoltaic System Losses When designing solar power cogeneration systems, the net energy output production must be calculated by taking into consideration losses associated with the totally integrated system. In general, losses occur due to the following design elements and environmental conditions:

- *PV dc nameplate derating*. This is a loss resulting from dc power output from modules that vary from 80 to 105 percent of the manufacturer's nameplate rating. Such losses may result from solar cell physical dimensions, interconnecting cell solder path bridge resistance, etc. The default value applied for such losses is 95 percent of the dc nameplate value or a multiplier value of 0.95.
- *Inverter and matching transformer losses.* These losses are a result of the conversion of dc to ac power. The efficiency of inverters used in solar power cogeneration

range from 88 to 96 percent. The mean value multiplier applied by the STC power rating is 92 percent, which translates into a multiplier value of 0.92.

- *PV module array interconnection mismatch.* As mentioned earlier, the dc output of manufactured PV modules do vary, and when the PV modules are connected in tandem, impedance mismatch results in power losses that may vary from 97 to 99 percent; hence, a median degradation multiplier of 0.98 is applied during solar array power output calculations.
- *Reverse diode losses.* These losses are attributed to the voltage drop across diodes that are used in each PV module to prevent reverse current flow into the unit. Diodes are unidirectional electronic check valves that pass current only in one direction and have intrinsic resistive characteristics; as a result they account for energy loss due to heat dissipation.
- *DC wiring losses.* A string of electrical wires that carry the dc output from PV module to PV module and to the inverters are subject to ohmic resistive losses. Alhough these losses could substantially be reduced by proper sizing of wires and conduits, nevertheless, they account for 97 to 99 percent of performance efficiency and are therefore assigned a multiplier value of 0.99.
- *AC wiring losses.* Similar to dc wiring, ac wiring from inverters to the switchgear or service power distribution hardware are also subject to voltage drop, conduit loss derating, and conduit solar exposure. Theses losses likewise could be substantially reduced by proper engineering design; however, a median loss multiplier value of 0.99 is generally applied to the calculations.
- *PV module dirt and soiling losses.* When PV module surfaces are exposed to dirt, dust, and snow, the efficiency of the performance can drop as much as 25 percent. Solar power installations in windy, desert, and high vehicular traffic areas should be cleaned periodically to maintain the optimum level of PV performance efficiency. PV modules supported by tilted platforms or inclined terrain, in addition to having a higher performance efficiency, are less susceptible to dirt collection and are relatively easier to clean and maintain. Likewise in northern locations during the winter, accumulated snow that blocks solar irradiance slides off the PV modules when PV arrays are tilted at an angle. Note that snow accumulation in northern parts of the country can reduce solar power output performance by as much as 70 to 80 percent. Also for soiling a derating factor of 0.95 is recommended.
- *System availability and mean time between failures (MTBF).* Solar power cogeneration configurations, whether grid connected or otherwise, are extremely reliable systems since the most important active components, namely, PV modules, are manufactured as hermetically sealed solid-state electronic devices with a life expectancy of over 40 years. Inverters are likewise solid-state power conversion devices that are guaranteed for at least 5 years by the manufacturers.

Since solar power systems do not make use of any moving mechanical devices, they are not subject to wear and tear like most energy-generating plants and equipment. The only downtime that may result from periodic equipment and module tests is essentially insignificant; however, a mean system availability multiplier of 0.98 is considered to be a safe derating factor. Other losses such as shading, PV degradation due to aging, and sun tracking are in general not taken into account.

With reference to the preceding, the overall calculated dc-to-ac losses amount to 0.77:

$$\text{DC-to-ac loss} = 0.95 \times 0.92 \times 0.98 \times 0.996 \times 0.98 \times 0.99 \times 0.95 \times 0.98 = 0.77$$

Array tilt angle loss   The optimum tilt angle for PV module performance is the latitude angle of the particular terrain. As discussed earlier in this chapter, irradiance at the latitude angle is perpendicular to the solar PV module. At this angle the annual solar power energy output from the PV module is at its optimum. An increased tilt angle above the latitude will increase power output production in wintertime; however, it will decrease in summertime. Likewise, decreasing the tilt angle from the latitude will increase power production in summertime.

The following table relates the tilt angle and roof pitch, which is the ratio of the vertical rise of the roof to its horizontal run.

ROOF PITCH	TILT ANGLE (DEGREES)
4/12	18.4
5/12	22.6
6/12	26.6
7/12	30.3
8/12	33.7
9/12	36.9
10/12	39.8
11/12	42.5
12/12	45.0

Photovoltaic array azimuth angle (0 to 360 degrees)   The azimuth angle is the angle measured clockwise from the true north of the direction facing the PV array. For fixed PV arrays, facing south, the azimuth angle is therefore 180 degrees clockwise from the north.

PV arrays that are mounted on sun-tracking platforms can move in either one-axis or two-axis rotation. In one-axis rotation the azimuth angle is rotated clockwise from the true north. In PV modules installed on platforms with two-axis rotation the azimuth angle does not come into play.

As a rule, for optimum energy output PV arrays in the northern hemisphere are mounted or secured in an azimuth angle of 180 degrees or tilted at an angle in a position north facing south direction and in southern hemisphere the azimuth angle is reversed installed in a tilted angle in a position south facing north.

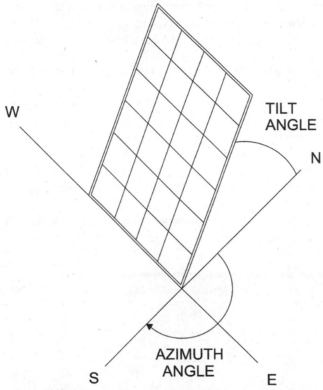

Figure 4.11   PV array facing south at a fixed tilt.

Figure 4.11 depicts PV array facing south at a fixed tilt and Figure 4.12 a single-axis PV array with a south-facing orientation. Figure 4.13 depicts a two-axis tracking array.

The following table shows the relationship of the azimuth angle and compass headings.

COMPASS HEADING	AZIMUTH ANGLE (DEGREES)
N	0 or 360
NE	45
E	90
SE	135
S	180
SW	225
W	270
NW	315

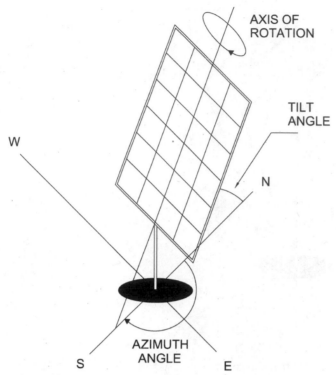

Figure 4.12   Single-axis PV array with a south-facing orientation.

Solar Power Design

SOLAR POWER REBATE APPLICATION PROCEDURE

Energy providers such as Nevada Power, the Los Angeles Department of Water and Power, Southern California Edison, and the Southern California Gas Company, which disburse renewable energy rebate funds, have established special design documentation submittal requirements that are mandatory for rebate qualification and approval. Initial rebate application forms require minimal design justification for sizing the solar power cogeneration. Information on the application forms is limited to the following:

■ Name and address of the owner and the solar power contractor
■ PV module manufacturer and model number (must be listed on CEC-approved equipment)
■ PTC watts of the PV module
■ Total number of PV modules
■ Total PV power output in PTC watts

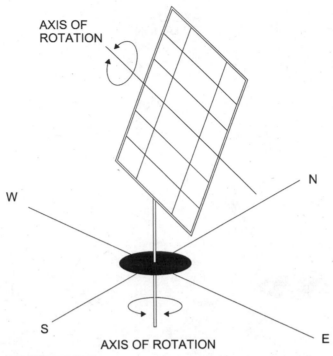

AXIS OF
ROTATION

N

W

S

E

AXIS OF ROTATION

Figure 4.13   Two-axis tracking array.

■ The inverter manufacturer (must be listed on CEC-approved equipment)
■ Inverter model number and number of units (total output capacity)
■ Inverter efficiency in percent
■ Maximum site electrical demand load

Upon completion of the parameters, project incentive and cost calculations are calculated using the following formula:

Total eligible rebate watts (TERW) = Total sum of PV PTC watts × Inverter efficiency

The total rebate amount is calculated by multiplying the TERW by the rebate per watt amount.

The following example demonstrates the eligible rebate amounts that can be expected from Southern California Gas Company. Let's assume that a solar power platform provides an unshaded area for a quantity of 530 SolarWorld modules (model SW 175 mono) and a properly sized inverter is selected from a manufacturer that has been listed under CEC-approved solar power equipment. The rebate application calculation will be as follows:

■ PV module: SolarWorld model SW 175 with dc-rated output of 175 W
■ PV module PTC power output rating: 158.3 W

- PV module area: 14 ft^2
- Approximate available unshaded platform space: 7500 ft^2
- Total number of PV modules to be installed: 530 units
- Total PTC output watts: 530 × 158.3 = 83,899 W
- Inverter unit capacity chosen: 100 kilowatts (kW)
- Inverter efficiency: 94.5%
- Total eligible rebate watts: 83,899 × 94.5% = 79,285 W
- Adjusted incentive rebate per watt: $2.50
- Total rebate eligibility: 79,285 × $2.50 = $198,212.50

Note that the maximum allowable rentable solar energy is regulated and capped by various rebate administrative agencies. For instance, Nevada Power caps the small commercial rebate to 30 kW per user meter, whereas the Los Angeles Department of Water and Power caps it to a maximum of 300 kW per user per meter and Southern California Gas and Southern California Edison limit the cap to 1000 kW per user address or meter per year.

It is suggested that the solar power design engineers before commencing their design should familiarize themselves with the specific requirements of energy service providers.

Additional documents that must be provided with the rebate application form are as follows:

- Electric power system single-line diagrams that show solar power PV arrays, dc combiner boxes, inverters, ac combiner boxes, conduit sizes, feeder cable sizes and associated voltage drops, the solar power fused service disconnect switch, the solar power meter, and the main service switchgear solar disconnect circuit breaker.
- Total building or project electrical demand load calculations (TBDL)
- Calculated percentage ratio of the total eligible rebate watts: (TERW/TBDL) × 100. This figure is required to confirm that the overall capacity of the solar power cogeneration does not exceed 125 percent of the total project or building electrical demand load.

In order to protect the client and avoid design error and omission liabilities, it is suggested that all documents and calculations be prepared by an experienced, qualified, registered electrical engineer.

Example of a solar power design Upon the approval and confirmation of the rebate by the energy service provider, the solar power engineer must undertake a comprehensive study of the entire solar power system and provide the following design documentation and paperwork to the agency for final incentive rebate approval. Note that times for submitting the required documents outlined here are mandatory, and they must be completed within a specified period of limited time.

To start the final solar power design, the electrical or solar power designer must conduct a thorough solar shading evaluation. If the solar platform is surrounded by trees, buildings, and tall objects such as power poles or hilled terrain, a comprehensive Solar Pathfinder field survey and analysis must be undertaken. In particular some agencies such as the Los Angeles Department of Water and Power base their incentive rebates on the Solar Pathfinder power performance analysis as described earlier in this chapter.

The following hypothetical shading study is applied to the solar power cogeneration system reflected in the preceding rebate application. Let us assume that the Solar Pathfinder diagram in Figure 4.14 represents the site shading diagram. Figure 4.14 depicts Pathfinder dome showing solar platform site shading and Figure 4.15 depicts Pathfinder showing marked up platform site shading area.

To determine the solar shading performance multiplier, we must first compile the monthly percentage totals and then calculate the yearly mean or average multiplier. In this particular example the solar power cogeneration platform is heavily shaded by trees and surrounding buildings, so we should expect a less-than-optimum energy output performance.

Upon completion of solar shading tabulation and establishment of the shading multiplier, the overall solar power output performance will be calculated by applying all power loss factors to the calculated total PTC.

Figure 4.14   Pathfinder dome showing solar platform site shading.

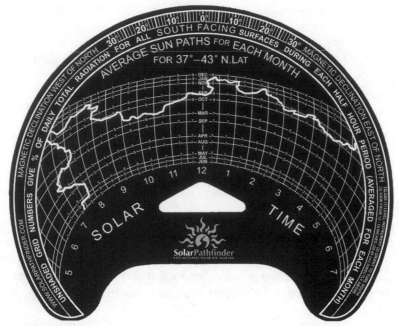

Figure 4.15   Pathfinder showing marked up platform site shading area.

The following shading tabulations are for the Solar Pathfinder (Figure 4.14) chart:

December: Totally shade; therefore, there are no percentage points = 0%

January: 7 = 7%

November: 8 + 7 = 15%

February: 8 + 7 + 7 + 7 + 6 = 35%

October: 8 + 7 + 7 + 6 + 6 + 5 + 4 + 3 + 2 + 1 = 49%

March: 2 + 3 + 4 + 6 + 6 + 7 + 7 + 7 + 7 + 7 + 6 + 6 + 5 + 4 + 3 + 2 + 2 + 1 + 1 = 88%

September: 2 + 3 + 4 + 6 + 6 + 7 + 7 + 7 + 7 + 7 + 6 + 6 + 5 + 4 + 3 + 2 + 2 + 1 + 1 + 1 = 89%

April: 2 + 3 + 4 + 6 + 6 + 7 + 7 + 7 + 7 + 7 + 6 + 6 + 5 + 4 + 3 + 2 + 2 + 1 + 1 + 1= 89%

August: 2 + 3 + 4 + 6 + 6 + 7 + 7 + 7 + 7 + 7 + 6 + 6 + 5 + 4 + 3 + 2 + 2 + 1 + 1 + 1 + 9 = 90%

May: 2 + 3 + 4 + 5 + 5 + 6 + 7 + 7 + 7 + 7 + 7 + 6 + 6 + 5 + 4 + 3 + 2 + 2 + 1 + 1 + 1 = 92%

July: 3 + 4 + 5 + 5 + 6 + 7 + 7 + 7 + 7 + 7 + 7 + 6 + 6 + 5 + 4 + 3 + 2 + 2 + 1 + 1 + 1 + 1 = 97%

June: 3 + 4 + 5 + 5 + 6 + 7 + 7 + 7 + 7 + 7 + 7 + 6 + 6 + 5 + 4 + 3 + 2 + 2 + 1 + 1 + 1 + 1 = 97%

$$\text{Mean shading multiplier} = (7 + 15 + 35 + 49 + 88 + 89 + 89 + 90 + 92 + 97 + 97)/ 12 = 54\%$$

Applying the dc-to-ac loss (0.77%) and mean shading multiplier to the calculated PTC, we obtain

$$\text{Net ac output watts} = (79{,}285 \text{ W PTC}) \times 0.77 \text{ (ac-to-dc loss)} \times 0.54 \text{ (shading multiplier)} = 32{,}966.7 \text{ W}$$

According to the preceding performance result, the effective energy production is reduced to such a degree that the return on investment of the solar power program will be hardly justified. To mitigate this problem the solar power platform must be reduced to the unshaded area.

The preceding solar power calculation should serve to caution design engineers to perform a comprehensive preliminary feasibility study prior to submitting the rebate application forms; otherwise, a substantial reduction in projected power production capacity could create many adverse consequences that may result in unnecessary budget overruns, project execution delays, rebate incentive application updates, and client dissatisfaction.

5

SOLAR POWER GENERATION
PROJECT IMPLEMENTATION

Introduction

In previous chapters we covered the basic concepts of solar power system design, reviewed various system configurations, and outlined all major system equipment and materials required to implement a solar power design. In this chapter the reader will become acquainted with a number of solar power installations that have been implemented throughout the United States and abroad. The broad range of projects reviewed include very small stand-alone pumping stations, residential installations, solar farm installations, large pumping stations, and a few significant commercial and institutional projects. All these projects incorporate the essential design concepts reviewed in Chapter 2. Prior to reviewing the solar power projects, the design engineers must keep in mind that each solar power system design presents unique challenges, requiring special integration and implementation, that may not have been encountered before and may recur in future designs.

Designing a Typical Residential Solar Power System

A typical residential solar power system configuration consists of solar photovoltaic (PV) panels; a collector fuse box; a dc disconnect switch; some lightning protection devices; a charge controller for a battery if required; an appropriately sized inverter; the required number of PV system support structures; and miscellaneous components, such as electrical conduits or wires and grounding hardware. Additional expenses associated with the solar power system will include installation labor and associated electrical installation permits.

Prior to designing the solar power system, the designer must calculate the residential power consumption demand load. Electrical power-consuming items in a household must be calculated according to the NEC-recommended procedure outlined in the following steps. The calculation is based on a 2000-ft^2 conventional single residential unit:

Step 1: Lighting load. Multiply the living space square area by 3 W: 2000 × 3 = 6000 W.

Step 2: Laundry load. Multiply 1500 W for each laundry appliance set, which consists of a clothes washer and dryer: 1500 × 1 = 1500 W.

Step 3: Small appliance load. Multiply kitchen appliance loads rated 1500 W by 2: 1500 W × 2 = 3000 W.

Step 4: Total lighting load. Total the sum of the loads calculated in steps 1 to 3: 6000 + 1500 + 3000 + 10,500 W.

Step 5: Lighting load derating. Use the first 3000 W of the summed-up load (step 4) and add 35 percent of the balance to it: 3000 + 2625 + 5625 W.

Step 6: Appliance loads. Assign the following load values (in watts) to kitchen appliances:

Dishwasher	1200
Microwave oven	1200
Refrigerator	1000
Kitchen hood	400
Sink garbage disposer	800
Total kitchen appliance load	4600

If the number of appliances equals 5 or more, then the total load must be multiplied by 75 percent, which in this case is 3450 W.

Step 7: Miscellaneous loads. Loads that are not subject to power discounts include air conditioning, Jacuzzi, pool, and sauna and must be totaled as per the equipment nameplate power ratings. In this example we will assume that the residence is equipped with a single five-tone packaged air-conditioning system rated at 17,000 W.

When totaling the load, the total energy consumption is 26,075 W:

17,000 (air conditioner) + 3450 (appliances) + 5625 (lighting power) = 26,075

At a 240-V entrance service this represents about 100 A of load. However, considering the average power usage, the realistic mean operating energy required discounts full-time power requirements by the air conditioning, laundry equipment, and kitchen appliances; hence, the norm used for sizing the power requirement for a residential unit boils down to a fraction of the previously calculated power. As a rule of thumb, an average power demand for a residential unit is established by equating 1000 to 1500 W per 1000 ft² of living space. Of course this figure must be augmented by considering the geographic location of the residence, number of habitants, occupancy time of the population within the dwelling unit, and so forth. As a rule, residential dwellings in hot climates and desert locations must take the air-conditioning load into consideration.

As a side note when calculating power demand for large residential areas, major power distribution companies only estimate 1000 to 1500 W of power per household, and this is how they determine their mean bulk electric power purchase blocks.

When using a battery backup, a 30 percent derating must be applied to the overall solar power generation output efficiency, which will augment the solar power system requirement by 2500 W.

In order to size the battery bank, one must decide how many hours the overall power demand must be sustained during the absence of sun or insolation. To figure out the ampere-hour (Ah) capacity of the battery storage system, the aggregate wattage worked out earlier must be divided by the voltage and then multiplied by the backup supply hours. For example, at 120 V ac, the amperes produced by the solar power system, which is stored in the battery bank, will be approximately 20 A. To maintain power backup for 6 hours, the battery system must be sized at about 160 Ah.

Example of Typical Solar Power System Design and Installation Plans for a Single Residential Unit

The following project represents a complete design and estimating procedure for a small single-family residential solar power system. In order to establish the requirements of a solar power system, the design engineer must establish the residential power demand calculations based on NEC design guidelines, as shown in the following.

REED RESIDENCE, PALM SPRINGS, CALIFORNIA

Electrical engineer consultant Vector Delta Design Group, Inc., 2325 Bonita Dr., Glendale, CA 91208

Solar power contractor Grant Electric, 16461 Sherman Way, Van Nuys, CA 91406

Project design criteria The residential power demand for a single-family dwelling involves specific limits of energy use allocations for area lighting, kitchen appliances, laundry, and air-conditioning systems. For example, the allowed maximum lighting power consumption is 3 W per square foot of habitable area. The laundry load allowed is 1500 W for the washer and dryer.

The first 3000 W of the total combined lighting and laundry loads are counted at 100 percent, and the balance is applied at 35 percent. The total appliance loads, when there are more than five appliances, are also derated by 25 percent. Air conditioning and other loads such as the pool, sauna, and Jacuzzi are applied at their 100 percent value.

The demand load calculation of the 1400-ft^2 residential dwelling shown in Figure 5.1 indicates a continuous demand load of about 3000 W/h. If it is assumed that the residence is fully occupied and is in use for 12 hours a day, the total daily demand load

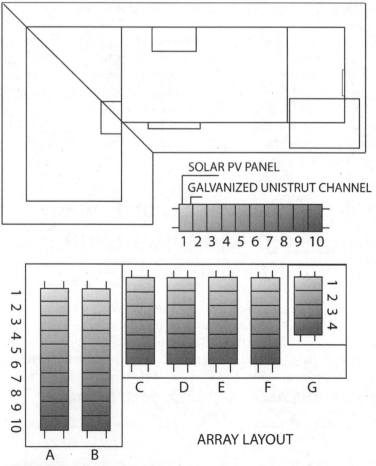

Figure 5.1   Roof-mount PV array layout, Reed residence, Palm Springs, California.

translates into 36,000 W/day. Figure 5.1 depicts a residential roof-mount PV array layout, in Palm Springs, California and Figure 5.2 the solar power system schematic wiring diagram.

Since the average daily insolation in southern California is about 5.5 hours, the approximate solar power system required to satisfy the daily demand load should be approximately 6000 W. Occupancies that are not fully inhabited throughout the day may require a somewhat smaller system.

In general, an average 8 hours of habitation time should be used for sizing the solar power system, which in this example would yield a total daily power demand of 24,000 Wh, which in turn translates into a 4000-W solar power system.

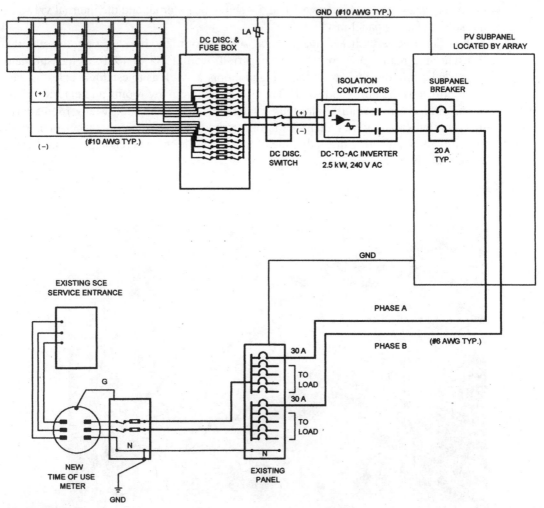

Figure 5.2   Solar power system schematic wiring diagram, Reed residence, Palm Springs, California.

THE HEAVENLY RETREAT PROJECT

The following project is an example of excellence in solar heating and power generation design by owners Joel Goldblatt and Leslie Danzinger. The designer of the project has taken maximum advantage of applying a sustainable energy systems design and has blended the technology with the natural setting of the environment.

The Heavenly Retreat is a unique, celestially aligned, passive and active solar home, with two built-in integrated greenhouses, located at a 9000-ft elevation in the northern New Mexico Rockies. The heavily forested land is located on the northwest face of an 11,600-ft peak facing Wheeler Peak, the highest point in New Mexico and a landmark in the southern Rockies. The property slopes toward the northwest, set amid many tall pines and fir trees. The south side of the property opens to a large garden with a solar calendar that marks the sunrises and sunsets throughout the year within a meditation circle and seating area. Figures 5.3 and 5.4 show the external and the internal views of a residential solar panel installation.

The designer established the passive solar design by first visiting the natural, forested area many times over a 3-year period, from 1985 to 1988, to measure and observe the motions and arc of the sun and moon, as well as to establish celestial cardinal points to the north and south. The design aligns many rooms and exterior walls toward Polaris in the north, and thus establishes the solar south. Many rooms have markers that cast shadows along the floors to show when the sun is at the solar south position, thus indicating the daily time of local solar noon and creating a silent rhythm within the living spaces.

Figure 5.3   **The Heavenly Retreat project (exterior view).** *Photo courtesy of Mr. Joel Goldblatt, New Mexico, United States.*

Figure 5.4   The Heavenly Retreat project (interior view). *Photo courtesy of Mr. Joel Goldblatt, New Mexico, United States.*

Using these studies of the sun, moon, and planetary alignments, as well as building with adobe, heavy beam, and recyclable product construction and insulation, the home is a tribute to its environment. Being embedded into the forest floor it rises with three terraced stories to a meditation room with a pyramid skylight and a geodesic dome garage. It produces its own solar electricity (3.6 kW) and solar thermal heating (180,000 Btu/h) for 10 zones of hydronic radiant floor heating. It includes three year-round producing greenhouses and an organically cultivated summer garden. Stepping outdoors from one of three patios, trails go off into the forest with over 50 acres of alpine forest and meadows, sharing the forest with deer, elk, and many other animals. The home was completed in 2002. It is being further refined and developed to serve as a spiritual retreat for guests, offering meditation and seminars in solar design and construction.

The owner, who is also a designer of the project, spent many years contemplating about the project and undertook special design efforts and planning to honor and respect the building's surrounding environment. The project was not designed to be just a sustainable energy design installation project but rather a holistic environmental design that could unite technology and ecology.

The sustainable energy system is configured by the deployment of 11 solar thermal panels with a control system that augments water heating by use of a supplemental propane boiler. The heated water is the primary heat source, which provides space heating as well as a domestic hot water (DHW) system. At present the solar thermal system accounts for about 60 to 70 percent of the heat, thus reducing propane use by about 40 to 50 percent.

The project was constructed in two phases. The first phase was completed from 1988 to 1994, and the second phase began in 1999 and was completed in 2002. The present net power output of the photovoltaic power system is rated at 3.6 kW ac. The solar power generation consists of two systems; one is installed in the garden, and the other in the kitchen above the counter, which also serves as a solar window for the adjacent interior greenhouse. The kitchen solar power panels are building-integrated photovoltaic (BIPV) panes custom-manufactured by Atlantis Energy. They are built with quadruple-pane tempered glass panels that sandwich Shell Solar Power Max cells by a lamination process; dc-to-ac power conversion is achieved by the use of two stacked Trace SW-4048 inverters and Outback MX-60 charge controllers.

The roof-mount solar BIPV panes are secured to the building structure at an angle that matches the pitch of the rest of the roof, which is about 32 degrees up from the horizon, in close proximity to the optimal angle matching the geographic latitude for an equinox alignment that favors the winter sun angle.

The battery backup system has a 1600-A/h backup capacity. The batteries alternate between a float status and net metering without the local utility—a rural electric cooperative.

The total system, including all underground infrastructure and the battery storage and monitoring equipment, but excluding the owner's installation time, cost about $80,000 or about $23/W, somewhat expensive compared to off-the-shelf type equipment, but it is a totally unique solar integrated solution in harmony with the indoor greenhouse environment.

Metering is dual-rate "time of use," and the utility consumption averages 1400 kWh/month. Even with recent electricity rate increases, the average monthly bill is $130/month. The owner also pays a $15 monthly fee to participate in the green power program offered by the local utility. The owner estimates that the monthly saving resulting from use of the solar power system amounts to at least $200/month. The owner's average propane bill now ranges from $80/month in the summer to about $250/month in the winter. Without the solar thermal system installation, his summer bill would range from $150/month and his winter bill would be over $600/month, an average of a 60 percent energy cost savings.

At present, the state of New Mexico does not have a cash rebate program, although as of 2006, with federal and newly passed state incentives the owner will be able to claim a 30 percent average tax credit from both federal and state governments.

Elsewhere in New Mexico, Public Service of New Mexico (PNM) has instituted a 13 cent/kWh production credit feed-in tariff within its service in the Santa Fe and Albuquerque areas.

Commercial Applications

The following plans are provided for illustrative purposes only. The actual design criteria and calculations may vary depending on the geographic location of the project, cost of labor, and materials, which can significantly vary from one project to another. The following projects were collaborations among the identified organizations.

TCA ARSHAG DICKRANIAN ARMENIAN SCHOOL PROJECT

Location Hollywood, California

Electrical engineer consultant Vector Delta Design Group, Inc., 2325 Bonita Dr., Glendale, CA 91208

Solar power contractor Grant Electric, 16461 Sherman Way, Van Nuys, CA 91406

Project design criteria The project described here is a 70-kW roof-mount solar power cogeneration system, which was completed in 2004. The design and estimating procedures of this project are similar to that of the residence in Palm Springs. Diagrams and pictures of this project are shown in Figures 5.5 through 5.11.

In order to establish the requirements of a solar power system, the design engineer must determine the commercial power demand calculations based on the NEC design guidelines. Power demand calculations for this project were calculated to be about 280 kWh. The solar power installed represents about 25 percent of the total demand load. Since the school is closed during summer, credited energy accumulated for 3 months is expected to augment the overall solar power cogeneration contribution to about 70 percent of the overall demand.

This project was commissioned in March 2005 and has been operating at optimum capacity, providing a substantial amount of the lighting and power requirements of the school. Fifty percent of the installed cost of the overall project was paid by CEC rebate

Figure 5.5 **Roof-mount solar power system, Arshag Dickranian School, Hollywood, California.** *Photo courtesy of Vector Delta Design Group, Inc.*

Figure 5.6   Inverter system, Arshag Dickranian School, Hollywood, California. *Photo courtesy of Vector Delta Design Group, Inc.*

funds. Figure 5.6 depicts the inverter system installation at Arshag Dickranian School, Hollywood, California.

BORON SOLAR POWER FARM COGENERATION SYSTEM PROJECT

Location   Boron, California

Electrical engineer consultant   Vector Delta Design Group, Inc., 2325 Bonita Dr., Glendale, CA 91208

Solar power contractor   Grant Electric, 16461 Sherman Way, Van Nuys, CA 91406

Project design criteria   The project described here is a 200-kW solar power farm cogeneration system, which was completed in 2003. The design and estimating procedures of this project are similar to the two projects already described.

In view of the vast project terrain, this project was constructed by use of relatively inexpensive, lower-efficiency film technology PV cells that have an estimated efficiency of about 8 percent. Frameless PV panels were secured on 2-in Unistrut channels, which were mounted on telephone poles that penetrated deep within the desert sand. The power produced from the solar farm is being used by the local Indian reservation. The project is shown in Figures 5.7 and 5.8.

WATER AND LIFE MUSEUM

Project location   Hemet, California

Architect   Lehrer Gangi Architects, 239 East Palm Ave., Burbank, CA 91502

Electrical and solar power consultants   Vector Delta Design Group, Inc., 2325 Bonita Dr., Glendale CA 91208

Electrical contractor   Morrow Meadows, 231 Benton Court, City of Industry, CA 91789

Figure 5.7   **Boron solar photovoltaic farm.** *Photo courtesy of Grant Electric.*

Figure 5.8　**Boron solar farm inverter system.** *Photo courtesy of Grant Electric.*

Project description　This project is located in Hemet, California, an hour-and-a-half drive from downtown Los Angeles. The project consists of a 150-acre campus with a Water Education Museum, sponsored by the Metropolitan Water District and the Water Education Board; an Archaeology and Paleontology Museum, sponsored by the City of Hemet; several lecture halls; a bookstore; a cafeteria; and two auditoriums. In this installation, PV panels are assembled on specially prefabricated sled-type support structures that do not require roof penetration. Roof-mount PV arrays are strapped together with connective ties to create large island platforms that can withstand 120-mi/h winds. A group of three PV assemblies with an output power capacity of about 6 kW are connected to a dedicated inverter. Each inverter assembly on the support incorporates overcurrent protective circuitry, fusing, and power collection busing terminals.

The inverter chosen for this project includes all technology features, such as anti-islanding, ac power isolation, voltage, and frequency synchronization required for grid connectivity. In addition, the inverters are also equipped with a wireless monitoring transmitter, which can relay various performance and fault-monitoring parameters to a centrally located data acquisition system. Figure 5.9 is a photograph of a section of the roof mount solar power cogeneration system and Figure 5.10 a section of roof mount inverter installation at Water and Life Museum, Hemet California.

Strategically located ac subpanels installed on rooftops accumulate the aggregated ac power outputs from the inverters. Outputs of subpanels are in turn accumulated

Figure 5.9   Water and Life Museum, rooftop solar power systems.
Photo courtesy of Vector Delta Design Group, Inc.

Figure 5.10   Water and Life Museum, rooftop solar power module
and inverter installation. *Photo courtesy of Vector Delta Design Group, Inc.*

Figure 5.11   **Water and Life Museum, distributed inverter systems.** *Photo courtesy of Vector Delta Design Group, Inc.*

by a main ac collector panel, the output of which is connected to a central collector distribution panel located within the vicinity of the main service switchgear. Grid connection of the central ac collector panel to the main service bus is accomplished by means of a fused disconnect switch and a net meter. Figure 5.11 is a photograph of distributed inverter systems at Water and Life Museum, Hemet, California.

The central supervisory system gathers and displays the following data:

- Project location (on globe coordinates, zoom in and out)
- Current and historic weather conditions
- Current positions of the sun and moon and the date and time (local and global)
- Power generation of the total system or individual buildings and inverters
- Historic power generation
- Solar power system environmental impact
- System graphic configuration data
- Educational PowerPoint presentations
- Temperature
- Wind velocity and direction
- Sun intensity
- Solar power output

- Inverter output
- Total system performance and malfunction
- Direct-current power production
- Alternating-current power production
- Accumulated daily, monthly, and yearly power production

Small-Scale Solar Power Pumping Systems

Pumping water with solar power is reliable and inexpensive and is achieved using a combination of a submersible pump and a solar cell panel that can be procured inexpensively. Engineers have spent years developing a water pumping system to meet the needs of ranchers, farmers, and homesteaders. These systems are reliable and affordable and can be set up by a person with no experience or very little mechanical or electrical know-how.

The solar submersible pump is probably the most efficient, economical, and trouble-free water pump. Figure 5.12 depicts a small submersed solar pumping system diagram. In some installations the procedure for installing a solar power pump simply

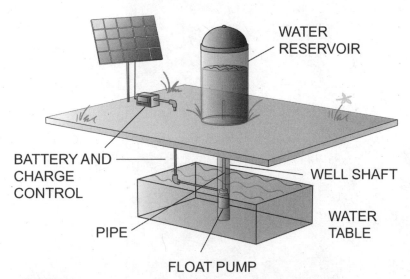

Figure 5.12   Water well solar power pumping diagram.

involves fastening a pipe to the pump and placing the unit in a water pond, lake, well, or river. The output of the solar panel is connected to the pump, and the panel is then pointed toward the sun and up comes the water. The pumps are generally lightweight and easily moved and are capable of yielding hundreds of gallons per day at distances of over 200 ft above the source. The pumps are of a rugged design and are capable of withstanding significant abuse without damage even if they are run in dry conditions for a short time.

In some instances the pumping systems can be equipped with a battery bank to store energy, in which case water can be pumped at any time, morning, noon, or night and on cloudy days. The system could also be equipped with simple float switch circuitry that will allow the pumps to operate on a demand basis. The solar pumping system just described requires very little maintenance. Figure 5.13 depicts a submersible solar water pump in a rural Philippine village.

Figure 5.13   **Submersible solar water pump in a rural Philippine village.** *Photo courtesy of WaterWorld & Power Corp.*

Figure 5.14   **Small solar water pump in a rural village in India.**
Photo courtesy of SolarWorld.

The batteries used for most systems are slightly different than ones used in cars. They are called deep-cycle batteries and are designed to be rechargeable and to provide a steady amount of power over a long period of time. Details about the design and application of battery systems are discussed in Chapter 3 of this book. In some farming operations solar-powered water systems pump the water into large holding tanks that serve as reserve storage supply during cloudy weather or at night.

In larger installations solar modules are usually installed on special ground- or pole-mounting structures. For added output efficiency, solar panels are installed on tracker-mounting structures that follow the sun like a sunflower. Figure 5.14 depicts a small solar water pump in a rural village in India.

Large-Capacity Solar Power Pumping Systems

A typical large-scale solar power pumping system, presented in Figure 5.15, is manufactured and system engineered by WorldWater & Power Corporation, whose headquarters are located in New Jersey. WorldWater & Power is an international solar engineering and water management company with a unique, high-powered solar technology that provides solutions to water supply and energy problems. The company has developed patented AquaMax solar electric systems capable of operating pumps and motors up to 600 horsepower (hp), which are used for irrigation, refrigeration and

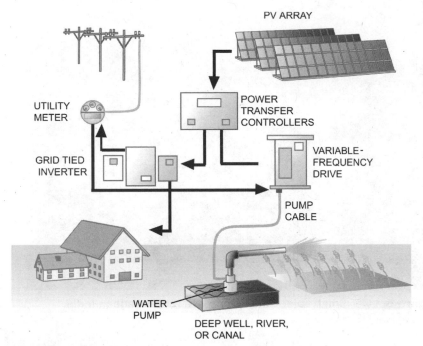

Figure 5.15   **Grid-connected solar pumping system diagram.**
Photo courtesy of WorldWater & Power Corporation.

cooling, and water utilities, making it the first company in the world to deliver main-stream solar electric pumping capacity.

THE AQUAMAX SYSTEM KEY FEATURES

In general, grid power normally provides the power to the pumps. With the AquaMax system, in the event of power loss, the system automatically and instantaneously switches power fully to the solar array. In keeping with the islanding provisions of the interconnection rules, when the power to the grid is off, the pump or motor keeps operating from solar power alone without interruption. This solar pumping system is the only one of its kind; other grid-tied solar power systems shut down when grid power is interrupted.

Power-blending technology   The AquaMax seamlessly blends dc power from the solar array and ac power from the grid to provide a variable-frequency ac signal to the pump or motor. This does two important things. It eliminates large power surges to the motor, and so reduces peak demand charges in the electric bill. It also increases the efficiency of the motor, so it uses less energy to operate. This power-blending technology also means that motors benefit from a "soft start" capability, which reduces wear and tear on the motor and extends its life.

Off-grid capability customers can elect to run a pump or motor off-grid on solar power alone. This may be useful if there is a time of day when, for example, running the pump or motor would incur a large demand charge that the customer wishes to avoid. The system makes operation still possible, while avoiding peak demand charges imposed by the utility.

The 2003 power outages in the Northeast and Midwest highlighted a critical application of this proprietary solar technology. The systems described here are capable of driving pumps or ac motors up to 600 hp as backup for grid power or in combination with the grid or other power sources, such as diesel generators. The systems can also be assembled for mobile or emergency use or used as part of a permanent power installation. In either case, they can provide invaluable power backup in emergency power outages and can operate independent of the electric grid, relying instead on the constant power of the sun.

Pump Operation Characteristics

The following discussion is presented to introduce design engineers to various issues related to pump and piping operational characteristics that affect power demand requirements. In general, the pumping and piping design should be trusted to experienced and qualified mechanical engineers.

Every cooling tower requires at least one pump to deliver water. Pump selection is based on the flow rate, total head, and ancillary issues such as type, mounting, motor enclosure, voltage, and efficiency.

The pumping volume is dictated by the manufacturer of the equipment. The total head (in feet) is calculated for the unique characteristics of each project as follows:

Total head = net vertical lift + pressure drop at cooling tower exit
+ pressure drop in piping to pump + pressure drop from pump to
destination storage compartment + pressure drop of storage
+ pressure drop through the distribution system + velocity pressure
necessary to cause the water to attain the required velocity

The total head is usually tabulated as the height of a vertical water column. Values expressed in pounds per square inch (lb/in^2) are converted to feet by the following formula:

$$\text{Head-feet} = \text{pounds per square inch} \times 2.31$$

The vertical lift is the distance the water must be lifted before it is let to fall. Typically, it is the distance between the operating level and the water inlet. The pressure drop in the piping to the pump consists of friction losses as the water passes through the pipe, fittings, and valves. The fittings and valves are converted to equivalent lengths of straight pipe (from a piping manual) and added to the actual run to get the equivalent length of suction piping.

Then the tabulated pressure drop from the piping manual for a specific length of pipe is compared to the length and pressure drop calculated by proportion:

$$\text{Pressure drop} = \text{pressure drop for specific length of pipe} \times \text{equivalent pipe length/pipe}$$

Typically, end suction pumps are selected and are of the close-coupled type (where the pump impeller fastens directly to the shaft and the pump housing bolts directly to the motor) for up to about 15 hp and the base-mounted type (where the separate pump and motor fasten to a base and are connected by a coupling) is used for larger sizes.

The static lift is typically the distance between the operating level in the cold water basin and the reservoir inlet near the top of the towers.

When selecting a pump, it is important to make sure the net available suction head exceeds the required net suction head. This ensures the application will not cause water to vaporize inside the pump causing a phenomenon called cavitation. Vaporization inside the pump occurs when small water particles essentially "boil" on the suction side of the pump. These "bubbles" collapse as they pass into the high-pressure side producing the classic "marbles sound" in the pump. If operated under this condition, pumps can be damaged.

Pumps are also required to operate under net positive suction head (NPSH) conditions, which means that the pump lift must be able to cope with the local barometric pressure and handle the friction losses in the suction line and vapor pressure of the water being pumped. Figure 5.16 depicts a large-scale solar pumping system in Imperial Valley California.

Figure 5.16   **Large-scale solar pumping system in Imperial Valley California.** *Photo courtesy of WorldWater & Power Corporation.*

Semitropic Open Field Single-Axis Tracking System PV Array—Technical Specifications

The following project was designed and built by Shell Solar and consists of a solar farm configured from 1152-kW solar array modules. The single axis solar power tracking system presented for discussion is one of the recent solar farm projects which was installed at Semitropic Water District (SWD), located at Wasco, California, approximately 150 miles northeast of Los Angeles. The 550 ft × 38 ft land provided by SWD is relatively flat, with no trees, and thus required minimal grading and brush clearing. As shown in Figure 5.17, PV arrays movement rotational axels are mounted in a north south orientation.

MECHANICAL DESCRIPTION

The PV array design for the single-axis tracking system was based upon the use of 7200 Shell Solar Industries model SQ-160 modules, which were assembled into 1800 panels.

Figure 5.17   Aerial view of solar farming system, Semitropic Water District, Wasco, California. *Photo courtesy of SolarWorld.*

Figure 5.18   **Large-scale single-axis solar farming system, Semitropic Water District, Wasco, California.** *Photo courtesy of SolarWorld.*

The proposed configuration of the array provides 60 rows approximately 170 ft long and spaced at 21 ft on center as shown in Figure 5.18. The proposed 21-ft row-to-row spacing extends the array's operational day and maximizes energy output by minimizing shadowing effects.

The north-south axis trackers utilize a 20-ton screw drive jack to provide 45-degree east to 45-degree west single-axis tracking to maximize the daily energy output from the array. The screw jacks are controlled by a square–D (or Allen-Bradley) programmable logic controller (PLC). A clock-based controller provides ±2 percent tracking accuracy for the flat plate PV arrays and allows backtracking to eliminate row-to-row shadowing.

The system is installed on top of 720 wooden utility-grade ground-embedded poles as foundations for the array structure. Each pole is 15 ft in length and is buried in the ground at a depth of approximately 7 to 8 ft. The panel support structure for the array utilizes square galvanized steel "torque" tubes that are free to rotate at ±45 degrees, which are in turn supported by galvanized steel bearing plates. The precise motion of these torque tubes is provided by screw jacks that are regulated by the controller system. Prewired solar panels are clamped directly to the steel structure with two panel clamps per panel.

The steel subassemblies form 60 rows consisting of 30 rows of two-pair matrix. These rows are then divided electrically to form five equal-size subsystems consisting of 360 prewired panels. Each panel is factory prewired with four Shell Solar SQ160 modules and delivered to the site in reusable shipping racks. The panels are also equipped with factory quick-disconnects to ease field wiring.

DC-to-ac power conversion is accomplished by use of five Xantrex model PV-225208 inverters that are centrally located on a 12 ft × 74 ft concrete pad placed adjacent to five step-up transformers, the collected output of which is connected to a low-voltage metering system. Centrally located power accumulation allows for shorter conductor runs to all five of the inverters.

ELECTRICAL DESCRIPTION

As described in the preceding, the arrays are electrically divided into five equal subsections consisting of 1440 SQ160 modules, dc circuit combiners, one 225-kW inverter, and a 225-kVA 208-12.47 kV step-up transformer. Thirty panels each containing four modules are used on each of the 60 single-axis tracking rows. Figure 5.19 depicts a solar tracking pilot insolation detection system, Semitropic Water District, Wasco, California.

The dc collectors feed underground current to each inverter's dc interface, which incorporate prefabricated fusing and a manually operated disconnect switch. The ac output of the inverters also includes manually operated disconnect switches that feed the low-voltage section of the step-up transformer. The low-voltage winding of the transformer includes a metering section that is fitted with an energy production meter. The meter includes a cellular modem that can be read remotely. The electronic meters are designed to store daily, weekly, and monthly energy production parameters.

The transformers step up the 208 V ac to 12.47 kV ac. The high-voltage output of each transformer includes fusing and a hot stick disconnect. All five transformers are loop-fed, and the final underground feed from the transformer pad extends 200 ft to the north section of the array where it terminates at a riser pole.

The Xantrex inverters used in this installation meet IEEE 929 and UL1741 standards and as such do not require any anti-islanding hardware.

Figure 5.19   Solar tracking pilot insolation detection system, Semitropic Water District, Wasco, California. *Courtesy of Vector Delta Design Group.*

ENERGY PERFORMANCE

Generally speaking, it is estimated that the annual energy production from a single-axis tracking system can be as much as 20 percent higher when compared with that of a comparable fixed-tilt system. In general, the single-axis tracking modeling software used in this project calculates energy production of a single north-south axis row of PV modules from sunrise to sunset (90 degrees east to 90 degrees west). The most popular software currently used for calculating solar array output performance such as PV Design Pro or PV watts use a 90-degrees east to 90-degrees west algorithm to calculate the maximum available annual energy. As discussed in earlier chapters, when calculating energy output performance, shadowing effects must be accounted for in the annual energy production model. Figure 5.20 depicts the tracking tilt actuator mechanism at Semitropic Water District, Wasco, California.

When tracking multiple rows of solar panels, the higher the tracking limit angle (in this case 90 degrees), the larger the shadow cast in the morning and afternoon hours. This shadowing will effectively shut down energy production from all the rows located behind the eastern-most row in the morning and the western-most row in the evening. This effect can be reduced by limiting the tracking limit angle to 45 degrees. From a practical standpoint, the linear actuators used in the most popular systems easily accommodate a 45-degree limit angle and are the hardware used in the proposed system. To further improve the energy performance of the system, a backtracking scheme is used in the morning and evening hours of each day to eliminate the row-to-row shadowing.

Figure 5.20   **Solar tracking tilt actuator mechanism, Semitropic Water District, Wasco, California.** *Photo courtesy of Vector Delta Design Group.*

Backtracking begins by adjusting the tilt angle of each row to 10 degrees east just before the sun rises in the morning. As the sun rises, each row begins tracking east just enough so that no row-to-row shading occurs. This backtracking continues until the tracker limit angle of 45 degrees is reached, at which time, the tracker controller waits until the sun catches up with the 45-degree tilt angle and then begins to follow the sun throughout the day. In the afternoon, the controller will repeat the backtracking scheme until the sun sets.

Shell Solar is including typical energy profiles for tracking arrays in December and June (the winter and summer solstices). These profiles illustrate the effects described in the preceding and the impact they have on the annual energy production of a multirow single-axis tracking system. See Appendix C.

This project was constructed by Shell Solar Industries for customer Semitropic Water District and placed in service in April 2005. Because of market conditions at the time, Shell SQ-85 modules were used in place of the SQ-160 modules. Then because of other project constraints (related to the state of California incentive funding program at the time) the project size was ultimately reduced from 11,520 kW to 979.2 kW.

In July 2006, Shell Solar Industries and its projects and technology including the Semitropic project and single-axis tracking technology was purchased by SolarWorld Industries.

Installation of a comparable system with the present, more efficient SolarWorld modules would require 6576 SW-175 modules = 1150.8 kW. The modules could be assembled into 1096 prewired panels. The basic elements of the single-axis tracking system design and array would remain the same; however, because of the reduced solar module surface area the system would require approximately 9 percent less space. Figure 5.21 depicts the solar power inverter installation, Semitropic Water District, Wasco, California.

Figure 5.21   Solar power inverter installation, Semitropic Water District, Wasco, California. *Photo courtesy of Vector Delta Design Group.*

6

ENERGY CONSERVATION

Introduction

Energy efficiency is an issue that affects all projects. Whether you are considering a renewable energy system for supplying electricity to your home or business, or just want to save money with your current electrical service supplier, the suggestions given in this chapter will help you reduce the amount of energy that you use.

If you are planning to invest in a renewable energy system, such as solar or wind, and your electricity is currently supplied from a local utility, increasing the energy efficiency of the project will help to conserve valuable nonrenewable resources and reduce the size and cost of the solar or wind energy system needed.

There are many ways to incorporate energy efficiency into a design. Most aspects of energy consumption, within a building, have more efficient options than traditional methods. In this chapter we will review the basic concepts of conventional electric power generation and distribution losses and provide some basic recommendations and suggestions about energy conservation measures, which could significantly increase the efficiency of energy use.

In addition to providing energy-saving suggestions, we will review automated lighting design and California Energy Commission Title 24 design compliance. In view of recent green building design measures and raised consciousness about energy conservation, we will review the U.S. Green Building Council's Leadership in Energy and Environmental Design (LEED).

General Energy-Saving Measures

The following recommendations involve simple yet very effective means of increasing energy use. By following the recommendations, energy use can be minimized noticeably without resorting to major capital investment.

LIGHTING

Providing lighting within a building can account for up to 30 percent of the energy used. There are several options for reducing this energy usage. The easiest method for reducing the energy used to provide lighting is to invest in compact fluorescent lights, as opposed to traditional incandescent lights. Compact fluorescent lights use approximately 75 percent less energy than typical incandescent lights. A 15-W compact fluorescent light will supply the same amount of light as a 60-W incandescent light, while using only 25 percent of the energy. Compact fluorescent lights also last significantly longer than incandescent lights with an expected lifetime of 10,000 hours on most models. Most compact fluorescent lights also come with a 1-year warranty.

Another option for saving money and energy related to lighting is to use torchieres. In recent years halogen torchieres have become relatively popular. However, they create extremely high levels of heat; approximately 90 percent of the energy used by a halogen lamp is emitted as heat, not light. Some halogen lamps generate enough heat to fry an egg on the top of the lamp. These lamps create a fire hazard due to the possibility of curtains touching the lamp and igniting or a lamp falling over and igniting carpet. Great alternatives to these types of lamps are compact fluorescent torchieres. Whereas a halogen torchiere used 4 hours per day will consume approximately 438 kWh in a year, a compact fluorescent torchieres used 4 hours per day will only consume 80 kWh in a year. If you currently pay $0.11 per kilowatt-hour, this would save you over $30 per year, just by changing one lamp.

APPLIANCES

There are many appliances used in buildings that require a significant amount of energy to operate. However, most of these appliances are available in highly efficient models.

Refrigerators   Conventional refrigerators are a major consumer of energy. It is possible to make a refrigerator more effective and efficient by keeping it full. In the event a refrigerator is not fully stocked with food, one must consider keeping jugs of water in it. When a refrigerator is full, the contents will retain the cold. If a refrigerator is old, then consideration should be given to investing in a new, highly efficient, star-rated model. There are refrigerators on the market that use less than 20 kWh per month. When you compare this to the 110 kWh used per month by a conventional refrigerator, you can save over $90 per year (based on $0.11/kWh).

Clothes washers   Washing machines are a large consumer of not only electricity but water as well. By using a horizontal-axis washing machine, also known as a front loader because the door is on the front of the machine, it is possible to save money from using less electricity, water, and detergent.

Front loaders have a more efficient spin cycle than top loaders, which further increases savings due to clothes requiring less time in the dryer. These are the types of machines typically found in Laundromats. The machines are more cost effective than conventional top loaders. Another option is to use a natural gas or propane washer and dryer, which is currently more cost effective than using electric models. If you are on a solar or wind energy system, gas and propane are options that will reduce the over-all electricity usage of your home.

Water heaters   Water heaters can be an overwhelming load for any renewable energy system, as well as a drain on the pocketbook for those using electricity from a local utility. The following are some suggestions to increase the efficiency of electric heaters.

■ Lower the thermostat to 120 to 130°F.
■ Fix any leaky faucets immediately.
■ Wrap your water heater with insulation.
■ Turn off the electricity to your indoor water heater if you will be out of town for 3 or more days.
■ Use a timer to turn off the water heater during the hours of the day when no one is at home.

If you are looking for a higher-efficiency water heater, you may want to consider using a "flash" or "tankless" water heater, which heats water on demand. This method of heating water is very effective and does not require excessive electricity to keep a tank of water hot. It also saves water because you do not have to leave the water running out of the tap while you wait for it to get hot. Propane or natural gas water heaters are another option for those who want to minimize their electricity demand as much as possible.

INSULATION AND WEATHERIZATION

Inadequate insulation and air leakage are leading causes of energy waste in many homes. By providing adequate insulation in your home, walls, ceilings, and floors will be warmer in the winter and cooler in the summer. Insulation can also help act as a sound absorber or barrier, keeping noise levels low within the home. The first step to improving the insulation of a building is to know the type of existing insulation.

To check the exterior insulation, simply switch off the circuit breaker to an outlet on the inside of an exterior wall. Then remove the electric outlet cover and check to see if there is insulation within the wall. If there is not, insulation can be added by an insulation contractor who can blow cellulose into the wall through small holes, which are then plugged. The geometry of attics will also determine the ease with which additional insulation can be added. Insulating an attic will significantly increase the ability to keep heat in during winter and out during summer.

One of the easiest ways to reduce energy bills and contribute to the comfort of your home or office space is by sealing air leaks around windows and doors. Temporary or permanent weather stripping can be used around windows and doors. Use caulk to seal other gaps that are less than $\frac{1}{4}$ inch wide and expanding foam for larger gaps. Storm windows and insulating drapes or curtains will also help improve the energy performance of existing windows.

HEATING AND COOLING

Every indoor space requires an adequate climate control system to maintain a comfortable environment. Most people live or work in areas where the outdoor temperature fluctuates beyond ideal living conditions. A traditional air-conditioning or heating system can be a tremendous load on a solar or wind energy system, as well as a drain on the pocketbook for those connected to the utility grid. However, by following some of the insulation and weatherization tips previously mentioned, it is possible to significantly reduce heating losses and reduce the size of the heating system.

The following heating and cooling tips will help further reduce heating and cooling losses and help your system work as efficiently as possible. These tips are designed to increase the efficiency of the heating and/or cooling system without drastic remodeling. Table 6.1 shows the energy distribution in residential and commercial projects.

TABLE 6.1   ENERGY DISTRIBUTION IN RESIDENTIAL AND COMMERCIAL PROJECTS

	% OF POWER USED
APARTMENT BUILDINGS	
Environmental control	70
Lighting, receptacles	15
Water heating	3
Laundry, elevator, miscellaneous	12
SINGLE RESIDENTIAL	
Environmental control	60
Lighting, receptacles	15
Water heating	3
Laundry, pool	22
HOTEL AND MOTELS	
Space heating	60
Air conditioning	10
Lighting	11
Refrigeration	4
Laundry, kitchen, restaurant, pool	15
Water heating, miscellaneous	5
RETAIL STORES	
Environmental control HVAC	30
Lighting	60
Elevator, security, parking	10

Heating   When considering the use of renewable energy systems, electric space and water heaters are not considered viable options. These require a significant quantity of electricity to operate at a time of the year when the least amount of solar radiation is available.

Forced-air heating systems also use inefficient fans to blow heated air into rooms that may not even be used during the day. They also allow for considerable leakage through poorly sealed ductwork. Ideally, an energy-independent home or office space with a passive solar design and quality insulation will not require heating or cooling. However, if the space requires a heating source, one should consider a heater that burns fuel to provide heat and does not require electricity. Some options to consider are woodstoves and gas or propane heaters.

Cooling   A conventional air-conditioning unit is an enormous electrical load on a renewable energy system and a costly appliance to use. As with heating, the ideal energy-independent home should be designed to not require an air-conditioning unit. However, since most homeowners considering renewable energy systems are not going to redesign their home or office space, an air-conditioning unit may be necessary.

If you adequately insulate your home or office space and plug any drafts or air leaks, air-conditioning units will have to run less, which thus reduces energy expenditure. Air-conditioning units must be used only when it is absolutely necessary.

Another option is to use an evaporative cooling system. Evaporative cooling is an energy-efficient alternative to traditional air-conditioning units. Evaporative cooling works by evaporating water into the airstream. An example of evaporative cooling is the chill you get when stepping out of a swimming pool and feeling a breeze. The chill you get is caused by the evaporation of the water from your body. Evaporative cooling uses this evaporation process to cool the air passing through a wetted medium.

Early civilizations used this method by doing something as simple as hanging wet cloth in a window to cool the incoming air. Evaporative cooling is an economical and energy-efficient solution for your cooling needs. With an evaporative cooling unit there is no compressor, condenser, chiller coils, or cooling towers. Therefore, the cost of acquiring and operating an evaporative cooling unit is considerably less than for a conventional air-conditioning unit, and maintenance costs are lower due to the units requiring simpler procedures and lower-skilled maintenance workers. Also, unlike conventional air-conditioning units, evaporative cooling does not release chlorofluorocarbons (CFCs) into the atmosphere.

By following these recommendations, it is possible to turn a home or office space into an energy-efficient environment.

Power Factor Correction

The intent of the following discussion is to familiarize the reader with the basic concepts of the power factor and its effect on energy consumption efficiency. Readers interested in a further understanding of reactive power concepts should refer to electrical engineering textbooks.

In large commercial or industrial complexes where large amounts of electric power are used for fluorescent lighting or heavy machinery, the efficiency of incoming power, which is dependent upon the maintenance of the smallest possible phase angle between the current and voltage, is usually widened, thus resulting in a significant waste of energy. The cosine of the phase angle between the current and voltage referred to as the power factor, is the multiplier that determines whether the electric energy is used at its maximum to deliver lighting or mechanical energy or is wasted as heat. Power (P) in electrical engineering is defined as the product of the voltage (V) and current (I) times the cosine of the phase angle, or $P \times V \times I \times$ cosine phase angle. When the phase angle between the current and voltage is zero, the cosine equals 1 and therefore $P = V \times I$, which represents the maximum power conversion or delivery.

The principal components of motors, transformers, and lighting ballasts are wound-copper coils referred to as inductance elements. A significant characteristic of inductors is that they have a tendency to shift the current and voltage phase angles, which results in power factors that are less than 1 and, hence, in reduced power efficiency. The performance of power usage, which is defined as the ratio of the output power to the maximum power, is therefore used as the figure of merit. The reduction of electric power efficiency resulting from reactive power is wasted energy that is lost as heat. In a reactive circuit, the phase angle between current and voltage shifts, thus, giving rise to reactive power that is manifested as unused power, which dissipates as heat.

Mitigation measures that can be used to minimize inductive power loss include the installation of phase-shifting capacitor devices that negate the phase angle created by induction coils. As a rule, the maximum power affordable for efficient use of electric power should be above 93 percent. In situations where the power factor measurements indicate a value of less than 87 percent, power losses can be minimized by the use of capacitor reactance.

A Few Words about Power Generation and Distribution Efficiency

It is interesting to observe that most of us, when using electric energy, are oblivious to the fact that the electric energy provided to our household, office, or workplace is mostly generated by extremely low efficiency conversion of fossil fuels such as coal, natural gas, and crude oils. In addition to producing substantial amounts of pollutants, electric plants when generating electric power operate with meager efficiency and deliver electricity to the end user with great loss. To illustrate the point let us review the energy production and delivery of a typical electric-generating station that uses fossil fuel.

By setting an arbitrary unit of 100 percent for the fossil fuel energy input into the boilers, we see that due to losses resulting from power plant machinery, such as turbines, generators, high-voltage transformers, transmission lines, and substations, the efficiency of delivered electric energy at the destination is no more than 20 to 25 percent. The efficiency of energy use is further reduced when the electric energy is used by motors, pumps, and a variety of equipment and appliances that have their own specific

TABLE 6.2   SOLAR POWER AND FOSSIL FUEL POWER GENERATION COMPARISON TABLE

	SOLAR ELECTRIC POWER	FOSSIL FUEL ELECTRIC POWER
Delivery efficiency	Above 90%	Less than 30%
Maintenance	Very minimal	Considerable
Transmission lines	None required	Very extensive
Equipment life span	25–45 years	Maximum of 25 years
Investment payback	8–14 years	20–25 years
Environmental impact	No pollution	Very high pollution index
Percent of total U.S. energy	Less than 1%	Over 75%
Reliability index	Very high	Good

performance losses. Table 6.2 depicts comparative losses between solar and fossil fuel power generation systems. As evidenced by this example, when comparing solar power generation with electric power generated by fossil fuel, the advantages of solar power generation in the long run become quite obvious. Figure 6.1 depicts transmission and distribution losses associated with electric fossil fuel power generation delivery losses.

A short-sighted assessment by various experts siding with conventional fossil power generation is that less burning of coal and crude oil to minimize or prevent global warming will increase the national expenditure to such a degree that governments will be prevented from meeting the society's needs for transportation, irrigation, heating,

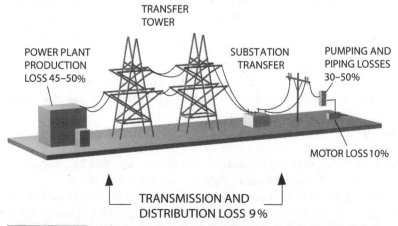

Figure 6.1   Transmission and distribution losses associated with electric fossil fuel power generation delivery losses.

and many other energy-dependent services. On the other hand, environmentalists argue that protection of nature and the prevention of global warming warrant the required expenditure to prevent inevitable climatic deterioration.

With advances in technology, the increased output efficiency of solar PV modules and the reduction in the cost of PV modules, which would result from mass production, will within the next decade make solar power installation quite economical. National policies should take into consideration that technologies aimed at reducing global warming could indeed be a major component of the gross national income and that savings from fossil fuel consumption could be much less than the expenditure for research and development of solar power and sustainable energy technologies.

In the recent past some industry leaders, such as DuPont, IBM, Alcan, NorskeCanada, and British Petroleum, have expended substantial capital toward the reduction of carbon dioxide and greenhouse gas emissions, which has resulted in billions of dollars of savings.

For example, British Petroleum has reduced carbon dioxide emissions by 10 percent in the past 10 years and as a result has cut $650 million of expenses. DuPont by reducing 72 percent of greenhouse gases has increased its production by 30 percent, which resulted in $2 billion of savings. The United States at present uses 47 percent less energy dollars than it did 30 years ago, which results in $1 billion per day of savings.

Computerized Lighting Control

In general, conventional interior lighting control is accomplished by means of hard-wired switches, dimmers, timers, lighting contactor relays, occupancy sensors, and photoelectric eyes that provide the means to turn various light fixtures on and off or to reduce luminescence by dimming.

The degree of interior lighting control in most instances is addressed by the state of California Title 24 energy regulations, which dictate specific design measures required to meet energy conservation strategies including:

■ Interior room illumination switching
■ Daylight illumination control or harvesting
■ Duration of illumination control by means of a preset timing schedule
■ Illumination level control specific to each space occupancy and task environment
■ Lighting zone system management
■ Exterior lighting control

Figures 6.2 through 6.10 depict various wiring diagrams and lighting control equipment used to increase illumination energy consumption efficiency.

In limited spaces such as small offices, commercial retail, and industrial environments (where floor spaces do not exceed 10,000 ft²), lighting control is undertaken by hardwiring of various switches, dimmers, occupancy sensors, and timers. However, in large environments, such as high-rise buildings and large commercial and industrial

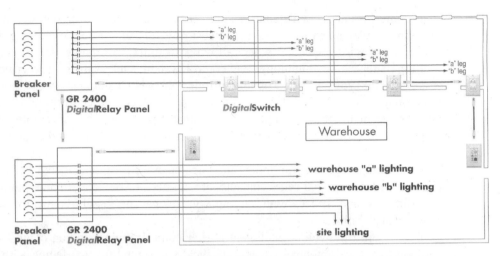

Figure 6.2   **A typical centralized lighting control wiring plan.** *Photo courtesy of LCD.*

environments, lighting control is accomplished by a computerized automation system that consists of a centralized control and display system that allows for total integration of all the preceding components.

A central lighting control system embeds specific software algorithms that allow for automated light control operations to be tailored to meet specific energy and automation management requirements unique to a special environment. An automated lighting control system, in addition to reducing energy waste to an absolute minimum, allows for total operator override and control from a central location.

Because of its inherent design, the centralized lighting control system offers indispensable advantages that cannot be accomplished by hard-wired systems. Some of these are as follows:

■ Remote manual or automatic on-off control of up to 2400 lighting groups within a predetermined zone.
■ Remote dimming of lighting within each zone.
■ Automatic sequencing control of individual groups of lights.
■ Sequencing and graded dimming or step activation of any group of lights.
■ Remote status monitoring of all lights within the overall complex.
■ Inrush current control for incandescent lights, which substantially prolongs the life expectancy of lamps.
■ Visual display of the entire system illumination throughout the complex by means of graphic interfaces.
■ Free-of-charge remote programming and maintenance of the central lighting and control unit from the equipment supplier's or manufacturer's headquarters.
■ Optional remote radio communication interfaces that allow for control of devices at remote locations without the use of conduits and cables.

In some instances, radio control applications can eliminate trenching and cable installation, which can offset the entire cost of a central control system. Contrary to conventional wiring schemes where all wires from fixtures merge into switches and lighting panels, an intelligent lighting control system, such as the one described here, makes use of Type 5 cable (a bundle of four-pair twisted shielded wires), which can interconnect up to 2400 lighting control elements. A central control and monitoring unit located in an office constantly communicates with a number of remotely located intelligent control boxes that perform the lighting control measures required by Title 24 and beyond.

Since remote lighting, dimming, and occupancy sensing is actuated by means of electronically controlled relay contacts, any number of devices such as pumps, outdoor fixtures, and devices with varying voltages can be readily controlled with the same master station.

In addition to providing intelligent master control, remote station control devices and intelligent wall-mount switches specifically designed for interfacing with intelligent remote devices provide local lighting and dimming control override. Moreover, a centralized lighting control system can readily provide required interlocks between heating, ventilation, and air-conditioning (HVAC) systems by means of intelligent thermostats. Figure 6.3 depicts a relay lighting control wiring plan.

Even though central intelligent lighting control systems, such as the one described here, add an initial cost component to the conventional wiring, in the long run the extended expectancy of lamps, lower maintenance cost, added security, and considerable savings resulting from energy conservation undoubtedly justify the added initial investment.

In fact, the most valuable feature of the system is flexibility of control and ease of system expansion and reconfiguration. The system is indispensable for applications such as the Water and Life Museum project discussed in Chapter 5.

The major cost components of a centralized lighting control system consist of the central- and remote-controlled hardware and dimmable fluorescent T8 ballasts. It is a well-established fact that centralized lighting control systems pay off in a matter of a few years and provide a substantial return on investment by the sheer savings on energy consumption. Needless to say, no measure of security can be achieved without central lighting control. Figure 6.4 depicts a centralized dimming and lighting control diagram.

The automated centralized lighting control system manufactured by Lighting Control & Design (LCD), and shown in Figure 6.4, provides typical control components used to achieve the energy conservation measures discussed earlier. Note that the lighting control components and systems presented in this chapter are also available from Lutron and several other companies.

Some of the major lighting system components available for system design and integration include centralized microprocessor-based lighting control relays that incorporate 32 to 64 addressable relay channels, 365-day programmable astronomical timers, telecommunication modems, mixed voltage output relays (120 or 277 V), manual override for each relay, and a linkup capability of more than 100 links to digital

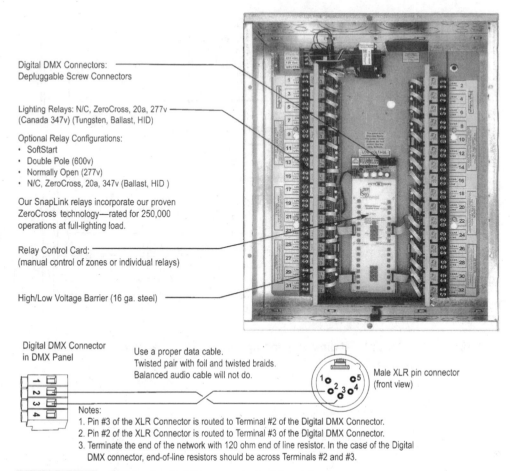

Digital DMX Connectors:
Depluggable Screw Connectors

Lighting Relays: N/C, ZeroCross, 20a, 277v
(Canada 347v) (Tungsten, Ballast, HID)

Optional Relay Configurations:
• SoftStart
• Double Pole (600v)
• Normally Open (277v)
• N/C, ZeroCross, 20a, 347v (Ballast, HID)

Our SnapLink relays incorporate our proven
ZeroCross technology—rated for 250,000
operations at full-lighting load.

Relay Control Card:
(manual control of zones or individual relays)

High/Low Voltage Barrier (16 ga. steel)

Digital DMX Connector
in DMX Panel

Use a proper data cable.
Twisted pair with foil and twisted braids.
Balanced audio cable will not do.

Male XLR pin connector
(front view)

Notes:
1. Pin #3 of the XLR Connector is routed to Terminal #2 of the Digital DMX Connector.
2. Pin #2 of the XLR Connector is routed to Terminal #3 of the Digital DMX Connector.
3. Terminate the end of the network with 120 ohm end of line resistor. In the case of the Digital
   DMX connector, end-of-line resistors should be across Terminals #2 and #3.

Figure 6.3   **DMX relay lighting control wiring plan.** *Photo courtesy of LCD.*

devices via category 5 patch cables and RJ45 connectors. Figure 6.5 depicts a centralized dimming circuit diagram.

The preceding systems also include smart breaker panels that use solenoid-operated thermal magnetic breakers that effectively provide overcurrent protection as well as lighting control. Overcurrent devices are usually available as single- or three-phase; with a current rating of 15, 20, and 30 A; and with an arc current interrupt capacity (AIC) of 14 kiloamperes (kA) @ 120/208 V and 65kA @ 277/480 V.

A microprocessor-based, current-limiting subbranch distribution panel provides lighting calculations for most energy regulated codes. For example, California's Title 24 energy compliance requirements dictate 45 W of linear power for track lighting, while the city of Seattle in the state of Washington requires 70 W/ft for the same track lighting system. The current limiting subpanel effectively provides a programmable circuit current-limiting capability that lowers or raises the voltampere (VA) rating requirement for track lighting circuits. The current-limiting capacity for a typical

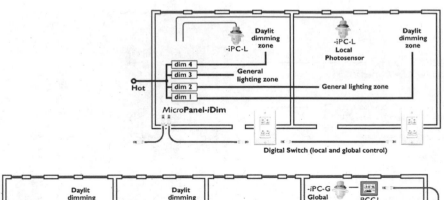

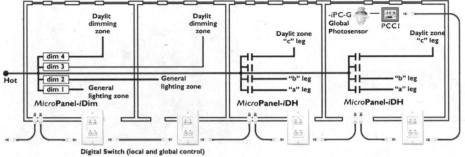

Figure 6.4   **Centralized dimming and lighting control diagram.** *Graphic courtesy of LCD.*

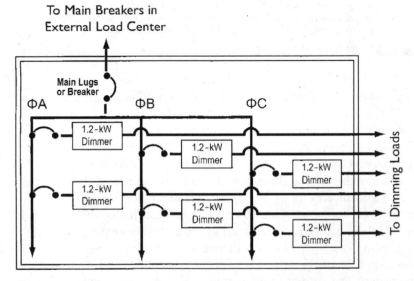

Dimming Cabinet with Distribution

Figure 6.5   **Centralized dimming circuit diagram.**

Graphic courtesy of LCD.

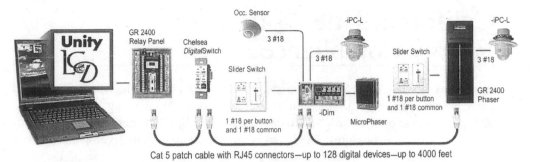

Figure 6.6   **Remote lighting control component configuration.** *Graphic courtesy of LCD.*

panel is 20 circuits, with each capable of limiting current from 1 to 15 A. Figure 6.6 depicts a remote lighting control component configuration.

Another useful lighting control device is a programmable zone lighting control panel, which is capable of the remote control of 512 uniquely addressable lighting control relays. Groups of relays can either be controlled individually, referred to as discrete mode, or can be controlled in groups, referred to as zone mode. Lighting relays in typical systems are extremely reliable and are designed to withstand 250,000 operations at full load capacity.

For limited area lighting control a compact microprocessor-based device, referred to as a micro control, provides a limited capability for controlling two to four switches and dimmable outputs. All microcontrolled devices are daisy chained and communicate with a central lighting command and control system.

A desktop personal computer with a monitor located in a central location (usually the security room) communicates with all the described lighting system panels and microcontrollers via twisted shielded category 5 communication cables. Wireless modem devices are also available as an alternative hard-wired system.

Other optional equipment and devices available for lighting control include digital astronomical time clocks, prefabricated connector cables, dimmer switches, lock-type switches, indoor and outdoor photosensor devices, and modems for remote communication. Figure 6.7 depicts a centralized light monitoring and control system.

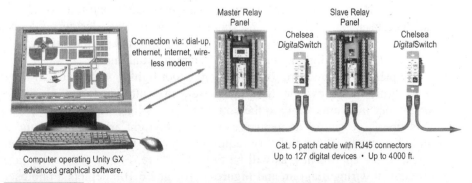

Figure 6.7   **Centralized light monitoring and control system.** *Graphic courtesy of LCD.*

California Title 24 Electric Energy Compliance

In response to the 2000 electricity crisis, the state of California legislature mandated the California Energy Commission (CEC) to update the existing indoor lighting energy conservation standards and to develop outdoor lighting energy efficiency compliant cost-effective measures. The intent of the legislature was to develop energy conservation standards that would reduce electricity system energy consumption.

Regulations for lighting have been enforced in California since 1977. However, the measures only addressed indoor lighting through control requirements and maximum allowable lighting power. Figure 6.8 depicts a local microprocessor-based control panel.

SCOPE AND APPLICATION

Earlier energy regulation standards only applied to interior and outdoor lighting of buildings that were air-conditioned, heated, or cooled. The updated standards, however, address lighting in non–air-conditioned buildings and also cover general site illumination and outdoor lighting. The standards include control requirements, as well as limits on installed lighting power, and also apply to internally and externally illuminated signs.

For detailed coverage of the energy control measures and regulations refer to the California Energy commission's standard publications.

Indoor Lighting Compliance

In this section we will review the requirements for indoor lighting design and installation, including controls. This discussion is addressed primarily to lighting designers, electrical engineers, and building department personnel responsible for lighting and electrical plan checking and inspection purposes.

Indoor lighting is perhaps the single largest consumer of energy (kilowatt-hours) in a commercial building, which amounts to approximately one-third of electric energy use. The principal purpose of the standards is to provide design guidelines for the effective reduction of energy use, without compromising the quality of lighting. Figure 6.12 depicts lighting energy use in a residential unit.

The primary mechanism for regulating indoor lighting energy under the standards is to limit the allowable lighting power, in watts, installed in the buildings. Mandatory measures apply to the entire building's lighting systems, and equipment consists of such items as manual switching, daylight area controls, and automatic shutoff controls. The mandatory requirements must be met either by prescriptive or performance approaches, as will be described here. Figure 6.9 depicts a photosensor control wiring diagram and Figures 6.10a and 6.10b depicts a photosensor control configurations.

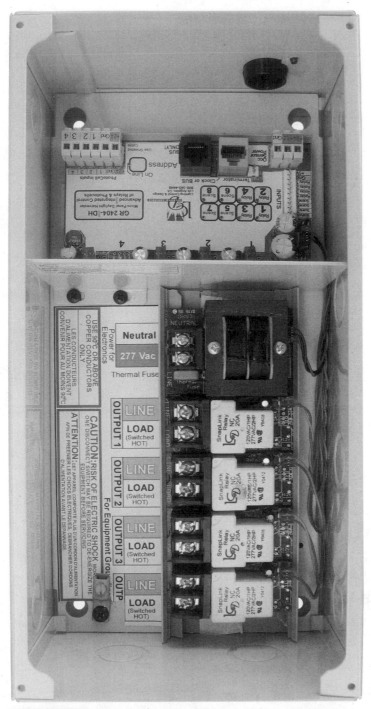

Figure 6.8   Local microprocessor-based control panel.

Photo courtesy of LCD.

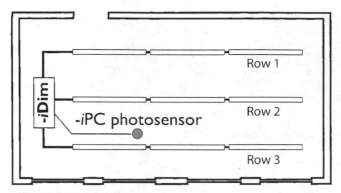

Figure 6.9   Photosensor control wiring diagram.
Graphic courtesy of LCD.

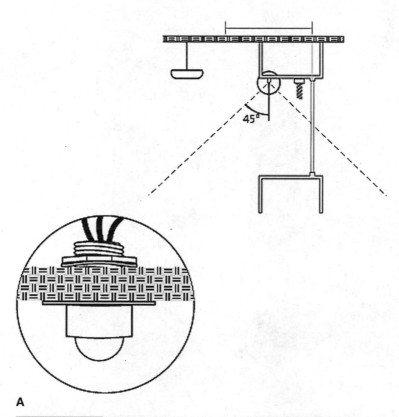

A

Figure 6.10   (A) Photosensor Omni directional control configuration scheme. (B) Photosensor omnidirectional control configuration scheme. *Graphics courtesy of LCD.*

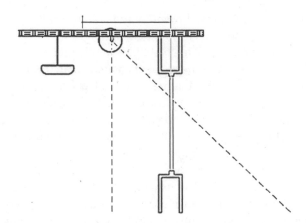

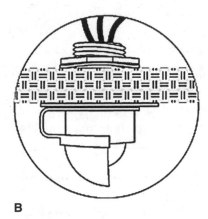

B

Figure 6.10 *(Continued)*

As a rule, allowed lighting power for a building is determined by one of the following five methods:

1 *Complete building method.* This method applies to situations when the entire building's lighting system is designed and permitted at one time. This means that at least 90 percent of the building has a single primary type of use, such as retail. In the case of wholesale stores, at least 70 percent of the building area must be used for merchandise sales functions. In some instances this method may be used for an entire tenant space in a multitenant building where a single lighting power value governs the entire building.

2 *Area category method.* This method is applicable for any permit situation, including tenant improvements. Lighting power values are assigned to each of the major function areas of a building, such as offices, lobbies, and corridors.

3 *Tailored method.* This method is applicable when additional flexibility is needed to accommodate special task lighting needs in specific task areas. Lighting power

allowances are determined room by room and task by task, with the area category method used for other areas in the building.

4 *Performance approach.* This method is applicable when the designer uses a CEC-certified computer program to demonstrate that the proposed building's energy consumption, including lighting power, meets the energy budget. This approach incorporates one of the three previous methods, which sets the appropriate allowed lighting power density used in calculating the building's custom energy budget. It may only be used to model the performance of lighting systems that are covered under the building permit application.

5 *Actual adjusted lighting power method.* This method is based on the total design wattage of lighting, less adjustments for any qualifying automatic lighting controls, such as occupant-sensing devices or automatic daylight controls. The actual adjusted lighting power must not exceed the allowed lighting power for the lighting system to comply.

LIGHTING TRADEOFFS

The intent of energy control measures is to essentially restrict the overall installed lighting power in the buildings, regardless of the compliance approach. Note that there is no general restriction regarding where or how general lighting power is used, which means that installed lighting could be greater in some areas and lower in others, provided that the total lighting energy wattage does not exceed the allowed lighting power.

A second type of lighting tradeoff, which is also permitted under the standards, is a tradeoff of performance between the lighting system and the envelope or mechanical systems. Such a tradeoff can only be made when permit applications are sought for those systems filed under performance compliance where a building with an envelope or mechanical system has a more efficient performance than the prescriptive efficiency energy budget, in which case more lighting power may be allowed. Figure 6.11 is a residential energy use distribution.

When a lighting power allowance is calculated using the previously referenced performance approach, the allowance is treated as if it is determined using one of the other compliance methods. Note that no tradeoffs are allowed between indoor lighting and outdoor lighting or lighting located in non–air-conditioned spaces.

MANDATORY MEASURES

Mandatory measures are compliance notes that must be included in the building design and on the engineering or Title 24 forms stating whether compliance is of the prescriptive or performance method building occupancy type.

The main purpose of mandatory features is to set requirements for manufacturers of building products, who must certify the performance of their products to the CEC. However, it is the designer's responsibility to specify products that meet these requirements.

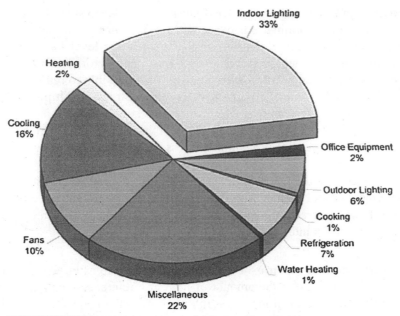

Indoor Lighting
33%

Heating
2%

Cooling
16%

Office Equipment
2%

Outdoor Lighting
6%

Cooking
1%

Refrigeration
7%

Fans
10%

Water Heating
1%

Miscellaneous
22%

Figure 6.11   **Residential energy use distribution.**
Chart courtesy of CEC.

LIGHTING EQUIPMENT CERTIFICATION

The mandatory requirements for lighting control devices include minimum specifications for features such as automatic time control switches, occupancy-sensing devices, automatic daylighting controls, and indoor photosensor devices. The majority of the requirements are currently part of standard design practice in California and are required for electrical plan checking and permitting.

Without exception all lighting control devices required by mandatory measures must be certified by the manufacturer before they can be installed in a building. The manufacturer must also certify the devices to the CEC. Upon certification, the device is listed in the Directory of Automatic Lighting Control Devices.

Automatic time switches   Automatic time switches, sometimes called time clocks, are programmable switches that are used to automatically shut off the lights according to preestablished schedules depending on the hours of operation of the building. The devices must be capable of storing weekday and weekend programs. In order to avoid the loss of programmed schedules, timers are required to incorporate backup power provision for at least 10 hours during power loss.

Occupancy-sensing devices   Occupancy-sensing devices provide the capability to automatically turn off all lights in an area for no more than 30 minutes after the area has been vacated. Sensor devices that use ultrasonic sensing must meet certain minimum health requirements and must have the built-in ability for sensitivity calibration to prevent false signals that may cause power to turn on and off.

Automatic daylight controls Daylighting controls consist of photosensors that compare actual illumination levels with a reference illumination level and gradually reduce the electric lighting until the reference level has been reached. These controls are also deployed for power adjustment factor (PAF) lighting credits in the day-lit areas adjacent to windows. It is also possible to reduce the general lighting power of the controlled area by separate control of multiple lamps or by step dimming.

Stepped dimming with a time delay prevents cycling of the lights, which is typically implemented by a time delay of 3 minutes or less before electric lighting is reduced or is increased.

Light control in daylight is accomplished by use of photodiode sensors. Note that this requirement cannot be met with devices that use photoconductive cells. In general, stepped switching control devices are designed to indicate the status of lights in controlled zones by an indicator.

Interior photosensor device Daylighting control systems in general use photosensor devices that measure the amount of light at a reference location. The photosensor provides light level illumination information to the controller, which in turn enables it to increase or decrease the area electric light level.

Photosensor devices must, as previously stated, be certified by the CEC. Devices having mechanical slide covers or other means that allow for adjusting or disabling of the photosensor are not permitted or certified.

Multilevel astronomical time switch controls Areas with skylights that permit daylight into a building area are required to be calculated by the prescriptive calculation method and to be controlled by mandatory automatic controls that must be installed to reduce electric lighting when sufficient daylight is available. Multilevel astronomical time switch controls or automatic multilevel daylight controls specially designed for general lighting control must meet the mandatory requirements for automatic controls when the particular zone has an area greater than 2500 ft^2.

The purpose of astronomical time switch controls is to turn off lights where sufficient daylight is available. Astronomical timers accomplish this requirement by keeping track of the time since sunrise and amount of time remaining before sunset. As a basic requirement, the control program must accommodate multilevel two-step control for each zone programmed to provide independently scheduled activation and deactivation of the lights at different times.

In the event of overly cloudy or overly bright days the astronomical timers are required to have manual override capability. Usually, the override switches in a zone are configured so that lights will revert to the off position within 2 hours, unless the time switch schedule is programmed to keep the lights on.

To comply with the power consumption regulation requirements, light control is not allowed to be greater than 35 percent of the total lighting load at the time of minimum light output. Device compliance also mandates that devices be designed to display the date and time, sunrise and sunset times, and switching times for each step of control.

To prevent a loss of settings due to a temporary loss of power, timers are required to have a 10-hour battery backup. Astronomical timers also are capable of storing the time of day and the longitude and latitude of a zone in nonvolatile memory.

Automatic multilevel daylight controls   Mandatory requirements stipulate that automatic multilevel daylight controls be used when the daylight area under skylights is greater than 2500 ft². In these circumstances the power consumption must not be greater than 35 percent of the minimum electric light output. This is achieved when the timer control automatically turns all its lights off or reduces the power by 30 percent.

Multilevel daylight control devices incorporate calibration and adjustment controls that are accessible to authorized personnel and are housed behind a switch plate cover, touch plate cover, or in an electrical box with a lock. They must have a minimum of two control steps so that electric lighting can be uniformly reduced. One of the control steps is intended to reduce lighting power from 100 percent to 70 and the second step to 50 percent of full rated power.

Fluorescent dimming controls, even though somewhat expensive, usually meet the minimum power requirements. Controls for high-intensity discharge (HID) lamps do not meet the power requirements at minimum dimming levels; however, a multistage HID lamp switching control can.

Outdoor astronomical time switch controls   Outdoor lighting control by means of astrological time switches is permitted if the device is designed to accommodate automatic multilevel switching of outdoor lighting. Such a control strategy allows all, half, or none of the outdoor lights to be controlled during different times of the day, for different days of the week, while ensuring that the lights are turned off during the daytime.

Incidentally, this feature is quite similar to the indoor multilevel astronomical control with the exception that this control scheme offers a less stringent offset requirement from sunrise or sunset. Mandatory certification for this device requires the controller to be capable of independently offsetting on-off settings for up to 120 minutes from sunrise or sunset.

INSTALLATION REQUIREMENTS

Automatic time switch control devices or occupant sensors for automatic daylight control must be installed in accordance with the manufacturer's instructions. They must also be installed so that the device controls only luminaires within daylit areas, which means that photosensors must either be mounted on the ceiling or installed in locations that are accessible only to authorized personnel. Requirements for specific items are as follows:

Certified ballasts and luminaires. All fluorescent lamp ballasts and luminaires are regulated by the Appliance Efficiency Regulations certified by the CEC and are listed in the efficiency database of these regulations.

Area controls. The best way to minimize energy waste and to increase efficiency is to turn off the lights when they are not in use. All lights must have switching or controls to allow them to be turned off when not needed.

Room switching. It is mandatory to provide lighting controls for each area enclosed by ceiling height partitions, which means that each room must have its own switches. Ganged switching of several rooms at once is not permitted. A switch may be manually or automatically operated or controlled by a central zone lighting or occupant-sensing system that meets the mandatory measure requirements.

Accessibility. It is mandatory to locate all switching devices in locations where personnel can see them when entering or leaving an area. In situations when the switching device cannot be located within view of the lights or area, the switch position and states must be annunciated or indicated on a central lighting panel.

Security or emergency. Lights within areas required to be lit continuously or for emergency egress are exempt from the switching requirements. However, the lighting level is limited to a maximum of 0.5 W/ft^2 along the path of egress. Security or emergency egress lights must be controlled by switches accessible only to authorized personnel.

Public areas. In public areas, such as building lobbies and concourses, switches are usually installed in areas only accessible to authorized personnel.

Outdoor Lighting and Signs

In response to the electricity crisis in 2000, the California legislature mandated the CEC to develop outdoor lighting energy efficiency standards that are technologically feasible and cost effective. The purpose of the legislature was to develop energy efficiency standards that could provide comprehensive energy conservation.

OUTDOOR ASTRONOMICAL TIME SWITCH CONTROLS

As briefly referenced earlier, outdoor lighting control by means of astrological time switches is permitted if the device is designed to accommodate automatic multilevel switching of outdoor lighting. Basically, such a control allows all, half, or none of the outdoor lights to be controlled during different times of the day, for different days of the week, while ensuring that the lights are turned off during the daytime.

Energy control measures for outdoor lighting and signs are intended to conserve energy and reduce winter peak electric demand. The standards also set design directives for minimum and maximum allowable power levels for large luminaires.

Permitted lighting power levels are based on Illuminating Engineering Society of North America (IESNA) recommendations, which are industry standard practices that have worldwide recognition. Note that outdoor lighting standards do not allow trade-offs between interior lighting, HVAC, building envelope, or water heating energy conformance requirements.

OUTDOOR LIGHTING ENERGY TRADEOFFS

Outdoor lighting tradeoffs are allowed only between the lighting applications with general site lighting illumination, which includes hardscape areas, building entrances without canopies, and outdoor sales lots. The requirements do not permit any trade-offs between outdoor lighting power allowances and interior lighting, HVAC, building envelope, or water heating. This includes decorative gas lighting; lighting for theatrical purposes, including stage, film, and video production; and emergency lighting powered by an emergency source as defined by the CEC.

SUMMARY OF MANDATORY MEASURES

The imposed mandatory measures on outdoor lighting include automatic controls that are designed to turn off outdoor lighting during daytime hours and during other times when it is not needed. The measures also require that all controls be certified by the manufacturer and listed in CEC directories. All luminaires with lamps larger than 175 W are required to have cutoff baffles so as to limit the light directed toward the ground. Luminaires with lamps larger than 60 W are also required to be high efficiency or controlled by a motion sensor.

The new CEC standards also limit the lighting power for general site illumination and for some specific outdoor lighting applications. General site illuminations specifically include lighting for parking lots, driveways, walkways, building entrances, sales lots, and other paved areas of a site. The measures also provide separate allowances for each of the previously referenced general site lighting applications and allow tradeoffs among these applications. In other words, a single aggregate outdoor lighting budget can be calculated for all the site applications together. Hardscape for automotive vehicular use, including parking lots; driveways and site roads; and pedestrian walkways, including plazas, sidewalks, and bikeways, are all considered general site lighting applications.

General site lighting also includes that for building entrances and facades such as outdoor sales lots, building facades, outdoor sales frontages, service station canopies, vehicle service station hardscape, other nonsales canopies, ornamental lightings, drive-up windows, guarded facilities, outdoor dining, and temporary outdoor lighting. Site lighting is also regulated by the Federal Aviation Regulation Standards.

General lighting standards also cover lighting of sports and athletic fields, children's playgrounds, industrial sites, automated teller machines (ATMs), public monuments, swimming pools or water features, tunnels, bridges, stairs, and ramps. Tradeoffs are not permitted for specific application lighting.

Allowable lighting power for both general site illumination and specific applications are governed by four separate outdoor lighting zone requirements, as will be described later. The lighting zones in general characterize ambient lighting intensities in the surrounding areas. For example, sites that have high ambient lighting levels have a larger allowance than sites with lower ambient lighting levels. The following are Title 24 CEC zone classifications:

Zone LZ1. Government assigned area

Zone LZ2. Rural areas as defined by the U.S. 2000 census

Zone LZ3. Urban areas as defined by the U.S. 2000 census

Zone LZ4. Currently not defined

SIGNS

Sign standards contain both prescriptive and performance approaches. Sign mandatory measures apply to both indoor and outdoor signs. Prescriptive requirements apply when the signs are illuminated with efficient lighting sources, such as electronic ballasts, while the performance requirement is applied when calculating the maximum power defined as a function of the sign surface area in watts per square foot.

INSTALLED POWER

The installed power for outdoor lighting applications is determined in accordance with specific mandatory measure calculation guidelines. Luminaire power for pin-based and high-intensity discharge lighting fixtures may be used as an alternative to determine the wattage of outdoor luminaires. Luminaires with screw-base sockets and lighting systems, which allow the addition or relocation of luminaires without modification to the wiring system, must follow the required guidelines. In commercial lighting systems no power credits are offered for automatic controls; however, the use of automatic lighting controls is mandatory.

MANDATORY MEASURES

Similar to indoor lighting, mandatory features and devices must be included in all outdoor lighting project documentation, whenever applicable. The mandatory measures also require the performance of equipment to be certified by the manufacturers and that fixtures rated 100 W or more must have high efficiency; otherwise they are required to be controlled by a motion sensor. Fixtures with lamps rated 175 W or more must incorporate directional baffles to direct the light toward the ground.

Fixture certification   Manufacturers of lighting control products are required to certify the performance of their products with the CEC. Lighting designers and engineers must assume responsibility to specify products that meet these requirements. As a rule, inspectors and code enforcement officials are also required to verify that the lighting controls specified carry CEC certification. The certification requirement applies to all lighting control equipment and devices such as photocontrols, astronomical time switches, and automatic controls.

Control devices are also required to have instructions for installation and startup calibration and must be installed in accordance with the manufacturer's directives. The control equipment and devices are required to have a visual or audio status signal that activates upon malfunction or failure.

Minimum lamp efficiency   All outdoor fixtures with lamps rated over 100 W must either have a lamp efficiency of at least 60 lumens per watt (lm/W) or must be controlled by a motion sensor. Lamp efficiencies are rated by the initial lamp lumens divided by the rated lamp power (W), without including auxiliary devices such as ballasts.

Fixtures that operate by mercury vapor principles and larger-wattage incandescent lamps do not meet these efficiency requirements. On the other hand, most linear fluorescent, metal halide, and high-pressure sodium lamps have lamp efficiencies greater than 60 lm/W and do comply with the requirements.

The minimum lamp efficiency does not apply to lighting regulated by a health or life safety statute, ordinance, or regulation, which includes, but is not limited to, emergency lighting. Also excluded are fixtures used around swimming pools; water features; searchlights or theme lighting used in theme parks, film, or live performances; temporary outdoor lighting; light-emitting diodes (LED); and neon and cold cathode lighting.

Cutoff luminaires   Outdoor luminaires with lamps rated greater than 175 W that are used in parking lots and other hardscapes, outdoor dining areas, and outdoor sales areas are required to be fitted with cutoff-type baffles or filters. They must also be specifically rated as "cutoff" in a photometric test report. A cutoff-type luminaire is defined as one where no more than 2.5 percent of the light output extends above the horizon 90 degrees or above the nadir and no more than 10 percent of the light output is at or above a vertical angle of 80 degrees above the nadir. The nadir is a point in the direction straight down, as would be indicated by a plumb line. Ninety degrees above the nadir is horizontal. Eighty degrees above the nadir is 10 degrees below horizontal.

Case study in the application of DC photovoltaic energy   The following is a case study of the application of solar power cogeneration dc electric energy without conversion to alternating current. The technology discussed is based on electronic lighting and rotating machinery control devices that operate with direct current. By eliminating the use of dc-to-ac conversion devices, solar power is applied without the losses associated with these devices.

To apply directly harvested dc power in electrical wiring, lighting devices and machinery must have specifically designed lighting ballasts and rotary machinery drive controls. As an example, Nextek Power Systems, a solar power technology development company located in New York State, has developed specially designed dc fluorescent ballasts and lighting control systems whereby the dc power harvested from photovoltaic modules is accumulated, controlled, and distributed from centrally located power routers.

A power router is an electronic dc energy control device that embodies the following functions:

- Routes all dc power harvested by photovoltaic modules as a primary source of energy and directs it either to the dc lighting control devices or channels the power to a bank of fast-discharge solar power battery banks for energy storage.
- In the event of loss of solar dc energy from the photovoltaic solar power system, it routes conventional electric energy as a secondary source of energy whereby the

alternating current is converted into direct current that provides energy for lighting and storage.

■ In the event of solar and electrical service power loss, the direct current stored in the battery banks is routed to the lighting control or machinery drives.

Nextek Power Systems recently designed and installed their high-efficiency renewable energy lighting system at a distribution center in Rochester, New York. The project, which is a LEED-rated gold facility, is equipped with a lighting system that utilizes dc fluorescent ballasts, roof-integrated solar panels, occupancy sensors, and daylight sensors for the highest possible efficiency. The building was designed by William McDonough and Partners of Charlottesville, Virginia.

The facility has 6600 ft^2 of office space and 33,000 ft^2 of warehouse space. The warehouse roof is equipped with skylights and 21 kW of amorphous panels by Solar Integrated Technologies that are bonded to the roof material. A canopy in the office area is also equipped with 2.1 kW of Sharp photovoltaic solar panels. Figure 6.12 depicts wiring diagram for NPS1000 module a direct PV output dc wiring with battery backup.

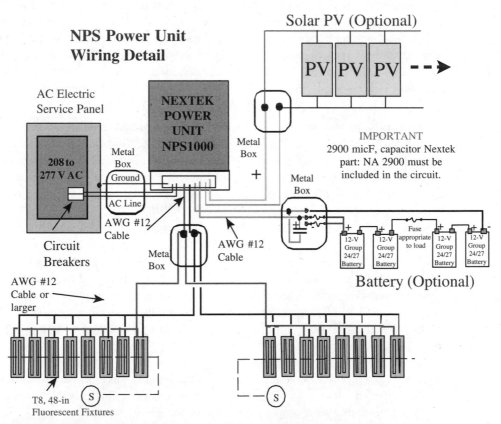

Figure 6.12   Wiring diagram for NPS1000 module a direct PV output dc wiring with battery backup. *Photo courtesy of Nextek Power Systems.*

The power from the solar panels is distributed in the following three ways: (1) 2.2 kW is dedicated to the lighting in the office; (2) 11.5 kW powers the lights in the warehouse; and (3) 11.5 kW is not needed by the lighting system, so it is inverted to alternating current and used elsewhere in the building or sold back to the utility.

The entire system consists of 35 Nextek model NPS1000 smart power routers. As mentioned previously, these devices take all the power from the solar panels and send it directly to the lighting without significant losses. Additional power, when needed at night or on cloudy days, is taken from the grid. Figure 6.13 depicts an interior lighting cluster control wiring plan.

In this project a number of NPS1000 lighting control power modules provide electric energy to 198 four-foot fluorescent fixtures that use two T8 energy-efficient lamps, illuminating the entire warehouse. Each of the fixtures is equipped with a high-efficiency dc ballast. Fixtures are controlled by a combination of manual switches, daylight sensors, and occupancy sensors located in various zones. Lamp fixtures with more than two lamps are controlled in a manner so that they can be dimmed by

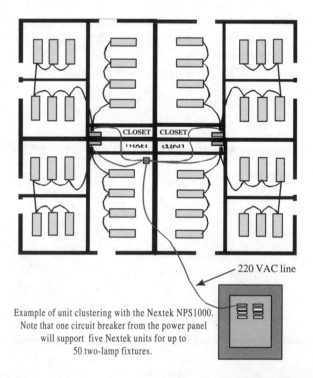

CLUSTERING ELECTRICAL LOADS
AROUND THE POWER MODULE

CLOSET   CLOSET

220 VAC line

Example of unit clustering with the Nextek NPS1000.
Note that one circuit breaker from the power panel
will support five Nextek units for up to
50 two-lamp fixtures.

Figure 6.13   **Interior lighting cluster control wiring.** *Photo courtesy of Nextek Power Systems.*

daylight sensors and occupancy sensors located throughout the area. The goal of the control architecture is to maintain a lighting nominal level with the use of maximum daylight whenever available

The strategy of the lighting control system is to provide optimum energy efficiency. This is achieved by a prioritizing scheme to make use of maximum daylight from the skylights, harvest the maximum power from the solar panels, and finally provide a fallback to grid power whenever daylight harvesting and solar power is not sufficient to maintain the lighting energy requirement.

A number of factors that contribute to the value of this system include the following:

- Using the electricity generated by the solar panels to power the lighting eliminates significant inverter losses and improves efficiency by as much as 20 percent.
- The low voltage control capability of the dc ballasts enables the innovative control system to be installed easily, without additional ac wiring.
- Roof-integrated solar panels reduce installation costs and allow the cost of the roof to be recovered using a 5-year accelerated depreciation formula.

Performance—Occupancy and Daylight Sensors

Most of the lighting comes on at 3 a.m. All lights are turned on from 6 a.m. to 6 p.m. The blue line shows the lighting load with the occupancy and daylight sensors controlling the lighting. Between March and mid-June 2005 about 20 to 30 percent savings have been achieved due to the controls.

An example of energy savings is shown in the following table:

Saved with controls	% kWh Saved	kWh saved $
$497.02	March—05 31%	4970
$345.80	April—05 22%	3458
$342.85	May—05 22%	3429
$324.75	June—05 30%	3248
$1,510.43	Total	15,104

The savings shown in the table are based on utility costs of $0.10/kWh. Savings could be significantly higher in areas such as southern California where electric energy tariffs are higher.

Web-Based Display Monitoring System

A special monitoring system developed by Nextek provides remote data acquisition and monitoring from a distant location over the Web. The data monitored and displayed include solar power generated, power used, and weather meteorological parameters. To access data from a remote location the client is provided with a special security password that allows the system performance parameters to be monitored from any location. In addition, the supervisory system also identifies anomalies in the system such as burned-out lamps and malfunctioning sensors.

Solar Power Facts

- Recent analysis by the U.S. Department of Energy showed that by the year 2025 half of new U.S. electricity generation could come from the sun.
- Based on current U.S. Department of Energy analysis, current known worldwide crude oil reserves at the present rate of consumption will last about 50 years.
- In the United States, 89 percent of the energy budget is based on fossil fuels.
- In 2000, the United States generated only 4 gigawatts (1 GW is 1000 MW) of solar power. By the year 2030, it is estimated to be 200 GW.
- A typical nuclear power plant generates about 1 GW of electric power, which is equal to 5 GW of solar power (daily power generation is limited to an average of 5 to 6 h/day).
- Global sales of solar power systems have been growing at a rate of 35 percent in the past few years.
- It is projected that by the year 2020, the United States will be producing about 7.2 GW of solar power.
- The shipment of U.S. solar power systems has fallen by 10 percent annually but has increased by 45 percent throughout Europe.
- The annual sale growth of PV systems globally has been 35 percent.
- The present cost of solar power modules on the average is $4.00/W. By 2030 it should be about $0.50/W.
- World production of solar power is 1 GW/year.
- Germany has a $0.75/W grid feed incentive, which will be valid for the next 20 years. The incentive is to be decreased by 5 percent per year.
- In the past few years, Germany has installed 130 MW of solar power per year.
- Japan has a 50 percent subsidy for solar power installations of 3- to 4-kW systems and has about 800 MW of grid-connected solar power systems. Solar power in Japan has been in effect since 1994.

- California, in 2007 under a California Solar Initiative (CSI) program, has set aside $2.2 billion for renewable energy, which will provide a scaled down rebate until year 2017.
- In the years 2015 through 2024, it is estimated that California could produce an estimated $40 billion of solar power sales.
- In the United States, 20 states have a solar rebate program. Nevada and Arizona have set aside a state budget for solar programs.
- Total U.S. production has been just about 18 percent of global production.
- For each megawatt of solar power, we employ 32 people.
- A solar power collector in the southwest United States, sized 100 × 100 miles, could produce sufficient electric power to satisfy the country's yearly energy needs.
- For every kilowatt of power produced by nuclear or fossil fuel plants, $\frac{1}{2}$ gallon of water is used for scrubbing, cleaning, and cooling. Solar power requires practically no water usage.
- Most sustainable energy technologies require less organizational infrastructure and less human power and therefore may result in human power resource reallocation.
- Significant impacts of solar power cogeneration are that it
    - Boosts economic development
    - Lowers the cost of peak power
    - Provides greater grid stability
    - Lowers air pollution
    - Lowers greenhouse gas emissions
    - Lowers water consumption and contamination
- A mere 6.7-mi/gal efficiency increase in cars driven in the United States could offset our share of imported Saudi oil.
- Types of solar power technology at present are:
    - Crystalline
    - Polycrystalline
    - Amorphous
    - Thin- and thick-film technologies
- Types of solar power technology in the future:
    - Plastic solar cells
    - Nonconstruction materials
    - Dye-synthesized cells

7

LEED—LEADERSHIP IN ENERGY AND ENVIRONMENTAL DESIGN

Energy Use and the Environment

Ever since the creation of tools, the formation of settlements, and the advent of progressive development technologies, humankind has consistently harvested the abundance of energy that has been accessible in various forms. Up until the eighteenth-century industrial revolution, energy forms used by humans were limited to river or stream water currents, tides, solar, wind, and to a very small degree geothermal energy, none of which had an adverse effect on the ecology.

Upon the discovery and harvesting of steam power and the development of steam-driven engines, humankind resorted to the use of fossil fuels and commenced the unnatural creation of air, soil, water, and atmospheric pollutants with increasing acceleration to a degree that fears about the sustenance of life on our planet under the prevailing pollution and waste management control has come into focus.

Since global material production is made possible by the use of electric power generated from the conversion of fossil fuels, continued growth of the human population and the inevitable demand for materials within the next couple of centuries, if not mitigated, will tax the global resources and this planet's capacity to sustain life as we know it.

To appreciate the extent of energy used in humanmade material production, we must simply observe that every object used in our lives from a simple nail to a super-computer is made using pollutant energy resources. The conversion of raw materials to finished products usually involves a large number of energy-consuming processes, but products made using recycled materials such as wood, plastics, water, paper, and metals require fewer process steps and therefore less pollutant energy.

In order to mitigate energy waste and promote energy conservation, the U.S. Department of Energy, Office of Building Technology, founded the U.S. Green Building Council. The Council was authorized to develop design standards that provide for improved environmental and economic performance in commercial buildings by the use of established or advanced industry standards, principles, practices, and materials. Note that the United States, with 5 percent of the world population, presently consumes 25 percent of the global energy resources.

The U.S. Green Building Council introduced the Leadership in Energy and Environmental Design (LEED) rating system and checklist. This system establishes qualification and rating standards that categorize construction projects with certified designations, such as silver, gold, and platinum. Depending on adherence to the number of points specified in the project checklist, a project may be bestowed recognition and potentially a set amount of financial contribution by state and federal agencies.

Essentially the LEED guidelines discussed in this chapter, in addition to providing design guidelines for energy conservation, are intended to safeguard the ecology and reduce environmental pollution resulting from construction projects. There are many ways to analyze the benefits of LEED building projects. In summary, green building design is about productivity. A number of studies, most notably a study by Greg Kats of Capital-E, have validated the productivity value.

There are also a number of factors that make up this analysis. The basic concept is that if employees are happy in their workspace, such as having an outside view and daylight in their office environment, and a healthy environmental quality, they become more productive.

State of California Green Building Action Plan

The following is adapted from the detailed direction that accompanies the California governor's executive order regarding the Green Building Action Plan, also referred to as Executive Order S-20-04. The original publication, which is a public domain document, can be found on the Californian Energy Commission's Web pages.

PUBLIC BUILDINGS

State buildings   All employees and all state entities under the governor's jurisdiction must immediately and expeditiously take all practical and cost-effective measures to implement the following goals specific to facilities owned, funded, or leased by the state.

Green buildings   The U.S. Green Building Council (USGBC) has developed green building rating systems that advance energy and material efficiency and sustainability, known as Leadership in Energy and Environmental Design for New Construction and Major Renovations (LEED-NC) and LEED Rating System for Existing Buildings (LEED-EB).

All new state buildings and major renovations of 10,000 ft^2 and over and subject to Title 24 must be designed, constructed, and certified by LEED-NC Silver or higher, as described later.

Life-cycle cost assessments, defined later in this section, must be used in determining cost-effective criteria. Building projects less than 10,000 ft² must use the same design standard, but certification is not required.

The California Sustainable Building Task Force (SBTF) in consultation with the Department of General Services (DGS), Department of Finance (DoF), and the California Energy Commission (CEC) is responsible for defining a life-cycle cost assessment methodology that must be used to evaluate the cost effectiveness of building design and construction decisions and their impact over a facility's life cycle.

Each new building or large renovation project initiated by the state is also subject to a clean on-site power generation. All existing state buildings over 50,000 ft² must meet LEED-EB standards by no later than 2015 to the maximum extent of cost effectiveness.

Energy efficiency All state-owned buildings must reduce the volume of energy purchased from the grid by at least 20 percent by 2015 as compared to a 2003 baseline. Alternatively, buildings that have already taken significant efficiency actions must achieve a minimum efficiency benchmark established by the CEC.

Consistent with the executive order, all state buildings are directed to investigate "demand response" programs administered by utilities, the California Power Authority, to take advantage of financial incentives in return for agreeing to reduce peak electrical loads when called upon, to the maximum extent cost effective for each facility.

All occupied state-owned buildings, beginning no later than July 2005, must use the energy-efficiency guidelines established by the CEC. All state buildings over 50,000 ft² must be retrocommissioned, and then recommissioned on a recurring 5-year cycle, or whenever major energy-consuming systems or controls are replaced. This is to ensure that energy and resource-consuming equipment is installed and operated at optimal efficiency. State facility leased spaces of 5000 ft² or more must also meet minimum U.S. EPA Energy Star standards guidelines.

Beginning in the year 2008, all electrical equipment, such as computers, printers, copiers, refrigerator units, and air-conditioning systems, that is purchased or operated by state buildings and state agencies must be Energy Star rated.

Financing and execution The consultation with the CEC, the State Treasurer's Office, the DGS, and financial institutions will facilitate lending mechanisms for resource efficiency projects. These mechanisms will include the use of the life-cycle cost methodology and will maximize the use of outside financing, loan programs, revenue bonds, municipal leases, and other financial instruments. Incentives for cost-effective projects will include cost sharing of at least 25 percent of the net savings with the operating department or agency.

Schools

New school construction The Division of State Architect (DSA), in consultation with the Office of Public School Construction and the CEC in California, was mandated to develop technical resources to enable schools to be built with energy-efficient resources. As a result of this effort, the state designated the Collaborative for High Performance Schools (CHPS) criteria as the recommended guideline. The CHPS is based on LEED and was developed specifically for kindergarten to grade 12 schools.

COMMERCIAL AND INSTITUTIONAL BUILDINGS

This section also includes private-sector buildings, state buildings, and schools. The California Public Utilities Commission (CPUC) is mandated to determine the level of ratepayer-supported energy efficiency and clean energy generation so as to contribute toward the 20 percent efficiency goal.

LEADERSHIP

Mission of green action team   The state of California has established an inter-agency team known as the Green Action Team, which is composed of the director of the Department of Finance and the secretaries of Business, Transportation, and Housing, with a mission to oversee and direct progress toward the goals of the Green Building Order.

LEED

LEED project sustainable building credits and prerequisites are based on LEED-NC2.1 New Construction. There are additional versions of LEED that have been adopted or are currently in development that address core or shell, commercial interiors, existing buildings, homes, and neighborhood development.

SUSTAINABLE SITES

Sustainable site prerequisite—construction activity pollution prevention The intent of this prerequisite is to control and reduce top erosion and reduce the adverse impact on the surrounding water and air quality.

Mitigation measures involve the prevention of the loss of topsoil during construction by means of a storm water system runoff as well as the prevention of soil displacement by gust wind. It also imposes measures to prevent sedimentation of storm sewer systems by sand dust and particulate matter.

Some suggested design measures to meet these requirements include deployment of strategies, such as temporary or permanent seeding, silt fencing, sediment trapping, and sedimentation basins that could trap particulate material.

Site selection, credit no. 1   The intent of this credit is to prevent and avoid development of a site that could have an adverse environmental impact on the project location surroundings.

Sites considered unsuitable for construction include prime farmlands; lands that are lower than 5 ft above the elevation of established 100-year flood areas, as defined by the Federal Emergency Management Agency (FEMA); lands that are designated habitats for endangered species; lands within 100 ft of any wetland; or a designated public parkland.

To meet site selection requirements, it is recommended that the sustainable project buildings have a reasonably minimal footprint to avoid site disruption. Favorable design practices must involve underground parking and neighbor-shared facilities.

The point weight granted for this measure is 1.

Development density and community connectivity, credit no. 2   The intent of this requirement is to preserve and protect green fields and animal habitats by means of increasing the urban density, which may also have a direct impact on the reduction of urban traffic and pollution.

A specific measure suggested includes project site selection within the vicinity of an urban area with high development density.

The point weight granted for this measure is 1.

Brownfield redevelopment, credit no. 3   The main intent of this credit is the use and development of projects on lands that have environmental contamination. To undertake development under this category, the Environmental Protection Agency (EPA) must provide a sustainable redevelopment remediation requirement permit.

Projects developed under Brownfield redevelopment are usually offered state, local, and federal tax incentives for site remediation and cleanup.

The point weight granted for each of the four measures is 1.

Alternative transportation, credit no. 4   The principal objective of this measure is to reduce traffic congestion and minimize air pollution. Measures recommended include locating the project site within $\frac{1}{2}$ mile of a commuter train, subway, or bus station; construction of a bicycle stand and shower facilities for 5 percent of building habitants; and installation of alternative liquid and gas fueling stations on the premises. An additional prequisite calls for a preferred parking facility for car pools and vans that serve 5 percent of the building occupants, which encourages transportation sharing.

The point weight granted for this measure is 1.

Site development, credit no. 5   The intent of this measure is to conserve habitats and promote biodiversity. Under this prerequisite, one point is provided for limiting earthwork and the destruction of vegetation beyond the project or building perimeter, 5 ft beyond walkways and roadway curbs, 25 ft beyond previously developed sites, and restoration of 50 percent of open areas by planting of native trees and shrubs.

Another point under this section is awarded for 25 percent reduction of a building footprint by what is allowed by local zoning ordinances.

Design mitigations for meeting the preceding goals involve underground parking facilities, ride-sharing among habitants, and restoring open spaces by landscape architecture planning that uses local trees and vegetation.

Storm water management, credit no. 6   The objective of this measure involves preventing the disruption of natural water flows by reducing storm water runoffs and promoting on-site water filtration that reduces contamination.

Essentially these requirements are subdivided into two categories. The first one deals with the reduction of the net rate and quantity of storm water runoff that is caused by the imperviousness of the ground, and the second relates to measures undertaken to remove up to 80 percent of the average annual suspended solids associated with the runoff.

Design mitigation measures include maintenance of natural storm water flows that include filtration to reduce sedimentation. Another technique used is construction of roof gardens that minimize surface imperviousness and allow for storage and reuse of storm water for nonpotable uses such as landscape irrigation and toilet and urinal flushing.

The point weight granted for each of the two categories discussed here is 1.

Heat island effect, credit no. 7   The intent of this requirement is to reduce the microclimatic thermal gradient difference between the project being developed and adjacent lands that have wildlife habitats. Design measures to be undertaken include shading provisions on site surfaces such as parking lots, plazas, and walkways. It is also recommended that site or building colors have a reflectance of at least 0.3 and that 50 percent of parking spaces be of the underground type.

Another design measure suggests use of Energy Star high-reflectance and high-emissivity roofing. To meet these requirements the project site must feature extensive landscaping. In addition to minimizing building footprints, it is also suggested that building rooftops have vegetated surfaces and that gardens and paved surfaces be of light-colored materials to reduce heat absorption.

The point weight granted for each of the two categories discussed here is 1.

Light pollution reduction, credit no. 8   Essentially, this requirement is intended to eliminate light trespass from the project site, minimize the so-called night sky access, and reduce the impact on nocturnal environments. This requirement becomes mandatory for projects that are within the vicinity of observatories.

To comply with these requirements, site lighting design must adhere to Illumination Engineering Society of North America (IESNA) requirements. In California, indoor and outdoor lighting design should comply with California Energy Commission (CEC) Title 24, 2005, requirements. Design measures to be undertaken involve the use of luminaires and lamp standards equipped with filtering baffles and low-angle spotlights that could prevent off-site horizontal and upward light spillage.

The point weight granted for this measure is 1.

WATER EFFICIENCY MEASURES

Water-efficient landscaping, credit no. 1   Basically, this measure is intended to minimize the use of potable water for landscape irrigation purposes.

One credit is awarded for the use of high-efficiency irrigation management control technology. A second credit is awarded for the construction of special reservoirs for the storage and use of rainwater for irrigation purposes.

Innovative water technologies, credit no. 2   The main purpose of this measure is to reduce the potable water demand by a minimum of 50 percent. Mitigation involves the use of gray water by construction of on-site natural or mechanical waste-water treatment systems that could be used for irrigation and toilet or urinal flushing. Consideration is also given to the use of waterless urinals and storm water usage.

The point weight granted for this measure is 1.

Water use reduction, credit no. 3   The intent of this measure is to reduce water usage within buildings and thereby minimize the burden of local municipal supply and water treatment. Design measures to meet this requirement involve the use of waterless urinals, high-efficiency toilet and bathroom fixtures, and nonpotable water for flushing toilets.

This measure provides one credit for design strategies that reduce building water usage by 20 percent and a second credit for reducing water use by 30 percent.

ENERGY AND ATMOSPHERE

Fundamental commissioning of building energy systems, prerequisite no. 1
This requirement is a prerequisite intended to verify intended project design goals and involves design review verification, commissioning, calibration, physical verification of installation, and functional performance tests, all of which are to be presented in a final commissioning report.

The point weight granted for this prerequisite is 1.

Minimum energy performance, prerequisite no. 2   The intent of this prerequisite is to establish a minimum energy efficiency standard for a building. Essentially, the basic building energy efficiency is principally controlled by mechanical engineering heating and air-conditioning design performance principles, which are outlined by ASHRAE/IESNA and local municipal or state codes. The engineering design procedure involves so-called building envelop calculations, which maximize energy performance. Building envelop computations are achieved by computer simulation models that quantify energy performance as compared to a baseline building. The point weight granted for this prerequisite is 1.

Fundamental refrigerant management, prerequisite no. 3   The intent of this measure is the reduction of ozone-depleting refrigerants used in HVAC systems. Mitigation involves replacement of old HVAC equipment with equipment that does not use CFC refrigerants. The point weight granted for this prerequisite is 1.

Optimize energy performance, credit no. 1   The principal intent of this measure is to increase levels of energy performance above the prerequisite standard in order to reduce environmental impacts associated with excessive energy use. The various

TABLE 7.1   NEW AND OLD BUILDING CREDIT POINT

% INCREASE IN ENERGY PERFORMANCE		
NEW BUILDINGS	EXISTING BUILDINGS	CREDIT POINTS
14	7	2
21	14	4
28	21	6
35	28	8
42	35	10

credit levels shown in Table 7.1 are intended to reduce the design energy budget for the regulated energy components described in the requirements of the ASHRAE/IESNA standard. The energy components include the building envelope, the hot-water system, and other regulated systems defined by ASHRAE standards.

Similarly to previous design measures, computer simulation and energy performance modeling software is used to quantify the energy performance as compared to a baseline building system.

On-site renewable energy, credit no. 2   The intent of this measure is to encourage the use of sustainable or renewable energy technologies such as solar photovoltaic cogeneration, solar power heating and air conditioning, fuel cells, wind energy, landfill gases, and geothermal and other technologies discussed in various chapters of this book.

The credit award system under this measure is based on a percentage of the total energy demand of the building. See Table 7.2.

Additional commissioning, credit no. 3   This is an enforcement measure to verify whether the designed building is constructed and performs within the expected or intended parameters. The credit verification stages include preliminary design documentation review, construction documentation review when construction is completed, selective submittal document review, establishment of commissioning documentation, and finally the postoccupancy review. Note that all these reviews must be conducted by an independent commissioning agency.

The point weight credit awarded for each of the four categories is 1.

TABLE 7.2   ENERGY-SAVING CREDIT AWARD

% TOTAL ENERGY SAVINGS	CREDIT POINTS
5	1
10	2
20	3

Enhanced refrigerant management, credit no. 4   This measure involves installation of HVAC, refrigeration, and fire suppression equipment that do not use hydrochlorofluorocarbon (HCFC) agents.

The point weight credit awarded for this category is 1.

Measurement and verification, credit no. 5   This requirement is intended to optimize building energy consumption and provide a measure of accountability. The design measures implemented include the following:

- Lighting system control, which may consist of occupancy sensors, photocells for control of daylight harvesting, and a wide variety of computerized systems that minimize the energy waste related to building illumination. A typical discussion of lighting control is covered under California Title 24 energy conservation measures, with which all building lighting designs must comply within the state.
- Compliance of constant and variable loads, which must comply with motor design efficiency regulations.
- Motor size regulation that enforces the use of variable-speed drives (VFD).
- Chiller efficiency regulation measures that meet variable-load situations.
- Cooling load regulations.
- Air and water economizer and heat recovery and recycling.
- Air circulation, volume distribution, and static pressure in HVAC applications.
- Boiler efficiency.
- Building energy efficiency management by means of centralized management and control equipment installation.
- Indoor and outdoor water consumption management.

Point weight credit awarded for each of the four categories is 1.

Green power, credit no. 6   This measure is intended to encourage the use and purchase of grid-connected renewable cogenerated energy derived from sustainable energy such as solar, wind, geothermal, and other technologies described throughout this book. A purchase-and-use agreement of this so-called green power is usually limited to a minimum of a 2-year contract. The cost of green energy use is considerably higher than that of regular energy. Purchasers of green energy, who participate in the program, are awarded a Green-e products certification.

MATERIAL AND RESOURCES

Storage and collection of recyclables, prerequisite no. 1   This prerequisite is a measure to promote construction material sorting and segregation for recycling and landfill deposition. Simply put, construction or demolition materials such as glass, iron, concrete, paper, aluminum, plastics, cardboard, and organic waste must be separated and stored in a dedicated location within the project for further recycling.

Building reuse, credit no. 1   The intent of this measure is to encourage the maximum use of structural components of an existing building that will serve to preserve and conserve a cultural identity, minimize waste, and reduce the environmental impact. Note that another significant objective of this measure is to reduce the use of newly manufactured material and associated transportation that ultimately result in energy use and environmental pollution.

Credit is given for implementation of the following measures:

- One credit for maintenance and reuse of 75 percent of the existing building.
- Two credits for maintenance of 100 percent of the existing building structure shell and the exterior skin (windows excluded).
- Three credits for 100 percent maintenance of the building shell and 50 percent of walls, floors, and the ceiling.

This simply means that the only replacements will be of electrical, mechanical, plumbing, and door and window systems, which essentially boils down to a remodeling project.

Construction waste management, credit no. 2   The principal purpose of this measure is to recycle a significant portion of the demolition and land-clearing materials, which calls for implementation of an on-site waste management plan. An interesting component of this measure is that the donation of materials to a charitable organization also constitutes waste management.

The two credits awarded under this measure include one point for recycling or salvaging a minimum of 50 percent by weight of demolition material and two points for salvage of 75 percent by weight of the construction and demolition debris and materials.

Material reuse, credit no. 3   This measure is intended to promote the use of recycled materials, thus, reducing the adverse environmental impact caused by manufacturing and transporting new products. For using recycled materials in a construction, one credit is given to the first 5 percent and a second point for a 10 percent total use. Recycled materials used could range from wall paneling, cabinetry, bricks, construction wood, and even furniture.

Recycled content, credit no. 4   The intent of this measure is to encourage the use of products that have been constructed from recycled material. One credit is given if 25 percent of the building material contains some sort of recycled material or if there is a minimum of 40 percent by weight use of so-called postindustrial material content. A second point is awarded for an additional 25 percent recycled material use.

Regional materials, credit no. 5   The intent of this measure is to maximize the use of locally manufactured products, which minimizes transportation and thereby reduces environmental pollution. One point is awarded if 20 percent of the material is manufactured within 500 miles of the project and another point is given if the total recycled material use reaches 50 percent. Materials used in addition to manufactured goods also include those that are harvested, such as rock and marble from quarries.

Rapidly renewable materials, credit no. 6   This is an interesting measure that encourages the use of rapidly renewable natural and manufactured building materials. Examples of natural materials include strawboards, woolen carpets, bamboo flooring, cotton-based insulation, and poplar wood. Manufactured products may consist of linoleum flooring, recycled glass, and concrete as an aggregate.

The point weight awarded for this measure is 1.

Certified wood, credit no. 7   The intent of this measure is to encourage the use of wood-based materials. One point is credited for the use of wood-based materials such as structural beams and framing and flooring materials that are certified by Forest Council Guidelines (FSC).

INDOOR ENVIRONMENTAL QUALITY

Minimum indoor air quality (IAQ) performance, prerequisite no. 1   This prerequisite is established to ensure indoor air quality performance to maintain the health and wellness of the occupants. One credit is awarded for adherence to ASHRAE building ventilation guidelines such as placement of HVAC intakes away from contaminated air pollutant sources such as chimneys, smoke stacks, and exhaust vents.

The point weight awarded for this measure is 1.

Environmental tobacco smoke (ETS) control, prerequisite no. 2   This is a prerequisite that mandates the provision of dedicated smoking areas within buildings, which can effectively capture and remove tobacco and cigarette smoke from the building. To comply with this requirement, designated smoking rooms must be enclosed and designed with impermeable walls and have a negative pressure (air being sucked in rather than being pushed out) compared to the surrounding quarters. Upon completion of construction, designated smoking rooms are tested by the use of a tracer gas method defined by ASHRAE standards, which impose a maximum of 1 percent tracer gas escape from the ETS area. This measure is readily achieved by installing a separate ventilation system that creates a slight negative room pressure.

The point weight awarded for this measure is 1.

Outdoor air delivery monitoring, credit no. 1   As the title implies, the intent of this measure is to provide an alarm monitoring and notification system for indoor and outdoor spaces. The maximum permitted carbon dioxide level is 530 parts per million. To comply with the measure, HVAC systems are required to be equipped with a carbon dioxide monitoring and annunciation system, which is usually a component of building automation systems.

The point weight awarded for this measure is 1.

Increased ventilation, credit no. 2   This measure is intended for HVAC designs to promote outdoor fresh air circulation for building occupants' health and comfort. A credit of one point is awarded for adherence to the ASHRAE guideline for naturally ventilated spaces where air distribution is achieved in a laminar flow pattern. Some HVAC design strategies used include displacement and low-velocity ventilation, plug flow or underfloor air delivery, and operable windows that allow natural air circulation.

Construction (IAQ) air quality management plan, credit no. 3   This measure applies to air quality management during renovation processes to ensure that occupants are prevented from exposure to moisture and air contaminants. One credit is awarded for installation of absorptive materials that prevent moisture damage and filtration media to prevent space contamination by particulates and airborne materials. A second point is awarded for a minimum of flushing out of the entire space, by displacement, with outside air for a period of 2 weeks prior to occupancy. At the end of the filtration period a series of test are performed to measure the air contaminants.

Low-emitting materials, credit no. 4   This measure is intended to reduce indoor air contaminants resulting from airborne particulates such as paints and sealants. Four specific areas of concern include the following: (1) adhesives, fillers, and sealants; (2) primers and paints; (3) carpet; and (4) composite wood and agrifiber products that contain urea-formaldehyde resins. Each of these product applications are controlled by various agencies such as the California Air Quality Management District, Green Seal Council, and Green Label Indoor Air Quality Test Program.
  The point weight awarded for each of the four measures is 1.

Indoor chemical and pollutant source control, credit no. 5   This is a measure to prevent air and water contamination by pollutants. Mitigation involves installation of air and water filtration systems that absorb chemical particulates entering a building. Rooms and areas such as document reproduction rooms, copy rooms, and blueprint quarters, which generate trace air pollutants, are equipped with dedicated air exhaust and ventilation systems that create negative pressure. Likewise, water circulation, plumbing, and liquid waste disposal are collected in an isolated container for special disposal. This measure is credited a single point.

Controllability of systems, credit no. 6   The essence of this measure is to provide localized distributed control for ventilation, air conditioning, and lighting. One point is awarded for autonomous control of lighting and control for each zone covering 200 ft^2 of area with a dedicated operable window within 15 ft of the perimeter wall. A second point is given for providing air and temperature control for 50 percent of the nonperimeter occupied area. Both of these measures are accomplished by centralized or local area lighting control and HVAC building control systems. These measures are intended to control lighting and air circulation. Each of the two measures is awarded one point.

Thermal comfort, credit no. 7   The intent of this measure is to provide environmental comfort for building occupants. One credit is awarded for thermal and humidity control for specified climate zones and another for the installation of a permanent central temperature and humidity monitoring and control system.

Daylight and views, credit no. 8   Simply stated, this measure promotes an architectural space design that allows for maximum outdoor views and interior sunlight exposure. One credit is awarded for spaces that harvest indirect daylight for 75 percent of spaces occupied for critical tasks. A second point is awarded for direct sight of vision glazing from 90 percent of normally occupied work spaces. Note that copy rooms, storage rooms, mechanical equipment rooms, and low-occupancy rooms do not fall into these categories. In other words 90 percent of the work space is required to have direct sight of a glazing window. Some architectural design measures taken to meet these requirements include building orientation, widening of building perimeter, deployment of high-performance glazing windows, and use of solar tubes.

INNOVATION AND DESIGN PROCESS

Innovation in design, credit no. 1   This measure is in fact a merit award for an innovative design that is not covered by LEED measures and in fact exceeds the required energy efficiency and environmental pollution performance milestone guidelines. The four credits awarded for innovation in design are: (1) identification of the design intent, (2) meeting requirements for compliance, (3) proposed document submittals that demonstrate compliance, and finally (4) a description of the design approach used to meet the objective.

LEED-accredited professional, credit no. 2   One point is credited to the project for a design team that has a member who has successfully completed the LEED accreditation examination.

Credit summary

Sustainable sites	10 points
Water efficiency	3 points
Energy and atmosphere	8 points
Material and resources	9 points
Indoor environmental quality	10 points
Innovation in design	2 points

The grand total is 42 points.

OPTIMIZED ENERGY PERFORMANCE SCORING POINTS

Additional LEED points are awarded for building efficiency levels, as shown in Table 7.3.
Project certification is based on the accumulated points, as shown in Table 7.4.

TABLE 7.3   LEED CERTIFICATION CATEGORIES AND ASSOCIATED POINTS

% INCREASE IN ENERGY PERFORMANCE		
NEW BUILDINGS	EXISTING BUILDINGS	POINTS
15	5	1
20	10	2
25	15	3
30	20	4
35	25	5
40	30	6
45	35	7
50	40	8
55	45	9
60	50	10

Los Angeles Audubon Nature Center—A LEED-Certified Platinum Project

The following project is the highest-ranked LEED-certified building by the U.S. Green Building Council within the United States. The pilot project, known as Debs Park Audubon Center, is a 282-acre urban wilderness that supports 138 species of birds. It is located in the center of the city of Los Angeles, California, and was commissioned on January 13, 2004. Based on the Building Rating System TM2.1, the project received a platinum rating, the highest possible. Figure 7.1 shows a view of the Los Angeles Audubon Center's roof-mount solar PV system.

The key to the success of the project lies in the design considerations given to all aspects of the LEED ranking criteria, which include sustainable building design parameters such as the use of renewable energy sources, water conservation, recycled

TABLE 7.4   LEED BUILDING EFFICIENCY POINTS

LEED certified	26–32 points
Silver level	33–38 points
Gold level	39–51 points
Platinum level	52–69 points

Figure 7.1   **Los Angeles Audubon Center.** *Photograph courtesy of LA Audubon Society. „ EHDD Architecture/Soltierra, LLC.*

building materials, and maintenance of native landscaping. The main office building of the project is entirely powered by an on-site solar power system that functions "off the grid." The building water purification system is designed such that it uses considerably less water for irrigation and bathrooms. To achieve the platinum rating the building design met 52 out of the total available 69 LEED energy conservation points outlined in Table 7.3.

The entire building, from the concrete foundation and rebars to the roof materials, was manufactured from recycled materials. For example, concrete-reinforcement rebars were manufactured from melted scrap metal and confiscated handguns. All wood materials used in the construction of the building and cabinetry were manufactured from wheat board, sunflower board, and Mexican agave plant fibers.

A 26-kW roof-mount photovoltaic system provides 100 percent of the center's electric power needs. A 10-ton solar thermal cooling system installed by SUN Utility Network Inc. provides a solar air-conditioning system believed to be the first of its kind in southern California. The HVAC system provides the total air-conditioning needs of the office building. The combination of the solar power and the solar thermal air-conditioning system renders the project completely self-sustainable requiring no power from the power grid. Figure 7.2 shows a roof-mount solar photovoltaic panel installation. The cost of this pilot project upon completion was estimated to be about $15.5 million. At present, the project houses a natural bird habitat, exhibits, an amphitheater, and a hummingbird garden. The park also has a network of many hiking trails enjoyed by local residents.

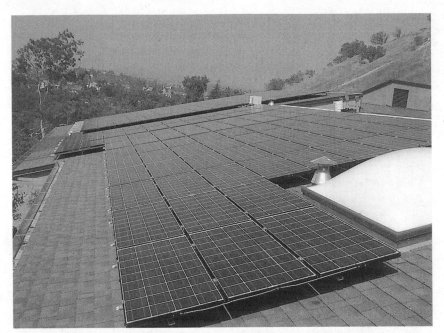

Figure 7.2   Los Angeles Audubon Center grid-independent solar power generation, system. *Photograph courtesy of LA Audubon Society.*

The thermal solar air-conditioning system, which is used only in a few countries, such as Germany, China, and Japan, utilizes an 800-ft^2 array of 408 Chinese-manufactured Sunda vacuum tube solar collectors. Each tube measures 78-in long and has a 4-in diameter, and each encloses a copper heat pipe and aluminum nitride plates that absorb solar radiation. Energy trapped from the sun's rays heats the low-pressure water that circulates within and is converted into a vapor that flows to a condenser section. A heat exchanger compartment heats up an incoming circulating water pipe through the manifold which allows for the transfer of thermal energy from the solar collector to a 1200-gal insulated high-temperature hot-water storage tank. When the water temperature reaches 180°F, the water is pumped to a 10-ton Yamazaki single-effect absorption chiller. A lithium bromide salt solution in the chiller boils and produces water vapor that is used as a refrigerant, which is subsequently condensed; its evaporation at a low pressure produces the cooling effect in the chiller. Figure 7.3 depicts the solar thermal heating and air-conditioning system diagram.

This system also provides space heating in winter and hot water throughout the year. Small circulating pumps used in the chiller are completely energized by the solar photovoltaic system. It is estimated that the solar thermal air-conditioning and heating system relieves the electric energy burden by as much as 15 kW. The cost of energy production at the Audubon Center is estimated to be $0.04/kW, which is substantially lower than the rates charged by the city of Los Angeles Department of Water and Power. Note that the only expense in solar energy cost is the minimal maintenance and

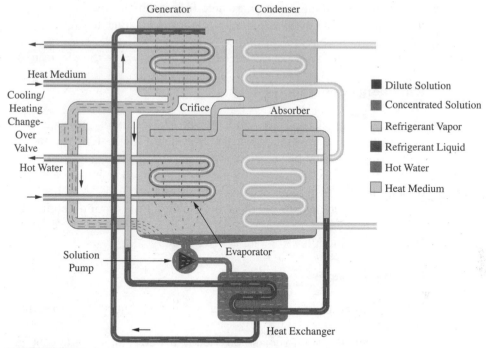

Figure 7.3   **Los Angeles Audubon Center, solar thermal heating and air-conditioning system.** *Graph courtesy of SUN Utility Network, Inc.*

investment cost, which will be paid off within a few years. The following are architectural and LEED design measures applied in the Los Angeles Audubon Center.

Architectural green design measures

- Exterior walls are ground-faced concrete blocks, exposed on the inside, insulated and stuccoed on the outside.
- Steel rebars have 97 percent recycled content.
- Twenty-five percent fly ash is used in cast-in-place concrete and 15 percent in grout for concrete blocks.
- More than 97 percent of construction debris is recycled.
- Aluminum-framed windows use 1-in-thick clear float glass with a low-emittance (low-E) coating.
- Plywood, redwood, and Douglas fir members for pergolas are certified by the Forestry Stewardship Council.
- Linoleum countertops are made from linseed oil and wood flour and feature natural jute tackable panels that are made of 100 percent recycled paper.
- Burlap-covered tackable panels are manufactured from 100 percent recycled paper.
- Batt insulation is formaldehyde-free mineral fiber with recycled content.
- Cabinets and wainscot are made of organic wheat boards and urea-formaldehyde-free medium-density fiberboard.

- Engineered structural members are urea-formaldehyde-free.
- Synthetic gypsum boards have 95 percent recycled content.
- Ceramic tiles have recycled content.
- Carpet is made of sisal fiber extracted from the leaves of the Mexican agave plant.

Green energy operating system

- One-hundred percent of the electric power for lighting is provided by an off-grid polycrystalline photovoltaic solar power system. The system also includes a 3- to ·5-day battery-backed power storage system.
- To balance the electric power provided by the sun, all lighting loads are connected or disconnected by a load-shedding control system.
- Heating and cooling are provided by a thermal absorption cooling and heating system.
- Windows open to allow for natural ventilation.

Green water system

- All the wastewater is treated on-site without a connection to the public sewer system.
- Storm water is kept on-site and diverted to a water-quality treatment basin before being released to help recharge groundwater.
- Two-stage, low-flow toilets are installed throughout the center.
- The building only uses 35 percent of the city water typically consumed by comparable structures.

TriCom Office Building

This project is a commercial and industrial-use type building, which has 23,300 ft^2 of building space. The TriCom Building incorporates three commercial uses of space, namely, executive office, showroom, and warehouse. The project was completed in 2003 at a cost of about $3.3 million. It was designed by Caldwell Architects and was constructed by Pozzo Construction. The solar power was designed and installed by Sun Utility, Solar Company.

The TriCom Building has received LEED certification by the U.S. Green Building Council, and it is the first certified building in Pasadena. It is a prototype for Pasadena and was undertaken with the partnerships of Pasadena Water and Power, a local landscape design school, and other entities.

The project site is in the expanded enterprise zone of the city of Pasadena, which was part of the redevelopment project. To promote alternative transportation, the building is located $\frac{1}{4}$ mile from nine bus lines and $\frac{3}{4}$ mile from the light rail. A bicycle rack and an electric vehicle charger are on site to encourage additional alternative modes of transportation.

The landscape is composed of drought-tolerant plants eliminating the need for a permanent irrigation system and thus conserving local and regional potable water resources. Significant water savings are realized by faucet aerators and dual-flush and low-gallons-per-flush (GPF) toilets.

A 31-kW roof-mount photovoltaic solar power cogeneration system, installed by Solar Webb Inc., provides over 50 percent of the building's demand for electric energy. The building construction includes efficient lighting controls, increased insulation, dual-glazed windows, and Energy Star–rated appliances that reduce energy consumption. Figure 7.4 shows the roof-mount solar PV system.

Approximately 80 percent of the building material, such as concrete blocks, the rebar, and the plants were manufactured or harvested locally, thereby minimizing the environmental pollution impacts that result from transportation.

In various areas of the building flooring, the cover is made from raw, renewable materials such as linseed oil and jute. The ceiling tiles are also made from renewable materials such as cork. Reused material consisted of marble and doors from hotels, used tiles from showcase houses, and lighting fixtures that augment the architectural character.

The carpet used is made of 50 percent recycled content, the ceiling tiles are made from 75 percent recycled content, and the aluminum building signage is made from 94 percent recycled content. Figure 7.5 shows the inverter system assembly.

To enhance environmental air quality, all adhesives, sealants, paints, and carpet systems contain little or no volatile organic compounds (VOCs), which makes the project comply with the most rigorous requirements for indoor air quality. Operable

Figure 7.4   **TriCom roof-mounted solar power cogeneration system.**
Photograph courtesy of Solar Integrated Technologies, Los Angeles, CA.

Figure 7.5   **TriCom inverter system assemblies.** *Photograph courtesy of Solar Integrated Technologies, Los Angeles, CA.*

windows also provide the effective delivery and mixing of fresh air and support the health, safety, and comfort of the building occupants.

Warehouse, Rochester, New York

This LEED gold-rated facility is equipped with a lighting system that utilizes dc fluorescent ballasts, roof-integrated solar panels, occupancy sensors, and daylight sensors for the highest possible efficiency. The building, including the innovative lighting design, was designed by William McDonough and Partners of Charlottesville, Virginia.

The facility has 6600 ft^2 of office space and 33,000 ft^2 of warehouse. The warehouse roof is equipped with skylights and 21 kW of solar panels bonded to the roof material (SR2001 amorphous panels by Solar Integrated Technologies). A canopy in the office area is equipped with 2.1 kW of Sharp panels.

The power from the solar panels is distributed in three ways:

- 2.2 kW is dedicated to the dc lighting in the office.
- 11.5 kW powers the dc lights in the warehouse.
- 11.5 kW is not needed by the lighting system, so it is inverted to alternating current and used elsewhere in the building or sold back to the utility.

The entire system consists of 35 NPS1000 Power Gateway modules from Nextek Power Systems in Hauppauge, New York. These devices take all the available power from the solar panels and send it directly to the lighting use without significant losses. Additional power, when needed at night or on cloudy days, is added from the grid.

In the office, six NPS1000 Power Gateway modules power 198 four-foot T-8 fluorescent lamps, illuminating most areas at 1.1 W/ft^2. Each of the fixtures is equipped with a single high-efficiency dc ballast for every two lamps. Most of the fixtures are controlled by a combination of manual switches, daylight sensors, and occupancy sensors in 13 zones.

In the warehouse area, 29 NPS1000 Power Gateway modules power 158 six-lamp T-8 fixtures. These fixtures have low, medium, and high settings for two-, four-, and six-lamp type fixtures, which are dimmed by daylight, and occupancy sensors are located throughout the area. The goal of the control architecture is to maintain a lighting level of 0.74 W/ft^2, using daylight when available, whenever the area is occupied. Figures 7.6 and 7.7 depict Nextek dc power system lighting controls.

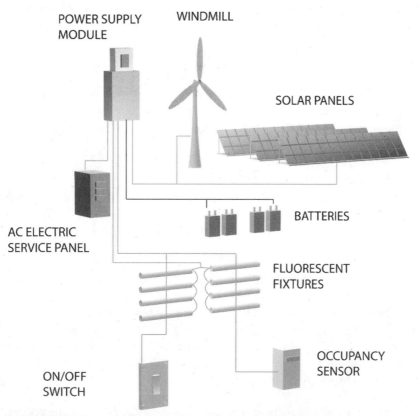

Figure 7.6   Nextek Power Systems solar power integration diagram.

SOLAR MODE

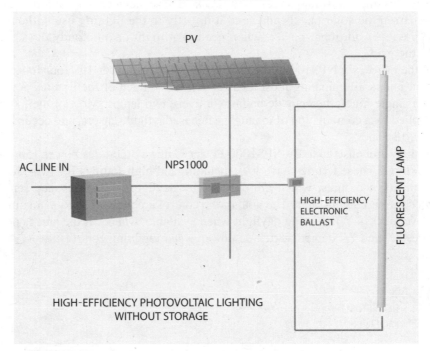

Figure 7.7   Nextek dc solar power lighting.

The logic of the lighting system is designed for optimum efficiency. Sources of light and power are prioritized as follows:

- First, daylight from the skylights is used.
- Second, if daylighting is not sufficient and the area is occupied, then power from the solar panels is added.
- Third, if daylight and solar power are not enough, then the additional power required is taken from the grid.

A number of factors contribute to the value of this system:

- Using the electricity generated by the solar panels to power the lighting eliminates inverter losses and improves efficiency by as much as 20 percent.
- The low-voltage control capability of the dc ballasts eliminates rectification losses and enables the innovative control system to be installed easily. Figure 7.8 depicts occupancy sensors efficiency chart

Roof-integrated solar panels reduce installation costs and allow the cost of the roof to be recovered using a 5-year accelerated depreciation formula.

RESULTS TO DATE SHOW THAT OCCUPANCY SENSORS AND DIMMING CONTROLS ACHIEVE UP TO 30% ENERGY SAVINGS

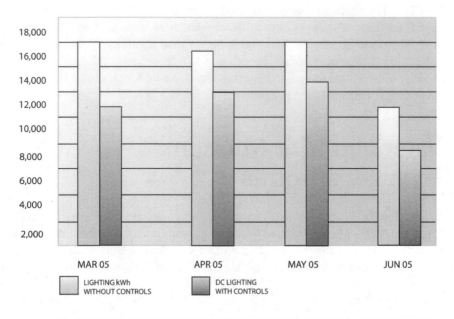

	kWh SAVED	$ SAVED WITH CONTROLS	% kWh OR $ SAVED
MARCH 05	4,970	497.02	31%
APRIL 05	3,458	345.80	22%
MAY 05	3,429	342.85	22%
JUNE 05	3,248	324.75	30% THRU 6/19
	15,104	1,510.43	

- ADJUSTMENTS MADE IN JUNE WILL ENSURE CLOSER TO 30% SAVINGS GOING FORWARD.
- IN OTHER PARTS OF THE UNITED STATES THE COST OF ENERGY IS OVER $0.20

Figure 7.8   **Occupancy sensors efficiency chart.**

WEB-BASED MONITORING

The Nextek data collection and monitoring system provides a Web-based display of power generated, power used, and weather. Additionally, the system also displays performance data and identifies anomalies in the system, such as burned-out lamps and sensors that are not operating properly.

VALUE

The Green Distribution Center (Green DC) has been successful in bringing more sustainable business practices into its facilities. The payback on its investment at the Rochester location, after rebates and accelerated depreciation, is approximately 12.6 years. The remaining system output will produce energy for 7.5 years, producing $60,000 in value at today's rates in Rochester. It is important to note that in areas where the avoided cost of peak power is higher than $0.10/kWh, this return of investment can drop to under 6 years, meaning that in those areas the facility would enjoy free peak power from the solar PV array for at least 14 years after the investment is returned. This equates to a $112,000 benefit at today's rates.

Water and Life Museum, Hemet, California

The Museum of Water and Life was a collaboration between architects Michael B. Lehrer FAIA and Mark Gangi AIA and electrical and solar power design consultant Vector Delta Design Group Inc. The project was designed as a sustainable campus, located within a recreational park at the entrance to Metropolitan's Diamond Valley Lake and is an example of a LEED-rated sustainable project.

The project consists of a large composite plan of six buildings, including the Center for Water Education Museum, Western Center of Archaeology and Paleontology Museum, Museum store, Museum café, and two conference room buildings, which are sited to produce a series of outdoor spaces from grand to intimate. A Grand Piazza, a Campus Way, and various courts strategically placed between buildings, give a civic sense to the campus.

The project was conceived as a LEED-designed campus in the platinum category, a designation reserved only for the most energy-efficient green buildings.

LEED DESIGN MEASURES

Solar Power Cogeneration   The Center for Water Education's solar system is one of the largest private solar installations in the western United States. The system, composed of 2925 solar panels, also includes custom-designed building-integrated photo-voltaic (BIPV) panels manufactured by Atlantis Energy Systems that cover the loggia. These panels are not only highly efficient, they are beautiful and add an architectural detail found nowhere else.

Part of the incentive to install such a large solar system was the generous rebate program provided by the California Energy Commision. At the time of purchase, The Center invested $4 million on the design and installation of the solar system.

Electrical engineering energy conservation design measures   In the process of designing an integrated electric and solar power system, special design measures were undertaken to significantly minimize the long-term operational cost of energy consumption. They consisted of significantly exceeding California Title 24 Energy Conformance minimum standards.

Lighting control automation   In order to exceed the earlier referenced energy economy standards, the electrical engineering design incorporated a wide variety of electronic sensing devices and timers to optimize daylight harvesting and zone lighting controls. Specifically, lighting control design measures incorporate the following:

- All buildings over 5000 ft^2 have been divided into lighting control zones that are controlled by a central computerized programmable astrologic timer. Each lighting zone is programmed to operate under varied timing cycles, which enables a substantial reduction in lighting power consumption.
- All campus lighting fixtures used throughout the project are high-efficiency fluorescent fixtures.
- All office lighting is controlled by occupancy sensors or photoelectric controls.
- All lights adjacent to windows are controlled by dedicated switches or photocells.
- Lighting levels in each room are kept below the minimum permitted levels of California Title 24 Energy Compliance levels.
- Outdoor banner projectors use light-emitting diode (LED) lamps, which use a minimal amount of electric power.
- All feeder conduits and branch circuit wires whenever warranted have been oversized to minimize voltage drop losses.

Solar power cogeneration design measures   One of the most significant energy-saving design features of the Water and Life Museum is the integration of photovoltaic solar power as an integral component of the architecture. A 540 kW dc solar power system consists of 2955 highly efficient Sharp Electronic solar power panels that cover the entirety of the rooftops of the two museum campuses, including the Water and Life Museum, Anthropology and Paleontology Museum, lecture hall, cafeteria, and gift shop buildings. Figure 7.9 depicts Water and Life Museum roof-mount solar power system.

At present, the power production capability of the aggregate solar power cogeneration system is designed to meet approximately 70 percent of the calculated demand load of the net meter which is effectively 48 percent above the 20 percent maximum qualification requirement for USGBC 3-point credit.

From the April 2006 commissioning date, until December 2006 while the project was under construction, the solar power system generated over 550 MWh of electric power, which translates into approximately 8 percent of the net invested capital. With 50 percent of the CEC solar power rebate received and the projected escalation of electric energy cost, the solar power cogeneration system investment is expected to recover its cost in less than 5 years.

In view of the 25-year guaranteed life span of the solar power panels and minimal maintenance cost of the system, the power cogeneration system in addition to saving

Figure 7.9   **Water and Life Museum roof-mount solar power system. Engineered by Vector Delta Design Group Inc.** *Photo courtesy of Fotoworks.*

significant amounts of energy will not contribute to atmospheric pollution because it will not be dumping millions of tons of carbon dioxide into the air.

Projected economic contribution   Considering maximum performance degradation over the life span of the solar power system, projected ac electric output power throughout the system life span under worst-case performance conditions is expected to be as follows:

Energy production

Daily energy production = 350 kWh (worst-case hourly energy output)
× 5.5 (hours of average daily insolation) = 1925 kWh

Yearly energy production = 1925 kWh × 365 days = 702,625 kWh

System lifetime (25 years) energy production = 702,625 kWh × 25 years
= 17,565,625 kWh or = 17.7 MWh

Assuming an optimistic rate of electric energy cost escalation over 25 years of system life at a mean value of $0.75/kWh, the projected saving contributions by the solar power system could be approximately $13,174,218.00. If we assume a rate of 1 percent of the solar power systems cost for general maintenance, the net energy saving over the life span of the system could be about $13 million.

Air pollution prevention   When generating electric power by brining fossil fuel, the resulting carbon dioxide production emitted into the atmosphere per kilowatt of electric energy ranges from 0.8 pounds for natural gas to 2.2 pounds for coal. The variation is dependent upon the type of fossil fuel such as natural gas, crude oil, or coal used.

In view of the preceding, the pollution abatement measure resulting from the use of the solar power, at the Museum of Water and Life over the life span of 25 years, is estimated to be 28 million pounds, which will be prevented from polluting the atmosphere.

As mentioned in Chapter 4, when generating electric energy by use of fossil fuels, power losses resulting from turbines, transformation, and transmission, which in some instances amount to as high as 70 percent, also contribute significantly toward generating considerable amounts of air and water pollutants.

Since solar power is produced on the site, it totally eliminates power generation and distribution losses; consequently, power production efficiency compared to conventional electric power plants is significantly higher. In other words, the cost of solar power compared to conventional electric energy, which is often generated hundreds or thousands of miles away and then transmitted, is significantly more efficient, cost effective, and less expensive when taking collateral expenses associated with state and federal pollution mitigation expenses into account. Figure 7.10 depicts Water and Life Museum building-integrated (BIPV) solar power system.

Solar power engineering design measures   Special electrical engineering design measures undertaken to maximize the solar power production output included the following:

- In order to maximize the available solar platform area, the PV modules were designed to cover the rooftops in a flat array formation. Losses resulting from the optimum tilt angle (about 11 percent) were significantly compensated for by a gain of more than 40 percent of surface area, which resulted in deployment of a much larger number of PV modules and less expensive support system platforms.
- Photovoltaic array string groupings were modularized to within 6-kW blocks. Each array block was assigned to a dedicated 6-kW highly efficient inverter. The distributed configuration of the arrays was intended to minimize the shading effect of a group of PV systems, which guarantees maximum independent performance by each array group.
- The inverters used are classified as the most efficient by the California Energy Commission (CEC-approved component list) and are designed for direct connection to the electrical grid.
- All ac solar power feeder conduits and cables were somewhat oversized to minimize voltage drop losses.
- In addition to direct power production, roof-mount solar panels provide significant roof shading, which extends the roof covering life by approximately 25 percent; they also keep the roofs cool, which increases the "R" insulation value. This in turn

Figure 7.10   Water and Life Museum building-integrated (BIPV) solar power system. Engineered by Vector Delta Design Group, Inc. *Photo courtesy of Fotoworks.*

leads to reduced air-conditioning system operation and therefore a notable amount of energy reduction.

■ An advanced computerized telemetry and monitoring system is designed to monitor and display real-time power output parameters from each building. Power output in kilowatt-hours, system power efficiency, accumulated power output statistics, barometric pressure, outdoor temperature, humidity, and many other vital operational parameters are instantaneously displayed on a number of screens on the supervisory console.

The Water and Life Museum provides a specially designed solar power interactive display on a large, flat plasma monitor whereby visitors to the museum are allowed to interact and request information regarding the solar power cogeneration system.

ADDITIONAL LEED CERTIFICATION DESIGN MEASURES

Half-flush/full-flush lavatories   All lavatories at The Center for Water Education have the option of a half flush or a full flush. This cuts back on water that is wasted when only a full flush is given as an option every time. A half flush uses half the amount of water as a full flush uses.

Waterless urinals   The Center's urinals are all waterless. This saves a significant amount of potable water.

Foam exterior   The Center's foam insulation cuts back on heater and air conditioner use and increases the R factor in the exterior walls, making it as energy-efficient as possible.

Lithocrete   The Center chose to use Lithocrete rather than concrete because it is a superior paving process that blends both "old world" paving finishes such as granite and stone with innovative paving finishes incorporating select, surface-seeded aggregates.

Carpet from recycled materials   All the carpets at The Center are made out of recycled materials making them environmentally safe.

LANDSCAPE

California-friendly plantings   The plants at The Center go from being ice-age native California plants on one side of the campus to modern-day native California plants on the other side of the campus. Native California plant species are an integral part of the design allowing the ground cover to blend into the adjacent Nature Preserve. These drought-tolerant, water-efficient plants are weaved in a unique and beautiful demonstration garden.

SMART controllers   Unlike most water sprinkler systems, The Center for Water Education's landscape features SMART controllers. Operated in conjunction with satellites, weather dictates the amount the system operates. SMART controllers are also available for residential installation.

Recycled water   State-of-the-art technology for irrigation, include the use of reclaimed as well as, and an exposed braided stream for water runoff are just a few of the features incorporated into the irrigation plan. The reclaimed water is distributed through purple colored pipes, which irrigate the root of the plants.

Grounds   Rock and decomposed granite enrich the color and water efficiency, which allows the ground cover to blend into the adjacent Nature Preserve. These elements are also perfect for residential installation.

Sustainable Sites   Many of the points listed in sustainable sites were not applicable to the project, such as urban redevelopment or brown field. The site selected was excess property from the dam construction. The project is located on recreation ground covering 1200 acres and which in the near future will become a recreational park that will include a golf course, a recreational lake, a swim and sports complex, and a series of bike trails, horse trails, and campgrounds. The museum complex is the gateway to the recreational grounds and is intended to become the civic center of the area.

The building is designed to accommodate bicycle storage for 5 percent of building occupants as well as shower and locker facilities. The project is designed to encourage alternative transportation to the site to reduce negative environmental impacts from automobile use. Because of the expected volume of pedestrian and bicycle visitors, vehicle parking spaces were reduced to provide adequate space for bicycle stands, which meets local zoning requirements.

The architecture of the grounds blends magnificent building shapes and open spaces and interpretive gardens throughout the campus. The small footprint of the building occupying the open space of the land is a significant attribute that qualifies the project for LEED points since it resulted in reduced site disturbance.

Braided streams weaving throughout the site provide a thematic story of water in southern California. The streams are also designed to mitigate storm water management for the site. The braided streams contain pervious surfaces conveying rainwater to the water table.

The parking grove of the project consists of shading trees and dual-colored asphalt. The remainder of the paving is light-colored, acid-washed concrete and Lithocrete of a light color. The roofs of the buildings are covered with a single-ply white membrane, which is shaded by solar panels. These light, shaded surfaces reduce the heat islanding effect.

The Museum of Water and Life is located within the radius of the Palomar Observatory, which has mandatory light pollution restrictions. As a result all lampposts are equipped with full cutoff fixtures and shutoff timing circuits.

Water Efficiency   The mission of The Center for Water Education is to transform its visitors into stewards of water. To this end, the campus is a showcase of water efficiency. The campus landscaping consists of California native plants. The irrigation systems deployed are state-of-the-art drip systems that use reclaimed water. Interpretive exhibits throughout the museum demonstrate irrigation technology ranging from that once used by Native Americans to the satellite-controlled technology. Each building is equipped with waterless urinals and dual-flush toilets.

Energy and Atmosphere   Energy savings begins with the design of an efficient envelope and then employs sophisticated mechanical systems. The Water and Life Museum is located in a climate that has a design load of 105°F in the summertime. The project's structures provide shading of the building envelope. High-performance glass and a variety of insulation types create the most efficient building envelope possible. The building exterior skin is constructed from three layers of perforated metal strips. The rooftop of all buildings within the campus is covered with high-efficiency photovoltaic solar power panels. The east elevation building has eight curtain walls that bridge the 10 towers. Each curtain wall is composed of 900 ft^2 of high-performance argon-filled glass. To compensate for heat radiation, a number of translucent megabanners are suspended in front of each curtain wall. The banners are located above the finish grade which preserves the beautiful views of the San Jacinto mountain range.

Mechanical System   The mechanical system uses a combination of radiant flooring, which is used for both heating and cooling, and forced-air units that run from the same chiller and boiler. The combination of the efficient envelope and the sophisticated mechanical system provides a project that is 38 percent more efficient than Title 24. This not only gains the project many LEED points in this category, but also provides a significant cost savings in the operation of these facilities.

Material and Resources   Lehrer Gangi Design Build specified local and recycled materials wherever possible, and during the construction phase kept a careful watch on how construction wastes were handled.

Low-emitting materials were selected. All contractors had restrictions regarding the types of materials that they were allowed to use. The mechanical engineering design deployed a three-dimensional model for the project, which tested the system's thermal comfort to comply with ASHRAE 55 requirements. The architectural building design ensured that 90 percent of spaces had views of natural light.

Innovation and Design Process   One of the significant points applied to the project design is the use of the buildings as a learning center for teaching sustainability to the visitors. Within the museum, exhibit spaces are uniquely devoted to solar cogeneration which is presented by means of interactive displays where visitors can observe real-time solar and meteorological statistical data on the display.

HEMET WATER AND LIFE MUSEUM LEED ENERGY CALCULATION

In order to qualify for LEED energy certification, upon completion of the final acceptance test and commissioning, the architect, the mechanical engineer, and the electrical engineer must provide projected or calculated energy savings for the project. The exercise involves calculation of energy consumption and savings resulted from LEED design measures. As reflected in the sample calculation that follows, each building within the project campus must account for energy consumption for space cooling, air handling, climate-control circulating pumps, domestic hot water, lighting, power receptacles, and process operations. All mechanical and electric energy calculations are accounted in British thermal units (kBtu) per square foot of the building.

Sample Energy Calculation   Vector Delta Design estimates the photovoltaic system will provide 930,750 kWh/yr. Reference: 3/17/06 EnergyPro 3.1 Analysis Performance Certificate of Compliance.

Building 1

ENERGY COMPONENT	(kBtu/ft^2-yr)	(kBtu/ft^2-yr)
Space heating	—	11.55
Space cooling	46.75	
Indoor fans	33.32	
Pumps & misc.	6.46	
Domestic hot water*	—	0.3
Lighting	22.93	
Receptacle	23.14	
Process	3.95	
Total	**136.55**	**11.85**

Building No. 1 total energy consumption distribution profile
Building 1 annual electricity usage:

$$136.55 \text{ kBtu/ft}^2\text{-yr} \times \text{kWh/3.413 kBtu} = 40.01 \text{ kWh/ft}^2\text{-yr}$$

$$\text{Conditioned floor area} = 16,773 \text{ ft}^2$$

$$40.01 \text{ kWh/ft}^2\text{-yr} \times 16,773 \text{ ft}^2 = 671,067.43 \text{ kWh/yr}$$

Local Hemet, CA, electricity cost: $0.43/kWh (per Vector Delta)

Bldg 1 annual electric energy cost: $0.43/kWh × 671,067.43 kWh/yr = $288,558.99/yr

Building 2

ENERGY COMPONENT	(kBtu/ft^2-yr)	(kBtu/ft^2-yr)
Space heating	—	5.03
Space cooling	27.22	
Indoor fans	29.71	
Pumps & misc.	4.60	
Domestic hot water*	—	0.3
Lighting	19.48	
Receptacle	16.15	
Process	6.34	
Total	**103.50**	**5.33**

Building No. 2 total energy consumption distribution profile
  Building 2 annual electricity usage:

$$103.5 \text{ kBtu/ft}^2\text{-yr} \times \text{kWh}/3.413 \text{ kBtu} = 30.32 \text{ kWh/ft}^2\text{-yr}$$

$$\text{Conditioned floor area} = 29,308 \text{ ft}^2$$

$$30.32 \text{ kWh/ft}^2\text{-yr} \times 29,308 \text{ ft}^2 = 888,771.75 \text{ kWh/yr}$$

$$\text{Local Hemet, CA, electricity cost: \$0.43/kWh}$$

Bldg 2 annual electric energy cost: $0.43/kWh \times 881,771.75 kWh/yr = $379,161.85/yr

Building 3

ENERGY COMPONENT	(kBtu/ft^2-yr)	(kBtu/ft^2-yr)
Space heating	—	8.33
Space cooling	35.24	
Indoor fans	19.84	
Pumps & misc.	5.01	
Domestic hot water*	—	0.3
Lighting	39.55	
Receptacle	16.95	
Process	—	
Total	**116.59**	**8.33**

Building No. 3 total energy consumption distribution profile
  Building 3 annual electricity usage:

$$116.59 \text{ kBtu/ft}^2\text{-yr} \times \text{kWh}/3.413 \text{ kBtu} = 34.16 \text{ kWh/ft}^2\text{-yr}$$

$$\text{Conditioned floor area} = 4248 \text{ ft}^2$$

$$34.16 \text{ kWh/ft}^2\text{-yr} \times 4248 \text{ ft}^2 = 145,114.06 \text{ kWh/yr}$$

$$\text{Local Hemet, CA, electricity cost: \$0.43/kWh}$$

Bldg 3 annual electric energy cost: $0.43/kWh \times 145,114.06 kWh/yr = $62,399.04/yr

Building 4

ENERGY COMPONENT	(kBtu/ft^2-yr)	(kBtu/ft^2-yr)
Space heating	—	32.04
Space cooling	55.95	
Indoor fans	33.81	
Pumps & misc.	9.44	
Domestic hot water*	—	0.3
Lighting	32.78	
Receptacle	10.30	
Process		
Total	142.28	32.34

Building No. 4 total energy consumption distribution profile
  Building 4 annual electricity usage:

$$142.28 \text{ kBtu/ft}^2\text{-yr} \times \text{kWh}/3.413 \text{ kBtu} = 41.68 \text{ kWh/ft}^2\text{-yr}$$

$$\text{Conditioned floor area} = 1748 \text{ ft}^2$$

$$41.68 \text{ kWh/ft}^2\text{-yr} \times 1748 \text{ ft}^2 = 72,870.03 \text{ kWh/yr}$$

$$\text{Local Hemet, CA, electricity cost: } \$0.43/\text{kWh}$$

Bldg 4 annual electric energy cost: $0.43/kWh × 72,870.03 kWh/yr = $31,334.11/yr

Building 5

ENERGY COMPONENT	(kBtu/ft^2-yr)	(kBtu/ft^2-yr)
Space heating	—	0.23
Space cooling	28.85	
Indoor fans	18.60	
Pumps & misc.	2.95	
Domestic hot water*	—	0.3
Lighting	32.50	
Receptacle	16.86	
Process		
Total	99.76	0.53

Building No. 5 total energy consumption distribution profile
Building 5 annual electricity usage:

$$99.76 \text{ kBtu/ft}^2\text{-yr} \times \text{kWh/3.413 kBtu} = 29.22 \text{ kWh/ft}^2\text{-yr}$$

$$\text{Conditioned floor area} = 1726 \text{ ft}^2$$

$$29.22 \text{ kWh/ft}^2\text{-yr} \times 1{,}726 \text{ ft}^2 = 50{,}449.97 \text{ kWh/yr}$$

$$\text{Local Hemet, CA, electricity cost: } \$0.43/\text{kWh}$$

Bldg 5 annual electric energy cost: $\$0.43/\text{kWh} \times 50{,}449.97 \text{ kWh/yr} = \$21{,}693.48/\text{yr}$

Building 8

ENERGY COMPONENT	(kBtu/ft²-yr)	(kBtu/ft²-yr)
Space heating	—	6.69
Space cooling	33.69	
Indoor fans	19.00	
Pumps & misc.	4.80	
Domestic hot water*	—	0.3
Lighting	44.14	
Receptacle	16.95	
Process	0.00	
Total	**118.58**	**6.99**

Building No. 8 total energy consumption distribution profile
Building 8 annual electricity usage:

$$118.58 \text{ kBtu/ft}^2\text{-yr} \times \text{kWh/3.413 kBtu} = 34.74 \text{ kWh/ft}^2\text{-yr}$$

$$\text{Conditioned floor area} = 4248 \text{ ft}^2$$

$$34.74 \text{ kWh/ft}^2\text{-yr} \times 4248 \text{ ft}^2 = 147{,}590.92 \text{ kWh/yr}$$

$$\text{Local Hemet, CA, electricity cost: } \$0.43/\text{kWh}$$

Bldg 8 annual electric energy cost: $\$0.43/\text{kWh} \times 147{,}590.92 \text{ kWh/yr} = \$63{,}464.09/\text{yr}$

Summary Analysis

Total electricity cost (Buildings 1, 2, 3, 4, 5, 8): $\$288{,}558.99/\text{yr} + \$379{,}161.85/\text{yr} + \$62{,}399.04/\text{yr} + \$31{,}334.11/\text{yr} + \$21{,}693.48/\text{yr} + \$63{,}464.09/\text{yr} = \$846{,}611.55/\text{yr}$

Average gas load density: [(11.85 kBtu/ft^2-yr × 16,773 ft^2) + (5.33 kBtu/ft^2-yr × 29,308 ft^2) + (8.63 kBtu/ft^2-yr × 4248 ft^2) + (32.33 kBtu/ft^2-yr × 1748 ft^2) + (0.53 kBtu/ft^2-yr × 1726 ft^2) + (6.99 kBtu/ft^2-yr × 4248)]/58,051 ft^2 = 8.24 kBtu/ft^2-yr

California default natural gas cost: \$0.00843/kBtu (taken from EIA 2003 Commercial Sector Average Energy Costs by State, Table 5: Default Energy Costs by State).

Total natural gas cost: 8.24 kBtu/ft^2-yr × 58,051 ft^2 × \$0.00843/kBtu = \$4032.40/yr

Total energy cost (Buildings 1, 2, 3, 4, 5, 8): \$846,611.55/yr + \$4032.40/yr = \$850,643.95/yr

Renewable energy cost contribution: \$0.43/kWh × 930,750 kWh/yr = \$400,222.50/yr

PV offset: \$400,222.50/\$850,643.95 = 0.47 or 47% which corresponds to 3 points.

Hearst Tower

Hearst Tower is the first building to receive a gold LEED certification from the U.S. Green Building Council. This towering architectural monument is located in New York around 57th Street and Eighth Avenue. The site was originally built in the 1920s and was a six-story office structure that served as the Hearst Corporation's headquarters. Construction began in 1927 and was completed in 1928 at a cost of \$2 million.

The original architecture consisted of a four-story building, set back above a two-story base. Its design consisted of columns and allegorical figures representing music, art, commerce, and industry. As it was an important heritage monument, in 1988 the building was designated as a landmark site by the New York Landmarks Preservation Commission.

ARCHITECTURE

In 2001 the Hearst organization commissioned Foster and Partners, Architects, and Cantor Seinuk, Structural Engineers, to design a new headquarters at the site of the existing building. The new headquarters is a 46-story, 600-ft-tall office tower with an area of 856,000 ft^2. One of the unique features of the architectural design was the requirement to preserve the six-story historical landmark facade. The old building had a 200- by 200-ft horseshoe-shaped footprint, which was totally excavated while keeping the landmark facade.

The new architectural design concept called for a tower with a 160- by 120-ft footprint. To maintain the historical facade, the design also called for a seven-story-high interior atrium.

The structural design utilizes a composite steel and concrete floor with a 40-ft interior column-free span for open office planning. The tower has been designed with two zones. The office zone starts 100 ft above street level at the 10th floor rising to the 44th-floor level. Below the 10th floor, the building houses the entrance and the lobby at the street level, the cafeteria, and an auditorium on the 3rd floor with an approximately 80-ft-high interior open space. At the seventh-floor elevation, the tower is connected to the existing landmark facade by a horizontal skylight system spanning approximately 40 ft from the tower columns to the existing facade.

STRUCTURAL DESIGN

The structural design is based upon a network of triangulated trusses of diagrid networks connecting all four faces of the tower, which has resulted in a highly efficient tube structure. The diagonal nodes are formed by the intersection of the diagonal and horizontal structural elements. Structurally the nodes act as hubs for redirecting the member forces. Figure 7.11 Figure is photograph of the old Hearst building and 7.12 depicts photograph of the new Hearst Tower in New York, NY.

The inherent lateral stiffness and strength of the diagrids provide a significant advantage for general structural stability under gravity, wind, and seismic loading conditions. As a result of the efficient structural system, the construction consumed 20 percent less steel in comparison to conventional moment frame structures. Figure 7.13 depicts digital structural members of the new Hearst tower.

Below the 10th-floor level, the structure is designed to respond to the large unbraced height by using a megacolumn system around the perimeter of the tower footprint, supporting the tower perimeter structure. The megacolumn is constructed of steel tube sections that are strategically filled with concrete.

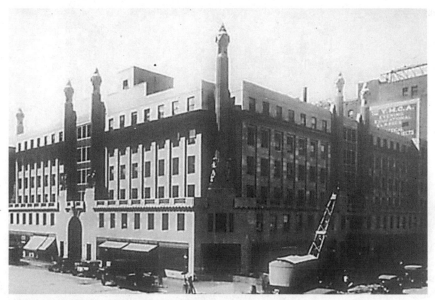

Figure 7.11   **Old Hearst Building.** *Photo courtesy of Hearst Corporation.*

Figure 7.12   **New Hearst Tower.** *Photograph courtesy of Hearst Corporation.*

Figure 7.13   **Digital structural members of new Hearst tower, New York, NY.** *Photo courtesy of Hearst Corporation.*

FACTS AND FIGURES

Gross area	856,000 ft^2
Typical floor	20,000 ft^2
Building height	597 ft
Number of stories	46
Client	Hearst Corporation
Architect	Foster and Partners
Associate architect	Adamson
Structural engineers	Seinuk Group
Lighting designer	George Sexton
Development manager	Tishman Speyer Properties

GREEN FACTS

For New York City, the benefits include a significant reduction in pollution and increased conservation of the city's vital resources, including water and electricity.

During construction the owners and the builder went to great lengths to collect and separate recyclable materials. As a result, about 85 percent of the original structure was recycled for future use. The architectural design uses an innovative diagrid system that has created a sense of a four-story triangle on the facade. The tower is the first North American high-rise structure that does not use horizontal steel beams. As a result of the structural design, the building has used 2000 tons less steel, a 20 percent savings over typical office buildings.

The computerized building automated lighting control system minimizes electric energy consumption based on the amount of natural light available at any given time by means of sensors and timers. The air-conditioning and space heating system uses high-efficiency equipment that utilizes outside air for cooling, and ventilation for 75 percent of the time throughout the year.

For water conservation, the building roof has been designed to collect rainwater, which reduces the amount of water dumped into the city's sewer system during rainfall. The harvested rainwater is stored in a 14,000-gal reclamation tank located in the basement of the tower. It is used to replace the water lost through evaporation in the office air-conditioning system. It is also used to irrigate indoor and outdoor plants and trees, thus reducing water usage by 50 percent. The interior paint used is of the low-VOC type, and furniture used is totally formaldehyde-free. All concrete surfaces are treated and sealed with low-toxicity sealants.

Statement by Cal/EPA Secretary Regarding Assembly Bill 32

The main objective of SB 32 is an aggressive measure to control emission of greenhouse gases.

The governor has set aggressive goals to reduce emissions to 1990 levels by 2020, which means that Californians must legislate regulations to create an emission reduction standard that allows for flexibility, using a combination of multi-sector market-based programs, incentive measures, best management practices and regulatory measures.

Present measures proposed are complex and affect the entire economy of California. Therefore, it's essential that the proposals have an economic safety valve to ensure that programs intended to reduce greenhouse gases do not have unintentional consequences that could be detrimental to the economy. The highlights of SB 32 and its intended goals include the following key components:

- Establishes a GHG cap to reduce emissions to 1990 levels by 2020, with enforceable benchmarks beginning in 2012 that includes a multi-sector market-based program to reduce GHG emissions in the most cost-effective manner.

- Requires mandatory reporting of GHG emissions for the largest sectors (oil and gas extraction, oil refining, electric power, cement manufacturing, and solid waste landfills).

- Includes a consolidated, accurate emissions inventory in order to determine 1990 GHG emissions baseline.

- Continues the California Climate Action Registry and gives credit to businesses that voluntarily joined the Registry.

- Creates an umbrella entity to develop and coordinate the implementation of a state GHG reduction plan to reduce GHG emissions to 1990 levels by 2020.

- Does not weaken existing environmental or public health requirements.

- Ensures public review and comment, with specific consideration for minority and low-income communities.

- Contains an economic safety valve to ensure the state GHG emission reduction plan protects public health and the environment, is technologically feasible, and is not detrimental to the California economy.

- Allows a one-year review period by the Legislature of the adopted state GHG emission reduction plan.

Conclusion

In summary, the main objective of LEED is a combination of energy-saving and environmental protection measures that are intended to minimize the adverse effects of construction and development. Some of the measures discussed in this chapter represent a significant financial cost impact, merits of which must be weighed and analyzed carefully.

8

CALIFORNIA SOLAR INITIATIVE
PROGRAM

The following is a summary highlight of the California Solar Incentive (CSI) program, a detailed version of which is available on the CSI Web page at www.csi.com.

Beginning January 1, 2007, the state of California introduced solar rebate funding for installation of photovoltaic power cogeneration that was authorized by the California Public Utilities Commission (CPUC) and the California Senate. The bill referenced as SB1 has allotted a budget of $2.167 billion that will be used over a 10-year period.

The rebate funding program known as the California Solar Initiative (CSI) awards rebates on the basis of performance, unlike earlier programs that allotted rebates based on the calculated projection of system energy output. The new rebate award system categorizes solar power installation into two incentive groups. An incentive program referred to as Performance-Based Incentive (PBI) addresses photovoltaic installation of 100 kW or larger and provides rebate dollars based on the solar power cogeneration's actual output over a 5-year period. Another rebate referred to as Expected Performance-Based Buydown (EPBB) is a one-time lump sum incentive payment program for solar power systems with performance capacities of less than 100 kW, which is a payment based on the system's expected future performance.

Three major utility providers that service various state territories distribute and administer the CSI funds. The three main service providers that administer the program in California are:

■ Pacific Gas and Electric (PG&E), which serves northern California
■ Southern California Edison (SCE), which serves mid-California
■ San Diego Regional Energy Office (SDREO)/San Diego Gas and Electric (SDG&E)

Note that municipal electric utility customers are not eligible to receive the CSI funds from the three administrator agencies.

Each of the preceding service providers administering the CSI program have Web pages that enable clients to access an online registration database and provide program handbooks, reservation forms, contract agreements, and all forms required by the CSI program. All CSI application and reservation forms are available on the CSI Web page.

The principal object of the CSI program is to ensure that 3000 MW of new solar energy facilities are installed throughout California by the year 2017.

CSI Fund Distribution

The CSI fund distribution administered by the three main agencies in California has a specific budget allotment that is proportioned according to the demographics of power demand and distribution. CSI budget allotment values are shown in Table 8.1.

The CSI budget shown in Table 8.1 can be divided into two customer segments, namely, residential and nonresidential. Table 8.2 shows the relative allocations of CSI solar power generation by customer sector.

The California Solar Initiative program budget also allocates $216 million to affordable or low-income housing projects.

CSI Power Generation Targets

In order to offset the high solar power installation costs and promote PV industry development, the CSI incentive program has devised a plan that encourages customer sectors to take immediate advantage of a rebate initiative that is intended to last for a limited duration of 10 years. The incentive program is currently planned to be reduced automatically over the duration of the 3000 MW of solar power reservation, in 10 step-down trigger levels that gradually distribute the power generation over 10 allotted steps. CSI megawatt power production targets are proportioned among the administrative agencies by residential and nonresidential customer sectors.

In each of the 10 steps, CSI applications are limited to the trigger levels. Table 8.3 shows the set trigger stages for the SCE and PG&E client sectors. Once the trigger level allotments are complete, the reservation process is halted and restarted at the next trigger level. In the event of a trigger level power surplus, the excess of energy allotment is transferred forward to the next trigger level.

TABLE 8.1 CSI PROGRAM BUDGET BY ADMINISTRATOR		
UTILITY	TOTAL BUDGET (%)	DOLLAR VALUE (MILLIONS)
PG&E	43.7%	$946
SCE	46%	$996
SDG&E/SDREO	10.3%	$223

TABLE 8.2   CSI POWER ALLOCATION BY CUSTOMER SECTOR

CUSTOMER SECTOR	POWER (MW)	PERCENT
Residential	557.5	33%
Nonresidential	1172.50	67
Total	1750	100

The CSI power production targets shown in Table 8.3 are based on the premise that solar power industry production output and client sector awareness will gradually be increased within the next decade and that the incentive program will eventually promote a viable industry that will be capable of providing tangible source or renewable energy base in California.

Incentive Payment Structure

As mentioned in the preceding, the CSI offers PBI and EPBB incentive programs, both of which are based on verifiable PV system output performance. EPBB incentive output characteristics are basically determined by factors such as the location of solar

TABLE 8.3   CSI SOLAR POWER PRODUCTION TARGETS BY UTILITY AND CUSTOMER SECTOR

TRIGGER STEP	ALLOTTED MW	PG&E (MW)		SCE (MW)		SDG&E (MW)	
		RES.	NONRES.	RES.	NONRES.	RES.	NONRES.
1	50						
2	70	10.1	20.5	10.6	21.6	2.4	4.8
3	100	14.4	29.3	15.2	30.8	3.4	6.9
4	130	18.7	38.1	19.7	40.1	4.4	9.0
5	160	23.1	46.8	24.3	49.3	5.4	11.1
6	190	27.4	55.6	28.8	58.6	6.5	13.1
7	215	31.0	62.9	32.6	66.3	7.3	14.8
8	250	36.1	73.2	38.0	77.1	8.5	17.3
9	285	41.1	83.4	43.3	87.8	9.7	19.7
10	350	50.5	102.5	53.1	107.9	11.9	24.2
Total allocation	1750	764.8		805.0		180.3	
Percent	100%	43.7%		46.0%		10.3%	

TABLE 8.4 CSI PBI AND EPBB TARGETED ENERGY PAYMENT AMOUNTS

TRIGGER STEP	ALLOTTED		EPBB PAYMENT/WATT($)			PBI PAYMENT/WATT ($)		
	MW	RES.	COMM.	GOV.	RES.	COMM.	GOV.
1	50	N/A	N/A	N/A	N/A	N/A	N/A
2	70	$2.5	$2.5	$3.25	$0.39	$0.39	$0.50
3	100	2.2	2.2	2.95	0.34	0.34	0.46
4	130	1.9	1.9	2.65	0.26	0.26	0.37
5	160	1.55	1.55	2.30	0.22	0.22	0.32
6	190	1.10	1.10	1.85	0.15	0.15	0.26
7	215	0.65	0.65	1.40	0.09	0.09	0.19
8	250	0.35	0.65	1.10	0.05	0.05	0.15
9	285	0.25	0.25	0.9	0.03	0.03	0.12
10	350	0.20	0.20	0.70	0.03	0.03	0.10

platforms, system size, shading conditions, tilt angle, and all the factors that were discussed in previous chapters. On the other hand, the PBI incentive is strictly based on a predetermined flat rate per kilowatt-hour output over a 5-year period. Incentive payment levels are reduced automatically over the duration of the program in 10 steps that are directly proportional to the megawatt volume reservation (see Table 8.4).

As seen from Table 8.4, rebate payments diminish as the targeted solar power program reaches its 3000-MW energy output. The main reasoning behind downscaling the incentive is the presumption that the solar power manufacturers will within the next decade be in a position to produce larger quantities of more efficient and less expensive photovoltaic modules. As a result of economies of scale, the state will no longer be required to extend special incentives to promote the photovoltaic industry by use of public funds.

Expected Performance-Based Buydown (EPBB)

As mentioned earlier, the EPBB is a one-time, upfront incentive that is based upon a photovoltaic power cogeneration system's estimated or predicted future performance. This program is targeted to minimize program administration works for relatively small systems that do not exceed 100 kWh. As a rule, factors that affect the computations of the estimated power performance are relatively simple and take into consideration such factors as panel count, PV module certified specifications, location of the solar platform, insolation, PV panel orientation, tilt angle, and shading losses. Each of these factors is entered into a predetermined equation that results in a buy-down incentive rate.

The EPBB program applies to all new projects other than systems that have building-integrated photovoltaics (BIPV). Figure 8.1a and 8.1b is a sample of Southern California Edison (SCE) CSI reservation request form. Table 8.2 is a sample of Southern California Edison CSI program cost breakdown. Figure 8.2 is Southern California Edison CSI program cost breakdown.

The EPBB one-time incentive payment calculation is based on the following formula:

$$\text{EPBB incentive payment} = \text{Incentive rate} \times \text{System rating (kW)} \times \text{Design factor.}$$

System rating (kW)

$$= \frac{\text{Number of PV modules} \times \text{CEC PTS value} \times \text{CEC inverter listed efficiency}}{1000}$$

We divided the preceding equation by 1000 to convert to kilowatts.

Special design requirements imposed are as follow:

■ All PV modules must be oriented between 180 and 270 degrees.
■ The optimal tilt for each compass direction shall be in the range of 180 and 270 degrees for optimized summer power output efficiency.
■ Derating factors associated with weather and shading analysis must be taken into account.
■ The system must be on an optimal reference and location.
■ The PV tilt must correspond to the local latitude.

Note that all residential solar power installations are also subject to the EPBB incentive payment formulation shown in Table 8.4.

Performance-Based Incentive (PBI)

As of January 1, 2007, this incentive applies to solar power system installations that have a power output equal to or exceed 100 kW. As of January 1, 2008, the base power output reference will be reduced to 50 kW and by January 1, 2010, to 30 kW. Each BPI payment is limited to a duration of 5 years following completion of the system acceptance test. Also included in the plan are custom-made building-integrated photovoltaic systems.

Host Customer

Any beneficiary of the CSI program is referred to as a host customer, including not only the electric utility customers but also retail electric distribution organizations such as PG&E, SCE, and SDG&E. Under the rules of the CSI program, an entity that applies for an incentive is referred to as an applicant, a host, or a system owner.

Southern California's

EDISON

California Solar Initiative Program
2007 Reservation Request Form and Preliminary Agreement

Reservation Number: _____ (Administrator Use only)

Return Reservation Request Form to:
Southern California Edison
California Solar Initiaive
213 Walnut Grove Ave. GO3, 3rd Floor B10
Rosemead, CA 91770

Instructions: Please confirm you are using the current Reservation Request From by going to your Program Administrator's website. Please refer to your CSI Program Handbook for instructions, and please include all required attachment with your submittal. Incomplete Reservation Request will be returned to the sender.

1. Host Customer

Customer Name: (Residential only)	
Company or Government: (Commmercial & Govt. only)	
Tax Payer ID:	
Contact Person Name:	Title:
Mailing Address:	(Suite/Apt. Street)
	(City. State, Zip)
Phone:	Secondary Phone
Email:	Public Entity: ☐ (Certification of AB107 Compliance must be attached)

2. Applicant if not Host Customer

Applicant Name:	

Installer: ☐ License Number: _____ Seller: ☐ License Number: _____

Contact Person Name:	Title:
Mailing Address:	(Suite/Apt. Street)
	(City. State, Zip)
Phone:	Secondary Phone
Email:	

3. System Owner (if not Host Customer)

Owner Name: (Residential only)	
Company or Government: (Commmercial & Govt. only)	
Contact Person Name:	Title:
Mailing Address:	(Suite/Apt. Street)
	(City. State, Zip)
Phone:	Secondary Phone
Email:	

4. Project Site Information

Project Type: ☐ Retrofit   ☐ New Construction or New Construction or Expanded

Incentive Type: ☐ Residential   ☐ Commercial   ☐ Government, Non-profit or Public

Site Address:	(Street)
	(City. State, Zip)
County:	
Est. Building Sq.Ft:	
Electric Utility Service Account Number:	Meter Number
TOU Account Number:	

5. Proposed Equipment Information

Array Number	PV Manufacturer	PV Model Number	PTC Rating (Watts/Unit)	Number of Units
1				
2				
3				
4				

Array Number	Inverter Manufacturer	Inverter Model Number	Inverter Eifficiancy	Number of Units
1				
2				
3				
4				

6. Project Incentive Calculation and Cost Information

Array Number	System Rating (SR) kW_{csc}	EPSS Design Factor $(DF)^2$	CSI System Size (SR × DF)	
1			0	kW
2			0	kW
3			0	kW

Figure 8.1   **(a) Southern California Edison CSI reservation request form page 1.**

Southern California

EDISON

California Solar Initiative Program

2007 Reservation Request Form and Preliminaray Agreement

Reservation Number: _____   (Administrator Use only)

4			0	KW

	Total CSI System Size:	0	KW
	Estimated Annual Production (kWh)[2]:		KWh

[1] System Rating (KW$_{CEC}$) PTCAC Rating
[2] Obtain from online tool at www.csi-epbb.com

Project System Size: ____0____ Watts

Eligible CSI System Size: ____0____ Watts   The lessor or 1,000,000 Watts or Project system Size

Total Eligible Project Cost: [_____]   #DIV/01   6a

Pro-rated Eligible Project Cost: ____#DIV/01____   #DIV/01   Simple Pro-ration to equale costs to Eligible CSI system size

Other Financial Incentives in Addition to CSI   List all other non-CSI incentive sources and amount being received for eligible project Cost ???. See CSI Handbook for more information.

Other incentive Source Name	Other Incentive Source Type	Amount	
			#DIV/01
			#DIV/01
			#DIV/01
Total Other Incentive Amount: $		–	#DIV/01	6b

Verify incentives Don't Exceed System Owner's Out of Pocket Expenses

Total Eligible Project Costs (6a) =	$	–	#DIV/01	6c
Total Other Incentive Amount (6b) =	$	–	#DIV/01	6d
CSI Rebate Amount (7a) =	$	–	#DIV/01	6e
System Owner Out of Pocket Expenses* =	[_____]		#DIV/01	6f

*Verification of Amount Shown required prior to CSI Incentive Payment

Line 6c-6d-6e-6f (if Total $'s is not equal to $1 zero please contact your Program Administrator) = $ –   #DIV/01

7. Requested Project Incentive

CSI System Size   [____0____] kW

CSI Incentive Amount:

EPBB (<100 kw or New Construction)
____$2.50____ $/watt

OR

PBI (>100 kW or opt in)
_____ $/kWh

Requested CSI Incentive = $	0	7a

Payment will be made to: ☐ Host Customer ☐ System Owner ☐ Third Party   _____

Please indicate Name of Third Party above

8. Checklist for Other Requrred Documents

This Checklist must be completed. Please refer to the latest version of the CSI Program Handbook for detailed instructions on eligibility and application requirements. The purpose of this checklist is to assist applicants in the completion of information materials required for review of Reservation Requests and to speed processing of the applications.

☐ Completed Reservation Request Application with Orignal Signature on CSI Program Contract
☐ Proof of Electric Utility Service for Site
☐ System Description Worksheet
☐ Electrical System Sizing Documentation (New or expanded load only)
☐ Application Fee for non-residential project over 10 kW (Requested CSI Incentive × 1%) =   [____0____]
☐ Certification of tax-exempt status and AB1407 compliance (Goverment, Non-profit, and Public Entities only)
☐ Documentation of an Energy Efficiency Audit (if you have not met Title 24 or other exemptions)
☐ Printout or EPBB Tool Calculation (www.csi-epbb.com)

Additional Required Documents for Residential Customers, Commercial <10 kW, or Govenment/Non-Profit/Public Entities < 10 kW:

☐ Copy of Executed Agreement of Solar System Purchase
☐ Copy of Executed Alternative System Ownership Agreement (if system owner is different from host customer)
☐ Copy of Application for Interconnection Agreement
☐ Host Customer Certificate of Insurance
☐ System Owner Certificate of Insurance (if different than Host Custmer)

[Please continue to program contract and provide original signatures on the following pages]

Figure 8.1   **(b) Southern California Edison CSI reservation request form page 2.**

⌐EDISON

California Solar Initiative
Project Cost Breakdown Worksheet

Reservation Number: _____

HOST CUSTOMER SITE ADDRESS (street): _____

(City, State Zip) _____

Original Submittal ☐
Revision ☐   dd/mm/yy
Final Submittal ☐

Instructions: Refer to the latest California Solar Initiative Program Handbook before completing and then submitting this form, along with the Proof of Project Advancement or Incentive Claim documentation.

Eligible Project Costs (refer to CSI Handbook for further definitions and examples):

Item No.	Eligible Cost Elements	Item Description	Cost of Item(s)
1	Planning & Feasibility Study Costs		$0.00
2	Engineering & Design Costs		$0.00
3	Permitting Costs (air quality, building permits, etc.)		$0.00
4	PV Equipment Costs (generator, ancillary equipment)		$0.00
	PV Modules		$0.00
	Inverter		$0.00
5	Construction & Installation Costs (labor & materials)		$0.00
6	Interconnection Costs - Electric (customer side of meter only)		$0.00
7	Warranty Cost (if not already included in Item 4)		$0.00
8	Maintenance Contract Cost (only if warranty is insufficient)		$0.00
9	Sales Tax		$0.00
10	Metering Costs		$0.00
	Hardware (Meter, Socket, CTs, PTs, etc.)		$0.00
	Communication (Hardware & Service Provider)		$0.00
	Power Monitoring (PMRS)/Meter Data Mgt (MDMA)		$0.00
11	Other Eligible Costs (Itemize Below)		$0.00
10.a			$0.00
10.b			$0.00
10.c			$0.00
	Total Eligible Project Costs:		$0

PV Module Costs	$0.00
Inverter Costs	$0.00
Metering Costs	$0.00
Balance of System (BOS) Costs	$0.00
Feasibility & Engineering Design Costs	$0.00
Installation Costs	$0.00
Fees/Permiting Costs	$0.00

Figure 8.2   Southern California Edison CSI program cost breakdown.

In general, host customers must have an outstanding account with a utility provider at a location of solar power cogeneration. In other words, the project in California must be located within the service territory of one of the three listed program administrators.

CSI provides a payment guarantee, called a reservation, that cannot be transferred by the owner; however, the system installer can be designated to act on behalf of the owner. Upon approval of the reservation, the host customer is considered as the system owner and retains sole rights to the reservation.

To proceed with the solar power program, the applicant or the owner must receive a written confirmation letter from the administrating agency and then apply for authorization for grid connectivity. In the event of project delays beyond the permitted period of fund reservation, the customer must reapply for another rebate to obtain authorization.

According to SCI regulations there are several categories of customers who do not qualify to receive the incentive. Customers exempted from the program are organizations that are in the business of power generation and distribution, publicly owned gas and electricity distribution utilities, or any entity that purchases electricity or natural gas for wholesale or retail purposes.

As a rule, the customer assumes full ownership upon reception of the incentive payment and technically becomes responsible for operation and maintenance of the overall solar power system.

Note that a CSI applicant is recognized as the entity that completes and submits the reservation forms and becomes the main contact person who must communicate with the program administrator throughout the duration of the project; however, the applicant may also designate an engineering organization or a system integrator, an equipment distributor, or even an equipment lessor to act as the designated applicant.

Solar Power Contractors and Equipment Sellers

Contractors in the state of California who specialize in solar power installation must hold an appropriate state contractor's license. In order to qualify as an installer by the program administrator, the solar power system integrator must provide the following information:

- Business name and address
- Principal's name or contact
- Business registration or license number
- Contractor's license number
- Contractor's bond (if applicable) and corporate limited liability entities
- Reseller's license number (if applicable)

All equipment such as PV modules, inverters, and meters sold by equipment sellers must be UL approved and certified by the California Energy Commission. All equipment provided must be new and have been tested for at least a period of 1 year. Use of refurbished equipment is not permitted. Note that experimental, field demonstrated, or proof-of-concept operation type equipment and material are not approved and do not qualify for a rebate incentive. All equipment used therefore must have UL certification and performance specifications that would allow program administrators to evaluate equipment performance.

According to CEC certification criteria, all grid-connected PV systems must carry a 10-year warranty and meet the following certification requirements:

- All PV modules must be certified to UL 1703 standards.
- All grid-connected solar watt-hour meters for systems under 10 kW must have an accuracy of ±5 percent. Watt-hour meters for systems over 10 kW must have a measurement accuracy of ±2 percent.
- All inverters must be certified to UL 1741 standards.

PV System Sizing Requirement

Note that the primary objective of solar power cogeneration is to produce a certain amount of electricity to offset a certain portion of the electrical demand load. Therefore power production of PV systems is set in a manner as not to exceed the actual energy consumption during the previous 12 months. The formula applied for establishing the maximum system capacity is:

Maximum system power output (kW)

$$= \frac{12 \text{ months of previous energy used (kWh)}}{(0.18 \times 8760 \text{ h/yr})}$$

The factor of $0.18 \times 8760 = 1577$ h/yr can be translated into an average of 4.32 h/day of solar power production, which essentially includes system performance and derating indexes applied in CEC photovoltaic system energy output calculations.

The maximum PV system size under the present CSI incentive program is limited to 1000 kW or 1 MW; however, if the preceding calculation limits permit, customers are allowed to install grid-connected systems of up to 5 MW, for which only 1 MW will be considered for receiving the incentive.

For new construction where the project has no history of previous energy consumption, an applicant must substantiate system power demand requirements by engineering system demand load calculations that will include present and future load growth projections. All calculations must be substantiated by corresponding equipment specifications, panel schedules, single-line diagrams, and building energy simulation programs such as eQUEST, EnergyPro, or DOE-2.

Energy Efficiency Audit

Recent rules enacted in January 2007 require that all existing residential and commercial customers when applying for a CSI rebate will be obligated to provide a certified energy efficiency audit for their existing building. The audit certification along with the solar PV rebate application forms must be provided to the program administrator for evaluation purposes.

An energy audit could be conducted either by calling an auditor or accessing a special Web page provided by each administrative entity. In some instances, energy audits could be waved if the applicants can provide a copy of an audit conducted in the past 3 years or provide proof of a California Title 24 Energy Certificate of Compliance, which is usually calculated by mechanical engineers. Projects that have a national LEED certification are also exempt from an energy audit.

Warranty and Performance Permanency Requirements

As mentioned previously, all major system components are required to have a minimum of 10 years of warranty by manufacturers and installers alike. All equipment including PV modules and inverters in the event of malfunction are required to be replaced at no cost to the client. System power output performance must include 15 percent power output degradation from the original rated projected performance for a period of 10 years.

To be eligible for CSI rebates, all solar power system installations must be permanently attached or secured to their platforms. PV modules supported by quick disconnect means or installed on wheeled platforms or trailers are not considered as legitimate stationary installations.

During the course of project installation, the owner or designated representative must maintain continuous communication with the program administrator and provide all required information regarding equipment specifications, warranties, the platform configuration, all design revisions and system modifications, updated construction schedules, and construction status on a regular basis.

In the event the location of PV panels are changed and panels are removed or relocated within the same project perimeters or service territory, the owner must inform the CSI administrator and establish a revised PBI payment period.

Insurance

At present the owner or the host customer of a system rated 30 kW or above and receiving CSI is required to carry a minimum level of general liability insurance. Installers also must carry worker's compensation and business auto insurance coverage.

Since U.S. government entities are self-insured, the program administrators will only require a proof of coverage.

Grid Interconnection and Metering Requirements

The main criterion for grid system integration is that the solar power cogeneration system must be permanently connected to the main electrical service network. As such, portable power generators are not considered eligible. In order to receive the incentive payment, the administrator must receive proof of grid interconnection. In order to receive additional incentives, customers whose power demands coincide with California's peak electricity demand become eligible to apply for time-of-use (TOU) tariffs that could increase their energy payback.

All meters installed must be physically located to allow the administrator's authorized agents to have easy access for test or inspection.

Inspection

All systems rated from 30 to 100 kW that have not adopted the PBI will be inspected by specially designated inspectors. In order to receive the incentive payment, the inspectors must verify the system operational performance, installation conformance to the application, eligibility criteria, and grid interconnection.

System owners who have opted for the Expected Performance-Based Buydown (EPBB) incentive must install the PV panels in the proper orientation and produce power that is reflected in the incentive application.

In the event of inspection failure, the owner will be advised by the administrator about shortcomings regarding material or compliance which must be mitigated within 60 days. Failure to correct the problem could result in cancellation of the application and a strike against the installer, applicant, seller, or any party deemed responsible. Entities identified as responsible for mitigating the problem that fail to do so after three attempts will be disqualified from participating in CSI programs for a period of 1 year.

CSI Incentive Limitations

A prerequisite for processing a CSI program application is that the project's total installed out-of-pocket expenses by the owner do not exceed the eligible costs. For this reason the owner or the applicant must prepare a detailed project cost breakdown that highlights only the relative embedded cost of the solar power system. A worksheet designed for this purpose is available on the CSI Web page.

It is important to note that clients are not permitted to receive incentives under other sources. In the event a project may be qualified to receive an additional incentive from another source for the same power cogenerating system, the first incentive amount will be discounted by the amount of the second incentive received. In essence the overall combined incentive amount must not exceed the total eligibility costs. At all times during the project construction, administrators reserve the right to conduct periodic spot checks and random audits to make certain that all payments received were made in accordance with CSI rules and regulations.

CSI Reservation Steps

The following summarized the necessary steps for a EPBB application:

1 The reservation form must be completed and submitted with the owner's or applicant's wet signature.
2 Proof of electric utility service or the account number for the project site must be shown on the application. In case of a new project the owner must procure a tentative service account number.
3 The system description worksheet available on the CSI Web page must be completed.
4 Electrical system sizing documents as discussed must be attached to the application form.
5 If the project is subject to tax-exemption form AB 1407 compliance for government and nonprofit organizations, it must be attached to the application.
6 For existing projects, an energy efficiency audit or Title 24 calculations must be submitted as well.
7 To calculate the EPBB use the CSI Web page calculator (www.csi-epbb.com).
8 Attach a copy of the executed purchase agreement from the solar system contractor or provider.
9 Attach a copy of the executed contract agreement if the system ownership is given to another party.
10 Attach a copy of the grid interconnection agreement if available; otherwise inform the administrator about steps taken to secure the agreement.

To submit a payment claim provide the following documents to the administrator. Figure 8.3 is a sample of Southern California Edison CSI system description form.

1 Wet-signed claim form available on the CSI Web page
2 Proof of authorization for grid integration
3 Copy of the building permit and final inspection sign-off
4 Proof of warranty from the installer and equipment suppliers
5 Final project cost breakdown
6 Final project cost affidavit

For projects categorized as BPI or nonresidential systems above 10 kW or larger, the owner must follow the following process:

California Solar Initiative (CSI) Program
System Description Worksheet

Southern California Edison
CSI Program
2131 Walnut Grove Ave
GO3, 3rd Floor, B-10
Rosemead, CA 91770

Section A.

Reservation Number (if unknown, leave blank): _____

Host Customer Name (per Reservation Request Form): _____

Site Address (Street Address, City, State, Zip): _____

Name of Person Completing This Form and Title: _____

Company Name: _____

Telephone Number: _____

Email Address: _____

Section B.

Please provide complete answers to all of the following questions regarding the proposed Project requesting a reservation under the CSI Program.

1. Is there any **existing generation** at this Site, including nonfunctioning and/or emergency back-up generation?

   ☐ Yes   ☐ No

   If yes, does any of this existing generation serve to export electricity for sale, either with an "over-the-fence", or other wholesale arrangement (e.g., qualifying facility contract)?

   ☐ Yes   ☐ No

2. Describe the **location of the major system components** on the site (e.g., PV system will be located on the *existing* administration building's roof) and attach a copy of a site or plot plan if available.

   _____

3. At what **stage of development** is this Project currently in and is **new building construction or major renovation** also involved?

   ☐ Conceptual planning   ☐ Signed contract for equipment purchase   ☐ System installed

   ☐ New construction   ☐ Major renovation/facility expansion

4. What is the estimated date of Project completion (all equipment being claimed as an eligible cost is installed, interconnected, permitted and operational)? Est. project completion date: ____/____/____

5. Have the necessary **interconnection application(s)** already been submitted to the serving local distribution company(s)?

   ☐ Yes   ☐ No (estimated date of application: ____/____/____)

6. If the Host Customer will not be the owner of the PV system after it is installed, describe the contractual arrangement between the System Owner and Host Customer, including duration of contract and cancellation policy (attach additional copy of contract).

   _____

7. Does the Host Customer named above have **legal ownership of the building and/or property** where the proposed PV system will be located?

   ☐ Yes   ☐ No

Last Printed 1/23/07   CSI_System_DescriptionNew2.doc

Figure 8.3   (a) Southern California Edison CSI system description form page 1.
(b) Southern California Edison CSI system description form page 2.

**California Solar Initiative (CSI) Program
System Description Worksheet**

Southern California Edison
CSI Program
2131 Walnut Grove Ave
GO3, 3rd Floor, B-10
Rosemead, CA 91770

If no, please indicate who does own the building and/or property, the contractual relationship with the Host Customer and the contract period. _____

_____

8. Is the Host Customer or Applicant aware of any plans to potentially **sell, transfer or relocate the proposed PV system and/or the buildings or property where the PV system will be located**, before the CSI required warranty period ends (i.e., ten years after the system is installed)?

☐ Yes   ☐ No

If yes, please describe _____

_____

9. Will the portion of the Project costs not covered by the CSI rebate require **outside financing** by the Host Customer?

☐ Yes   ☐ No

If yes, please indicate the planned source of financing:

_____ Seller of System, _____ Government Agency, _____ Commercial Lender, _____ Undecided

10. Are there other financial incentives with regards to this Project? Examples of other incentives include but are not limited to: other program rebates, grants and gifted equipment.

☐ Yes   ☐ No

If yes, please describe and attach copies of agreement(s) involved. _____

11. Are there any **post-sale agreements or contracts** anticipated as part of either the sale or lease of this system between either the seller/installer, the system owner and/or the Host Customer which go into effect after the initial sale is made (e.g., cash payment(s) to Host Customer for agreeing to allow seller/installer to use system in a sales or promotion campaign, cash payment(s) to Host Customer if system does not perform to a certain level, anticipated sale of Renewable Energy Credits (RECs) or Green Credits, forgiven loans, performance payments, etc.)?

☐ Yes   ☐ No

If yes, please describe and attach copies of agreement(s) involved.

_____

_____

I hereby certify that the answers provided to the questions above are complete and accurate to the best of my knowledge. I am signing on behalf of the:

☐ **Applicant**   ☐ **Host Customer**   ☐ **System Owner**

Signature: _____

Company Name _____

Date: _____

Figure 8.3   *(Continued)*

1 The reservation form must be completed and submitted with the owner's or applicant's wet signature.

2 Proof of electric utility service or the account number for the project site must be shown on the application. In case of a new project the owner must procure a tentative service account number.

3 The system description worksheet available on the CSI Web page must be completed.

4 Electrical system sizing documents as discussed previously must be attached to the application form.

5 Attach an application fee (1 percent of the requested CSI incentive amount).

6 If the project is subject to tax-exemption form AB1407 compliance for government and nonprofit organizations, it must be attached to the application.

7 For existing projects an energy efficiency audit or Title 24 calculations must be submitted as well.

8 Forward the printout of the calculated PBI. Use the CSI Web page calculator.

9 Attach a copy of the executed purchase agreement from the solar system contractor or provider.

10 Attach a copy of the executed contract agreement if the system ownership is given to another party.

11 Attach a copy of the grid interconnection agreement if available; otherwise inform the administrator about the steps taken to secure the agreement.

12 Provide the following documents:
   a Completed proof of project milestone
   b Host customer certificate of insurance
   c System owner's certificate of insurance (if different from the host)
   d Copy of the project cost breakdown worksheet
   e Copy of an alternative system ownership such as a lease or buy agreement
   f A copy of the RFP or solicitation document if the customer is a government or nonprofit organization or a public entity

To submit the claim for the incentive documents to the administrator the owner or the contractor must provide the following:

1 Wet-signed claim form available on the CSI Web page
2 Proof of authorization for grid integration
3 Copy of the building permit and final inspection sign-off
4 Proof of warranty from the installer and equipment suppliers
5 Final project cost breakdown
6 Final project cost affidavit

In the event of incomplete document submittal, the administrators will allow the applicant 20 days to provide missing documentation or information. The information provided must be in a written form mailed by the U.S. postal system. Faxes or hand-delivered systems are not allowed.

All changes to the reservation must be undertaken by a formal letter that provides a legitimate justification for the delay. Requests to extend the reservation expiration date are capped to a maximum of 180 calendar days. Written time extension requests must explicitly highlight circumstances that were beyond the control of the reservation holder such as the permitting process, manufacturing delays and extended delivery of PV modules or critical equipment, and acts of nature. All correspondence associated with the delay must be transmitted with the letter.

Incentive Payments

Upon completion of final field acceptance and submission of the earlier referenced documents, for EPBB projects the program administrator will within a period of 30 days issue complete payment. For PBI programs the first incentive payment is commenced and issued within 30 days of the first scheduled performance reading from the wattmeter. All payments are made to the host customer or the designated agent.

In some instances a host could request the administrator to assign the all payments to a third party. For payment reassignment the host must complete a special set of forms provided by the administrator.

The EPBB one-time lump-sum payment calculation is based on the following formula:

EPBB incentive payment = Final CSI system size × Reserved EPBB incentive rate

PBI payments for PV systems of 100 kW or greater are made on a monthly basis over a period of 5 years. Payment is based on the actual electric energy output of the photovoltaic system. If chosen by the owner, systems less than 100 kW could also be paid on the PBI incentive basis.

The PBI payment calculation is based on the following formula:

Monthly PBI incentive payment = Reserved incentive rate × Measured kWh output

In the event of a change in the PV system, the original reservation request forms must be updated and the incentive amount recalculated. Figure 8.4 depicts CSI Performance Based Initiative (PBI) and Figure 8.5 is CSI Estimated Performance-Based Buydown Initiative rebate structures.

An Example of the Procedure for Calculating the California Solar Incentive Rebate

The following example is provided to assist the reader with details of CSI reservation calculation requirements. To apply for reservation, all of reservation forms along with project site information and solar power projected performance calculations must be

CSI INCENTIVE STRUCTURE

YEAR	EPBB MEGAWATTS	RES. ($)	COMM. ($)	GOV. ($)
	50			
1	70	$ 2.50	$ 2.50	$ 3.25
2	100	2.20	2.20	2.95
3	130	1.90	1.90	2.65
4	160	1.55	1.55	2.30
5	190	1.10	1.10	1.85
6	215	0.65	0.65	1.40
7	250	0.35	0.65	1.10
8	285	0.25	0.25	0.90
9	350	0.20	0.20	0.70

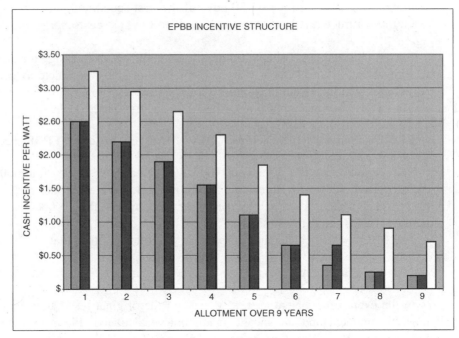

COLUMN 1—RESIDENTIAL
COLUMN 2—COMMERCIAL
COLUMN 3—GOVERNMENT/NONPROFIT

Figure 8.4   **CSI Performance-Based Initiative rebate.**

submitted to the SCE program manager. Regardless of the system size and classification as EPBB or PBI, the supportive documentation for calculating the solar power system cogeneration remains identical.

In order to commence reservation calculations, the designer must undertake the following preliminary design measures:

CSI INCENTIVE STRUCTURE

YEAR	PBI MEGAWATTS	RES. ($)	COMM. ($)	GOV. ($)
	50			
1	70	$ 0.39	$ 0.39	$ 0.50
2	100	0.34	0.34	0.46
3	130	0.26	0.26	0.37
4	160	0.22	0.22	0.32
5	190	0.15	0.15	0.26
6	215	0.09	0.09	0.19
7	250	0.05	0.05	0.15
8	285	0.03	0.03	0.12
9	350	0.03	0.03	0.10

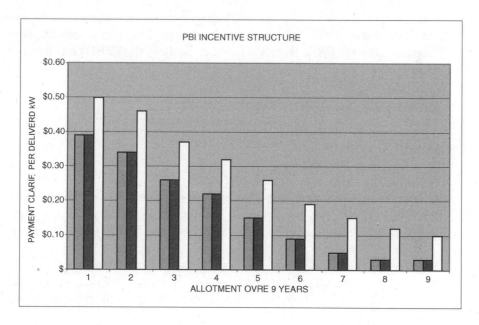

COLUMN 1—RESIDENTIAL
COLUMN 2—COMMERCIAL
COLUMN 3—GOVERNMENT/NONPROFIT

Figure 8.5   **CSI Estimated Performance-Based Buydown Initiative rebate.**

- Outline the solar power cogeneration system's net unobstructed platform area.
- Use a rule of thumb to determine watts per square feet of a particular PV module intended for use. For example, a PV module area of 2.5 ft × 5 ft = 12.5 ft^2 that produces 158 W PTC will have an approximate power output of 14 W/ft^2.
- By dividing the available PV platform area by 12.5 ft^2 we could determine the number of panels required.

In order to complete the CSI reservation forms referenced earlier, the designer must also determine the type, model, quantity, and efficiency of the CEC-approved inverter.

For this example let us assume that we are planning for a ground-mount solar farm with an output capacity of 1 MW. The area available for the project is 6 acres, which is adequate for installing a single-axis solar tracking system. Our solar power module and inverter are chosen from approved CEC listed equipment.

Solar power system components used are:

- *PV module.* SolarWorlds Powermax, 175p, unit W dc = 175, PTC = 158.3 W ac, 6680 units required.
- *Inverter.* Xantrex Technology, PV225S-480P, 225 kW, efficiency 94.5 percent.

Prior to completing the CSI reservation form, the designer must use the CSI EPBB calculator (available at www.csi-epbb.com) to determine the rebate for systems that are smaller than 100 kW or larger. Even though BPI calculations are automatically determined by the CSI reservation form spreadsheet, the EPBB calculation determines a Design Factor number required by the form.

EPBB CALCULATION PROCEDURE

To conduct the EPBB calculation the designer must enter the following data in the blank field areas:

- Project area zip code, for example, 92596
- Project address information
- Customer type such as residential, commercial, government, or nonprofit
- PV module manufacturer, model, associated module dc and PTC rating, and unit count
- Inverter manufacturer, model, output rating, and percent efficiency
- Shading information, such as minimal
- Array tilt, such as 30 degrees. For maximum efficiency the tilt angle should be close to latitude.
- Array azimuth in degrees, which determines orientation. For northern hemisphere PV modules installations use 180 degrees.

After the preceding data have been entered, the CSI calculator will output the following results:

- Optimal tilt angle at proposed azimuth.
- Annual kilowatt-hour output at optimal tilt facing south.
- Summer month output from May to October.
- CEC ac rating—a comment will indicate if the system is greater than 100 kW.
- Design correction factor—required for calculating the CSI reservation form.
- Geographic correction.
- Design factor.

- Incentive rate in dollars per watt.
- Rebate incentive cash amount if system is qualified as EPBB.

CALIFORNIA SOLAR INITIATIVE RESERVATION FORM CALCULATIONS

Like the EPBB form referenced earlier, the CSI reservation is also a Web page spreadsheet that can be accessed at the CSI Web page. Data required to complete the solar equipment are the same as the ones used for EPBB, except that the EPBB Design Factor derived from the earlier calculation must be inserted in the project incentive calculations.

The California Solar Initiative Program Reservation form consists of the following six major information input data fields:

1 Host customer. The information required includes the customer's name, business class and company information, and taxpayer identification; and a contact person's name, title, mailing address, telephone, fax, and e-mail.
2 Applicant information if the system procurer is not the host customer.
3 System owner information.
4 Project site information—same as that for EPBB. System platform information such as available building or ground area must be specified. In this file the designer must also provide the electrical utility service account and meter numbers if available. For new projects there should be a letter attached to the reservation form indicating the account procurement status.
5 PV and inverter hardware information identical to the one used in calculating EPBB.
6 Project incentive calculation. In this field the designer must enter the system rating in kilowatts (CEC) and Design Factor data obtained from the EPBB calculation.

When these entries are completed, the CSI reservation spreadsheet will automatically calculate the project system's power output size in watts. In a field designated Total Eligible Project Cost, the designer must insert the projected cost of the system, which automatically produces a per-watt installed cost, CSI rebate amount, and system owner out-of-pocket expenses.

The following CSI reservation request calculation is based on the same hardware information used in the preceding EPBB calculation. The data entry information and calculation steps are as follows:

- Platform—single-axis tracker
- Shading—none
- Insolation for zip code 92596, San Bernardino, California = 5.63 average h/day
- PV module—SolarWorld model SW175 mono/P
- DC watts—175
- PTC—158.2
- PV count—6680
- Total power output—1057 kW ac

- Calculated CSI system size by the spreadsheet—1032 kW ac
- Inverter—Xantrex Technology Model PV225S-480P
- Power output capacity—225 kW ac
- Efficiency—94.5 percent
- Resulting output = 1057 × 94.5% = 999 kW
- CSI EPBB Design Factor = 0.975
- CSI system size = 999 × 0.975 = 975 kWh
- PV system daily output = 957 × 5.63 (insolation) = 5491 kWh/day
- Annual system output = 5491 × 365 (days) = 2,004,196 kWh/yr

Assuming an incentive class for a government or nonprofit organization for year 2007, the allocated performance output per kilowatt-hour = $0.513.

- Total incentive over 5 years = 2,004,196 × $0.513 = $1,028,152.55
- Projected installed cost = $8,500,000.00 ($8.50/W)
- System owner's out-of-pocket cost = $3,489,510.00
- Application fee of 1 percent of incentive amount = $50,105.00

Equipment Distributors

Eligible manufacturers and companies who sell system equipment must provide the CEC with the following information on the equipment seller information form (CEC-1038 R4). For all CEC approved equipment see Appendix C. Figure 8.6 is a sample of Southern California Edison CSI final project cost affidavit.

- Business name, address, phone, fax, and e-mail address
- Owner or principal contact
- Business license number
- Contractor license number (if applicable)
- Proof of good standing in the records of the California secretary of state, as required for corporate and limited liability entities
- Reseller's license number

Special Funding for Affordable Housing Projects

California Assembly Bill 58 mandates the CEC to establish an additional rebate for systems installed on affordable housing projects. These projects are entitled to qualify for an extra 25 percent rebate above the standard rebate level, provided that the total amount rebated does not exceed 75 percent of the system cost. The eligibility criteria for qualifying are as follows:

All communications under Affidavit shall be forwarded directly to:
Southern California Edison
California Solar Incentive Program
2131 Walnut Grove Ave
GO3 , 3rd Floor B-10
Rosemead, CA 91770

FINAL PROJECT COST AFFIDAVIT

By signing this affidavit ("Affidavit"), _____ ("Host Customer") and
_____ ("System Owner", if different than the "Host Customer"), jointly referred to
as "Parties", with respect to the Photovoltaic (PV) system project ("Project") at
_____ (site address), which is partially funded by the _____
("Program Administrator") California Solar Incentive ("CSI") Program under Reservation Number
_____, each certify and declare under penalty of perjury under the laws of the State of
California that each of the statements in the paragraphs below are complete, true and correct.

Parties attest that the statements in the following paragraphs are true.

1. At the time the incentive payment is made, _____ ("System Owner") is the
   owner of the PV system which comprise the Project and all the statements below are true and
   correct:

    o The information provided on the following forms is true and accurate:

        • Final Incentive Calculation Worksheet

        • Final Incentive Claim Form

        • Final Project Cost Breakdown Worksheet,

    o System Owner incurred all costs referenced on the Project Cost Breakdown Worksheet,

    o Project is operating as intended according to Contract,

    o There are no post-sale agreements or agreements which go into effect after the initial sale is
       made which allow the seller or installer to use the PV system which comprise the Project in a
       sales or promotion campaign, or provide for cash payment(s) to Host Customer if the PV system
       does not perform to a certain level.

    o Project is financed or paid for in full except for an amount, which does not exceed the amount of
       incentive funding to be provided by Contract.

2. Costs to Project as referenced below are defined in the Incentive Claim Form and are identical to the
   costs submitted by Parties to Program Administrator in the Final Project Cost Breakdown
   Spreadsheet (attached).

    Total Eligible Project Cost: $ _____

    Total Incentive Amount: $ _____

Each of the undersigned certifies under penalty of perjury that the foregoing is true and correct and that
each is duly authorized to sign this Affidavit.

[HOST CUSTOMER]	**[SYSTEM OWNER]**
Signature: _____	Signature: _____
Name Printed: _____	Name Printed: _____
Title: _____	Title: _____
Date: _____	Date: _____

Last Printed 1/23/2007   DRAFTCSI_Project_Cost_Affidavit.doc

Figure 8.6   Southern California Edison CSI final project cost affidavit.

The affordable housing project must adhere to California health and safety codes. The property must expressly limit residency to extremely low, very low, lower, or moderate income persons and must be regulated by the California Department of Housing and Community Development.

Each residential unit (apartments, multifamily homes, etc.) must have individual electric utility meters.

The housing project must be at least 10 percent more energy efficient than current standards specified.

Special Funding for Public and Charter Schools

A special amendment to the CEC mandate, enacted in February 4, 2004, established a Solar Schools Program to provide a higher level of funding for public and charter schools to encourage the installation of photovoltaic generating systems at more school sites. At present the California Department of Finance has allocated a total of $2.25 million for this purpose. To qualify for the additional funds, the schools must meet the following criteria:

Public or charter schools must provide instruction for kindergarten or any of the grades 1 through 12.

The schools must have installed high-efficiency fluorescent lighting in at least 80 percent of classrooms.

The schools must agree to establish a curriculum tie-in plan to educate students about the benefits of solar energy and energy conservation.

Principal Types of Municipal Lease

There are two types of municipal bonds. One type is referred to as a "tax-exempt municipal lease," which has been available for many years and is used primarily for the purchase of equipment and machinery that has a life expectancy of 7 years or less. The second type is generally known as an "energy efficiency lease" or a "power purchase agreement" and is used most often on equipment being installed for energy efficiency purposes and is used where the equipment has a life expectancy of greater than 7 years. Most often this type of lease applies to equipment classified for use as a renewable energy cogeneration, such as solar PV and solar thermal systems. The other common type of application that can take advantage of municipal lease plans includes energy efficiency improvement of devices such as lighting fixtures, insulation, variable-frequency motors, central plants, emergency backup systems, energy management systems, and structural building retrofits.

The leases can carry a purchase option at the end of the lease period for an amount ranging from $1.00 to fair market value and frequently have options to renew the lease at the end of the lease term for a lesser payment over the original payment.

TAX-EXEMPT MUNICIPAL LEASE

A tax-exempt municipal lease is a special kind of financial instrument that essentially allows government entities to acquire new equipment under extremely attractive terms with streamlined documentation. The lease term is usually for less than 7 years. Some of the most notable benefits are:

- Lower rates than conventional loans or commercial leases.
- Lease-to-own. There is no residual and no buyout.
- Easier application, such as same-day approvals.
- No "opinion of counsel" required for amounts under $100,000.
- No underwriting costs associated with the lease.

ENTITIES THAT QUALIFY FOR A MUNICIPAL LEASE

Virtually any state, county, or city municipal government and its agencies, such as law enforcement, public safety, fire, rescue, emergency medical services, water port authorities, school districts, community colleges, state universities, hospitals, and 501 organizations qualify for municipal leases. Equipment that can be leased under a municipal lease includes essential-use equipment and remediation equipment such as vehicles, land, or buildings. Some specific examples are listed here:

- Renewable energy systems
- Cogeneration systems
- Emergency backup systems
- Microcomputers and mainframe computers
- Police vehicles
- Networks and communication equipment
- Fire trucks
- Emergency management service equipment
- Rescue construction equipment such as aircraft helicopters
- Training simulators
- Asphalt paving equipment
- Jail and court computer-aided design (CAD) software
- All-terrain vehicles
- Energy management and solid waste disposal equipment
- Turf management and golf course maintenance equipment
- School buses

- Water treatment systems
- Modular classrooms, portable building systems, and school furniture such as copiers, fax machines, closed-circuit television surveillance equipment
- Snow and ice removal equipment
- Sewer maintenance

The transaction must be statutorily permissible under local, state, and federal laws and must involve something essential to the operation of the project.

DIFFERENCE BETWEEN A TAX-EXEMPT MUNICIPAL LEASE AND A COMMERCIAL LEASE

Municipal leases are special financial vehicles that provide the benefit of exempting banks and investors from federal income tax, allowing for interest rates that are generally far below conventional bank financing or commercial lease rates. Most commercial leases are structured as rental agreements with either nominal or fair-market-value purchase options.

Borrowing money or using state bonds is strictly prohibited in all states, since county and municipal governments are not allowed to incur new debts that will obligate payments that extend over multiyear budget periods. As a rule, state and municipal government budgets are formally voted into law; as such there is no legal authority to bind the government entities to make future payments.

As a result, most governmental entities are not allowed to sign municipal lease agreements without the inclusion of nonappropriation language. Most governments, when using municipal lease instruments, consider obligations as current expenses and do not characterize them as long-term debt obligations.

The only exceptions are bond issues or general obligations, which are the primary vehicles used to bind government entities to a stream of future payments. General obligation bonds are contractual commitments to make repayments. The government bond issuer guarantees to make funds available for repayment, including raising taxes if necessary. In the event, when adequate sums are not available in the general fund, "revenue" bond repayments are tied directly to specific streams of tax revenue. Bond issues are very complicated legal documents that are expensive and time consuming and in general have a direct impact on the taxpayers and require voter approval. Hence, bonds are exclusively used for very large building projects such as creating infrastructure like sewers and roads.

Municipal leases automatically include a nonappropriation clause; as such they are readily approved without counsel. Nonappropriation language effectively relieves the government entity of its obligation in the event funds are not appropriated in any subsequent period, for any legal reason.

Municipal leases can be prepaid at any time without a prepayment penalty. In general, a lease amortization table included with a lease contract shows the interest principal and payoff amount for each period of the lease. There is no contractual penalty, and a payoff schedule can be prepared in advance. It should also be noted that equipment and installations can be leased.

Lease payments are structured to provide a permanent reduction in utility costs when used for the acquisition of renewable energy or cogeneration systems. A flexible leasing structure allows the municipal borrower to level out capital expenditures from year to year. Competitive leasing rates of up to 100 percent financing are available with structured payments to meet revenues that could allow the municipality to acquire the equipment without having current fund appropriation.

The advantages of a municipal lease program include

- Enhanced cash flow financing allows municipalities or districts to spread the cost of an acquisition over several fiscal periods leaving more cash on hand.
- A lease program is a hedge against inflation since the cost of purchased equipment is figured at the time of the lease and the equipment can be acquired at current prices.
- Flexible lease terms structured over the useful life span of the equipment can allow financing of as much as 100 percent of the acquisition.
- Low-rate interest on a municipal lease contract is exempt from federal taxation, there are no fees, and rates are often comparable to bond rates.
- Full ownership at the end of the lease most often includes an optional purchase clause of $1.00 for complete ownership.

Because of budgetary shortfalls, leasing is becoming a standard way for cities, counties, states, schools, and other municipal entities to get the equipment they need today without spending their entire annual budget to acquire it.

Municipal leases are different from standard commercial leases because of the mandatory nonappropriation clause, which states that the entity is only committing to funds through the end of the current fiscal year, even if the entity is signing a multi-year contract.

Electric Energy Cost Increase

In the past several decades the steadily increasing cost of electric energy production has been an issue that has dominated global economics and geopolitical politics, affected our public policies, become a significant factor in the gross national product equation, created numerous international conflicts, and made more headlines in newsprint and television than any other subject. Electric energy production not only affects the vitality of international economics, but it is one of the principal factors that determines standards of living, health, and general well-being of the countries that produce it in abundance.

Every facet of our economy one way or another is connected to the cost of electric energy production. Since a large portion of global electric energy production is based on fossil-fuel-fired electric turbines, the price of energy production is therefore determined by the cost of coal, crude oil, or natural gas commodities. As discussed in Chapter 1, the consequences of burning fossil fuel have significantly contributed to global warming and have adversely impacted the terrestrial ecology and our life style. In order to mitigate the devastating effects of fossil fuels, the international community has in the recent past taken steps to minimize the excess use of fossil fuels in electric energy production.

The state of California Assembly has recently introduced an act known as AB32 that mandates the entire state industry by year 2020 to reduce their carbon dioxide CO_2 footprint to 1990 levels.

California Assembly Bill 32

The following is a summary of California Assembly Bill AB 32, a complete text of which can be accessed at www.environmentcalifornia.org/html/AB32-finalbill.pdf. The legislated act addresses health and safety codes relating to air pollution. Under this act, also known as the Global Warming Solutions Act of 2006, the law mandates the State Air Resources Board, the State Energy Resources Conservation and Development Commission (Energy Commission), and the California Climate Action Registry to assume responsibilities with regard to the control of emissions of greenhouse gases. The Secretary for Environmental Protection is also mandated to coordinate emission reductions of greenhouse gases and climate change activity in state government.

To implement the bill, the state board is required to adopt regulations to require the reporting and verification of statewide greenhouse gas emissions and to monitor and enforce compliance with this program, as specified. The bill requires the state board to adopt a statewide greenhouse gas emissions limit equivalent to statewide greenhouse gas emissions levels in 1990 to be achieved by 2020, as specified. The bill would require the state board to adopt rules and regulations in an open public process to achieve the maximum technologically feasible and cost-effective greenhouse gas emission reductions, as specified. It also authorizes the state board to adopt market-based compliance mechanisms. Additionally the bill requires the state board to monitor compliance with and enforce any rule, regulation, order, emissions limitation, emissions reduction measure, or market-based compliance mechanism adopted by the state board, pursuant to specified provisions of existing law. The bill authorizes the state board to adopt a schedule of fees to be paid by regulated sources of greenhouse gas emissions, as specified. Because bill AB 32 requires the state board to establish emissions limits and other requirements, the violation of which would be a crime, this bill would create a state-mandated local program.

The California constitution requires the state to reimburse local agencies and school districts for certain costs mandated by the state. Statutory provisions also establish procedures for making that reimbursement.

IMPACT OF CALIFORNIA ASSEMBLY BILL 32

At present, California produces 50 percent of its electric energy mainly through coal, natural gas turbines, and nuclear power–generating stations. Hydroelectric and nuclear power represent a small percentage of state local electric power. The state of California therefore imports a significant amount of its electric energy from outside energy providers. Much of this outside electric energy is produced by coal- and gas-fired turbines, hydroelectric power, and nuclear power–generating stations.

By mandating reduction of greenhouse gas production, California will within the near future abstain from the purchase and import of electric energy from sources that use coal-based electric turbines. Most significantly all coal-based electric power-generating stations within California will be mandated to use less polluting natural gas turbines.

With reference to Figure 8.7, the cost of electric energy production historically has been going up at an accelerated rate. As indicated in the figure, the average annual cost of electric energy in California has escalated at an average of 4.18 percent. However, because of the enactment of AB 32 and other factors, it is expected that in the near future the rate of electric energy cost will increase at a higher rate.

Factors that affect the electric energy cost escalation in California, and soon will affect that of other states, are as follows:

- The cost of natural gas has recently gone up by 13 percent.
- Natural gas production has decreased over the last decade.
- No new natural gas refineries have been built in the United States.
- Within 3 years all electric power–generating utility companies in the state of California using coal-fired turbines must be converted to natural gas.
- The demand for natural gas within the near future will cause the prices to increase by 20 to 25 percent, which will inevitably be passed on to the consumer.
- At present only 5 percent of electric power produced in California is generated by the use of natural gas; this percentage within the near future is expected to increase to 50 to 60 percent.
- By year 2012 Hoover Dam, which presently provides inexpensive hydroelectric power to the states of California, Arizona, and Nevada, will most likely be privatized, which will inevitably result in a significant increase in the cost of electric energy.
- Since the cost of natural gas electric energy production is higher than that of hydroelectric power, due to market forces, the cost of electric power will most likely be equalized to a higher level.

In addition to the preceding, risk factors that may also have an effect on energy cost escalation include geopolitical unrest in the Middle East and Venezuela, international terrorism, and an accelerated demand for fossil fuels by Asian countries such as India and China that have achieved a gross national product (GNP) of 8 percent and over. It is therefore not unreasonable to predict that for the foreseeable future the annual cost increase in electric power would range between 6 to 8 percent.

In view of the preceding energy cost escalation, at present the initial capital investment required by solar power programs appears to be fully justified.

Average Retail Prices of Electricity 1973–2006

Year	National Residential	National Commercial	California	California Annual Rate Increase (%)
1973	2.50	2.40		
1975	3.50	3.50		
1980	5.40	5.50	5.58	
1985	7.39	7.27	8.28	9.68
1990	7.83	7.34	9.20	2.22
1995	8.40	7.69	9.94	1.61
1996	8.36	7.64	9.98	0.40
1997	8.43	7.59	10.06	0.80
1998	8.26	7.41	10.09	0.30
1999	8.16	7.26	10.11	0.20
2000	8.24	7.43	10.42	3.07
2001	8.58	7.92	13.30	27.64
2002	8.44	7.89	13.41	0.83
2003	8.72	8.03	13.92	3.80
2004	8.95	8.17	11.30	-18.82
2005	9.45	8.67	11.41	0.97
2006	10.46	9.39		

California 25-Year Average Annual Price Increase: 4.18%

Source: U.S. Department of Energy.
*Nominal Cents per kilowatt, Including Taxes.

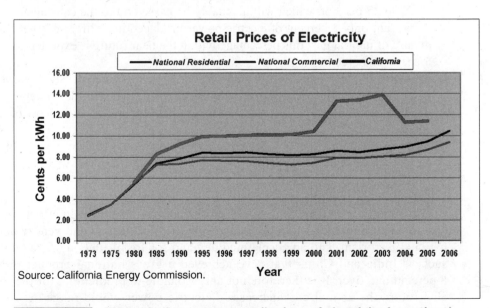

Source: California Energy Commission.

Figure 8.7   The table lists average retail prices of electricity for national residential use, national commercial use, and the state of California for the years 1973 to 2006. These values are also compared graphically.

Example of Energy Cost Increase in Solar Power Financial Analysis

The following economic analysis is a hypothetical example of a 1-MW (PTC) solar power installation under the present California Solar Initiative (CSI) rebate program. The analysis is based upon the following assumptions:

- The project execution will be completed within the required rebate energy category schedule using the PBI incentive.
- The project is classified as a government or nonprofit organization.
- The rebate amount paid for a period of 5 years will be $0.46/kWh.
- The prevailing electric energy cost at the end of the 5 years (year 2012) will be $0.14/kWh.
- The solar power system will be mounted on single-axis tracking platforms.
- The CSI EPBB design factor (calculated and derived from the CSI Web page) is 0.976.
- The location of the solar power installation is in San Bernardino County, California, with a latitude of 34.1 and an insolation of 5.84 hours.
- The life expectancy of the solar power installation is 30 years.
- The overall solar power system maintenance cost is negligible.

This example serves to indicate the impact of the electric energy cost increase 5 years after expiration of the 5-year rebate period.

Table 8.5 shows the result of annual power production produced by a quantity of 6840 PV modules with 175 dc rating that have a CEC PTC rating of 158.3 W ac and a quantity of five inverters rated at 225 kW with 94.5 percent efficiency.

Table 8.6 shows monthly insolation hours and kilowatt-hour production of the solar power system for each day during the month and the respective energy value contribution. Based on an average insolation value of 5.84 hours and an annual production of 2,128,978 kWh of energy and an energy incentive payment of $0.46/kWh, the total yearly energy value amounts to $978,329.97. Assuming an installation cost of $8.00/W per PTC, the total installed cost of the project for 1,082,772 W of energy amounts to $8,662,176.00.

Table 8.7 is a compounded table that shows energy cost increases from the present value of $0.12/kWh escalated by 6 and 8 percent. The lower part of the table indicates the amount of air pollution emission prevention by a 1-MW solar power cogeneration system.

Table 8.8 represents the energy cost contribution of the solar power system over the system's life span of 30 years, taking into account an electric energy cost escalation (after the 5 years of the guaranteed rebate contribution period) at an energy cost of $0.15/kWh.

TABLE 8.5 ENERGY OUTPUT CALCULATIONS FOR A 1-MW SOLAR POWER PLANT

SOLAR PHOTOVOLTAIC POWER OUTPUT CALCULATION

PLATFORM — SINGLE-AXIS TRACKER   Energy cost/kWh $ 0.14

SHADING — NONE

INSOLATION — 5.81

SAN BERNARDINO, CA 92569

PV MODULE—SOLARWORLD	MODEL	DC WATTS	PTC	PV COUNT	TOTAL POWER (kW)	CSI SYST. SIZE (kWh)
PV MODULE	SW175 MONO/P	175	158.3	6840	1083	1057
INVERTER	MODEL	POWER (kW)	UNITS	EFFICIENCY	OUTPUT	
XANTREX	PV225S-480P	225	5	94.5%	1063	

POWER OUTPUT — 1023

CSI EPBB DESIGN FACTOR — 0.976

CSI SYSEM SIZE (kWh) — 999

AVERAGE INSOLATION — 5.81

DAILY OUTPUT POWER (kWh) — 5.802

ANNUAL POWER PRODUCTION — 21,17,813

TABLE 8.6 ENERGY OUTPUT CALCULATIONS FOR A 1-MW SOLAR POWER PLANT

Incentive per kWh	$ 0.46			

Energy Calculation—Year 2007

Months	Daily Insolation Hours	Daily AC output (kWh)	Monthly AC output (kWh)	Energy Value
January	4.74	4,733.66	146,743.43	$ 67,501.98
February	5.24	5,232.99	146,523.73	$ 67,400.92
March	5.92	5,912.08	183,274.50	$ 84,306.27
April	6.32	6,311.55	189,346.37	$ 87,099.33
May	6.55	6,541.24	202,778.37	$ 93,278.05
June	6.56	6,551.22	196,536.73	$ 90,406.90
July	6.55	6,541.24	202,778.37	$ 93,278.05
August	6.73	6,721.00	208,350.91	$ 95,841.42
September	6.29	6,281.59	188,447.57	$ 86,685.88
October	5.51	5,502.63	170,581.50	$ 78,467.49
November	5.14	5,133.12	153,993.72	$ 70,837.11
December	4.51	4,503.97	139,622.97	$ 64,226.57
Total power value year 2007	5.84		2,128,978.19	$ 979,329.97
Installed cost per PTC watts		$8.00		
Total installed PTC watts		1,082,772		
Total installed cost		$8,662,176.00		

245

TABLE 8.7   PRESENT BASE ENERGY COST ESCALATION AT RATES OF 6 TO 8 PERCENT OVER 5 YEARS

		6%	8%
CURRENT kWh		$ 0.12	$ 0.12
	YEAR 1	0.14	0.14
	YEAR 2	0.13	0.13
	YEAR 3	0.13	0.14
	YEAR 4	0.14	0.15
	YEAR 5	0.15	0.16

SOLAR GAS POLLUTION EMISSION REDUCTION

	POUNDS	POUNDS	POUNDS
1 kWh ELECTRICITY	1.2	0.0011	0.00062
TOTAL YEARLY EMISSIONS	10394611.2	9528.3936	5370.54912
TOTAL EMISSIONS IN 30 YEARS	311,838,336	285,852	161,116

TABLE 8.8 NET ENERGY VALUE CONTRIBUTIONS OVER THE PV SYSTEM'S LIFE SPAN OF 30 YEARS

ENERGY MEAN VALUE OVER 30 YEARS	4.50%	$0.15 5.00%	5 YR INCENT. 6.00%	$0.46 7.00%	8.00%	ENERGY UNIT COST REBATE
Applied annual enegy cost escal.						$0.46
Annual % energy cost escalation	104.50%	105.00%	106.00%	107.00%	108.00%	
Energy cost increase multiplier						
Operational life cycle						MEDIAN COST $0.15
Year 1	$979,330	$979,330	$979,330	$979,330	$979,330	
Year 2	$979,330	$979,330	$979,330	$979,330	$979,330	
Year 3	$979,330	$979,330	$979,330	$979,330	$979,330	
Year 4	$979,330	$979,330	$979,330	$979,330	$979,330	
Year 5	$979,330	$979,330	$979,330	$979,330	$979,330	
Year 6	$338,947	$340,569	$343,812	$347,056	$350,299	
Year 7	$354,200	$357,597	$364,441	$371,350	$378,323	
Year 8	$370,138	$375,477	$386,307	$397,344	$408,589	
Year 9	$386,795	$394,251	$409,486	$425,158	$441,276	
Year 10	$404,200	$413,963	$434,055	$454,919	$476,578	
Year 11	$422,390	$434,662	$460,098	$486,764	$514,704	
Year 12	$441,397	$456,395	$487,704	$520,837	$555,881	
Year 13	$461,260	$479,214	$516,966	$557,296	$600,351	
Year 14	$482,017	$503,175	$547,984	$596,306	$648,379	
Year 15	$503,707	$528,334	$580,863	$638,048	$700,250	
Year 16	$526,374	$554,750	$615,715	$682,711	$756,270	
Year 17	$550,061	$582,488	$652,658	$730,501	$816,771	
Year 18	$574,814	$611,612	$691,818	$781,636	$882,113	
Year 19	$600,680	$642,193	$733,327	$836,350	$952,682	
Year 20	$627,711	$674,303	$777,326	$894,895	$1,028,897	
Year 21	$655,958	$708,018	$823,966	$957,538	$1,111,208	
Year 22	$685,476	$743,419	$873,404	$1,024,565	$1,200,105	
Year 23	$716,323	$780,590	$925,808	$1,096,285	$1,296,113	
Year 24	$748,557	$819,619	$981,357	$1,173,025	$1,399,802	
Year 25	$782,242	$860,600	$1,040,238	$1,255,136	$1,511,787	
Year 26	$817,443	$903,630	$1,102,652	$1,342,996	$1,632,730	
Year 27	$854,228	$948,812	$1,168,811	$1,437,006	$1,763,348	
Year 28	$892,668	$996,252	$1,238,940	$1,537,596	$1,904,416	
Year 29	$932,838	$1,046,065	$1,313,276	$1,645,228	$2,056,769	
Year 30	$974,816	$1,098,368	$1,392,073	$1,760,394	$2,221,310	
Energy Value 30 years	**$20,001,890**	**$21,151,004**	**$23,759,737**	**$26,847,588**	**$30,505,602**	
Energy cost savings in five years	**$1,270,688**	**$1,270,688**	**$1,270,688**	**$1,270,688**	**$1,270,688**	
Total Energy Value 30 years	$21,272,578	$22,421,692	$25,030,425	$28,118,276	$31,776,289	
Less installed cost	$8,120,790	$8,120,790	$8,120,790	$8,120,790	$8,120,790	
Adjusted Energy Value 30 years	**$13,151,788**	**$14,300,902**	**$16,909,635**	**$19,997,486**	**$23,655,499**	
Net payback in 5 years	$6,167,338	$6,167,338	$6,167,338	$6,167,338	$6,167,338	

9

ECONOMICS OF SOLAR
POWER SYSTEMS

Introduction

Perhaps the most important task of a solar power engineer is to conduct preliminary engineering and financial feasibility studies, which are necessary for establishing an actual project design. The essence of the feasibility study is to evaluate and estimate the power generation and cost of installation for the life span of the project. The feasibility study is conducted as a first step in determining the limitations of the solar project's power production and return on investment, without expending a substantial amount of engineering and labor effort. The steps needed to conduct the preliminary engineering and financial study are presented in this chapter.

Preliminary Engineering Design

Conduct a field survey of the existing roof or mounting area. For new projects, review the available roof-mount area and mounting landscape. Care must be taken to ensure that there are no mechanical, construction, or natural structures that could cast a shadow on the solar panels. Shade from trees and sap drops could create an unwanted loss of energy production. One of the solar PV modules in a chain, when shaded, could act as a resistive element that will alter the current and voltage output of the whole array.

Always consult with the architect to ensure that installation of solar panels will not interfere with the roof-mount solar window, vents, and air-conditioning unit ductwork. The architect must also take into consideration roof penetrations, installed weight, anchoring, and seismic requirements.

After establishing solar power area clearances, the solar power designer must prepare a set of electronic templates representing standard array configuration assemblies. Solar array templates then can be used to establish a desirable output of dc power. Note that, when laying blocks of PV arrays, consideration must be given to the desirable tilt inclination to avoid cross shadowing. In some instances, the designer must also consider trading solar power output efficiency to maximize the power output production. As mentioned in Chapter 2, the most desirable mounting position for a PV module to realize maximum solar insolation is the latitude minus 10 degrees. For example, the optimum tilt angle in New York is 39 degrees, whereas in Los Angeles it is about 25 to 27 degrees. The sun exposures caused by various insolation tilts over the course of the year in Los Angeles are shown in Figures 9.1 through 9.5. To avoid cross shading, the adjacent profiles of two solar rows of arrays can be determined. Simple trigonometry can be used to determine the geometry of the tilt by the angle of the associated sine (shading height) and cosine (tandem array separation space) of the support structure incline. Note that flatly laid solar PV arrays may incur about a 9 to 11 percent power loss, but the number of installed panels could exceed 30 to 40 percent on the same mounting space.

An important design criterion when laying out solar arrays is grouping the proper number of PV modules that would provide the adequate series-connected voltages and current required by inverter specifications. Most inverters allow certain margins for dc inputs that are specific to the make and model of the manufactured unit. Inverter power capacities may vary from a few hundred to many thousands of watts. When designing a solar power system, the designer should choose the specific PV and inverter makes and models in advance, thereby establishing the basis of the overall configuration.

It is not uncommon to have different sizes of solar power arrays and matching inverters on the same installation. In fact, in some instances, the designer may, for unavoidable

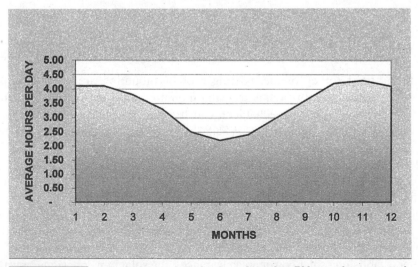

Figure 9.1   Insolation graph for Los Angeles PV panels mounted at a 0-degree tilt angle.

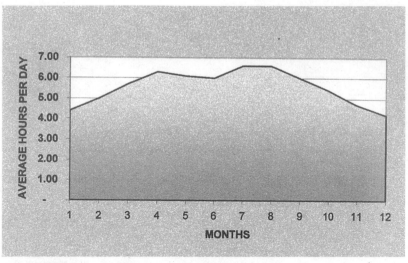

Figure 9.2   Insolation graph for Los Angeles PV panels mounted at a 15-degree tilt angle.

occurrences of shading, decide to minimize the size of the array as much as possible, thus limiting the number of PV units in the array, which may require a small-size power capacity inverter. The most essential factor that must be taken into consideration is that all inverters used in the solar power system must be completely compatible.

When laying out the PV arrays, care should be taken to allow sufficient access to array clusters for maintenance and cleaning purposes. In order to avoid deterioration of power output, solar arrays must be washed and rinsed periodically. Adequately

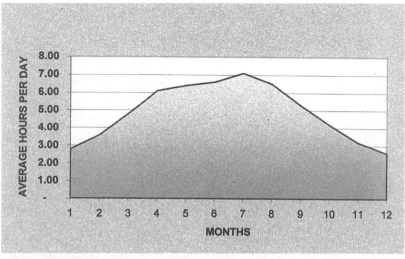

Figure 9.3   Insolation graph for Los Angeles PV panels mounted at a 33-degree tilt angle.

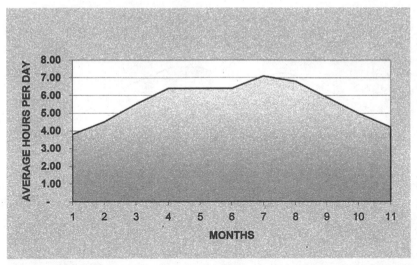

Figure 9.4   Insolation graph for Los Angeles PV panels mounted at a 15-degree tilt angle.

spaced hose bibs should be installed on rooftops to facilitate flushing of the PV units in the evening time only, when the power output is below the margin of shock hazard.

After completing the PV layout, the designer should count the total number of solar power system components, and by using a rule of thumb must arrive at a unity cost estimate such as dollars per watt of power. That will make it possible to better approximate the total cost of the project. In general, net power output from power PV arrays, when converted to ac power, must be subjected to a number of factors that can degrade the output efficiency of the system.

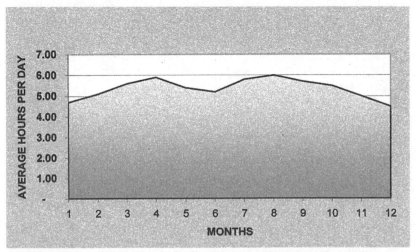

Figure 9.5   Insolation graph for Los Angeles PV panels mounted at 90-degree tilt angle.

The California Energy Commission (CEC) rates each manufacturer-approved PV unit by a special power output performance factor referred to as the power test condition (PTC). This figure of merit is derived for each manufacturer and PV unit model by extensive performance testing under various climatic conditions. These tests are performed in a specially certified laboratory environment. Design parameters that affect the system efficiency are as follows:

- Geographic location (PV units work more efficiently under sunny but cool temperatures)
- Latitude and longitude
- Associated yearly average insolation
- Temperature variations
- Building orientation (north, south, etc.)
- Roof or support structure tilt
- Inverter efficiency
- Isolation transformer efficiency
- DC and ac wiring losses resulting from the density of wires in conduits
- Solar power exposure
- Long wire and cable runs
- Poor, loose, or corroded wire connections
- AC power transmission losses to the isolation transformers
- Poor maintenance and dust and grime collection on the PV modules

Meteorological Data

When the design is planned for floor-mount solar power systems, designers must investigate natural calamities such as extreme wind gusts, periodic or seasonal flooding, and snow precipitation. For meteorological data contact the NASA Surface Meteorology and Solar Energy Data Set Web site at http://eosweb.larc.nasa.gov/sse/. To search for meteorological information on this Web site, the inquirer must provide the latitude and longitude for each geographic location. For example, to obtain data for Los Angeles, California, at latitude 34.09 and longitude 118.4, the statistical data provided will include the following recorded information for each month of the year for the past 10 years:

- Average daily radiation on horizontal surface [kWh/(m² · day)]
- Average temperature (°C)
- Average wind speed (m/s)

To obtain longitude and latitude information for a geographic area refer to the Web site www.census.gov/cgi-bin/gazetter. For complete listings of latitude and longitude data please refer to Appendix A. The following page lists a few examples for North American metropolitan area locations.

Los Angeles, California	34.09 N/118.40 W
Toronto, Canada	43.67 N/–79.38 W
Palm Springs, California	33.7 N/116.52 W
San Diego, California	32.82 N/117.10 W

To obtain ground surface site insolation measurements refer to the Web site http://eosweb.lac.nasa.gov/sse.

A certified registered structural engineer must design all solar power installation platforms and footings. Upon completing and integrating the preliminary design parameters previously discussed, the design engineer must conduct a feasibility analysis of the solar power cogeneration project. Some of the essential cost components of a solar power system required for final analysis are:

- Solar PV module (dollars per dc watts)
- Support structure hardware
- Electric devices, such as inverters, isolation transformers, and lightning protection devices; and hardware such as electric conduits, cables, and grounding wire

Additional costs may include:

- Material transport and storage
- Possible federal taxes and state sales taxes
- Labor wages (prevailing or nonprevailing) and site supervision (project management)
- Engineering design, which includes electrical, architectural, and structural disciplines
- Construction drawings and reproduction
- Permit fees
- Maintenance training manuals and instructor time
- Maintenance, casualty insurance, and warranties
- Spare parts and components
- Testing and commissioning
- Overhead and profit
- Construction bond and liability insurance
- Mobilization cost, site office, and utility expenses
- Liquidated damages

Energy Cost Factor

Upon completion of the preliminary engineering study and solar power generation potential, the designer must evaluate the present costs and project the future costs of the electric energy for the entire life span of the solar power system. To determine the present value of the electric energy cost for an existing building, the designer must evaluate the actual electric bills for the past 2 years. Note that the general cost per kilowatt-hour of energy provided by service distributors consists of an average of numerous charges such as commissioning, decommissioning, bulk purchase, and other miscellaneous cost items that generally appear on electric bills (that vary seasonally) but go unnoticed by consumers.

The most significant of the charges, which is in fact a penalty, is classified as *peak hour energy*. This charge occurs when the consumer's power demand exceeds the established boundaries of energy consumption as stipulated in tariff agreements. In order to maintain a stable power supply and cost for a unit of energy (a kilowatt-hour), service distributors, such as Southern California Electric (SCE) and other power-generating entities, generally negotiate a long-term agreement whereby the providers guarantee distributors a set bulk of energy for a fixed sum. Since energy providers have a limited power generation capacity, limits are set as to the amount of power that is to be distributed for the duration of the contract. A service provider such as SCE uses statistics and demographics of the territories served to project power consumption demands, which then form the baseline for the energy purchase agreement. When energy consumption exceeds the projected demand, it becomes subject to much costlier tariffs, which are generally referred to as the *peak bulk energy rate*.

Project Cost Analysis

As indicated in the preliminary solar power cogeneration study, the average installed cost per watt of electric energy is approximately $9 as shown in Figure 9.6a to c. The unit cost encompasses all turnkey cost components, such as engineering design documentation, solar power components, PV support structures, electric hardware, inverters, integration labor, and labor force training. Structures in that cost include roof-mount support frames and simple carport canopies, only. Special architectural monuments if required may necessitate some incremental cost adjustment. As per the CEC, all solar power cogeneration program rebate applications applied for before December of 2002 were subject to a 50 percent subsidy. At present, rebate allotments are strictly dependent on the amount of funding available at the time of application and are granted on a first come, first serve basis.

MAINTENANCE AND OPERATIONAL COSTS

As mentioned earlier, solar power systems have a near-zero maintenance requirement. This is due to solid-state technology, lamination techniques, and the total absence of mechanical or moving parts. However to prevent marginal degradation in output performance from dust accumulation, solar arrays require a biyearly rinsing with a water hose.

Figure 9.6 is a detailed estimate, designed by the author, for a solar power project for the Water and Life Museum located in Hemet, California. As discussed in Chapter 3 the project consists of two museum campuses with a total of seven buildings, each constructed with roof-mount solar power PV systems.

Economic Analysis:

INITIAL COSTS & CREDITS

Engineering Rate — $150.00 per hour

	Hours	Rate	Total	%
Site Investigation	40	$ 150.00	$ 6,000.00	46.88%
Preliminary Design Coordination	24	$ 150.00	$ 3,600.00	28.13%
Report Preparation	8	$ 150.00	$ 1,200.00	9.38%
Travel & Accommodations	1	$ 2,000.00	$ 2,000.00	15.63%
Other			$ -	0.00%
Sub Total			$ 12,800.00	100.00%

DEVELOPMENT

Permits & Rebate Applications	8	$ 150.00	$ 1,200.00	5.41%
Project Management	120	$ 150.00	$ 18,000.00	81.08%
Travel Expenses	1	$ 2,000.00	$ 2,000.00	9.01%
Other	1	$ 1,000.00	$ 1,000.00	4.50%
Sub Total			$ 22,200.00	100.00%

ENGINEERING

PV Systems Design	90	$ 150.00	$ 13,500.00	10%
Architectural Design	90	$ 150.00	$ 13,500.00	10%
Structural Design	90	$ 150.00	$ 13,500.00	10%
Electrical Design	420	$ 150.00	$ 63,000.00	48%
Tenders & Contracting	48	$ 150.00	$ 7,200.00	5%
Construction Supervision	94	$ 150.00	$ 14,100.00	11%
Training Mauals	48	$ 150.00	$ 7,200.00	5%
Sub Total			$ 132,000.00	100%

RENEWABLE ENERGY EQUIPMENT

PV Modules (per kWh-DC)	255	$ 3,900.00	$ 994,500.00	92%
Transportation	1	$ 5,000.00	$ 5,000.00	0%
Other			$ -	0%
Tax (Equipment Only)	8.25%		$ 82,046.25	8%
Sub Total			$ 1,081,546.25	100%

INSTALLATION EQUIPMENT

PV Module Support Structure (per kWh)	255	$ 500.00	$ 127,500.00	18%
Inverter (per kWh)	320	$ 488.00	$ 156,160.00	22%
Electrical Materials (per kW)	320	$ 250.00	$ 80,000.00	11%
System Installation Labor (per kWh)	320	$ 1,000.00	$ 320,000.00	45%
Transportaiton	1	$ 3,000.00	$ 3,000.00	0%
Other	0		$ -	0%
Tax (Equipment Only)	8.25%		$ 30,001.95	4%
Sub Total			$ 716,661.95	100%

MISCELLANEOUS

Training	48	$ 150.00	$ 7,200.00	100%
Contingencies (If applicable)	0		-	0%
Sub Total			$ 7,200.00	100%

OVERHEAD AND PROFIT	18%		$ 355,033.48	100%
TOTAL PROJECT COST ESTIMATE:			$ 2,327,441.68	100%

(a)

Figure 9.6 Material and equipment estimate for Water Education Museum.

Projected cost escalation		15 YEARS	20 YEARS	25 YEARS
Energy cost escalation rate	$	39,232.55	$ 41,586.51	$ 44,081.70
Inflation rate	$	18,521.63	$ 19,077.28	$ 19,649.60
Total yearly cost escalation	$	57,754.19	$ 60,663.79	$ 63,731.30

Solar power energy generation

		15 YEARS	20 YEARS	25 YEARS
Energy generation capacity (kWh AC)		212	212	212
Average daily solar irradiance (h)		5.5	5.5	5.5
Yearly energy generation (kWh)		425,037	425,037	425,037
5 Year Inflation Rate per kWh	$	0.25	$ 0.29	$ 0.33
Yearly cost savings by PV modules	$	106,801.15	$ 122,821.32	$ 141,244.52
Yearly aggregate savings	$	164,555	$ 183,485	$ 204,976

(b)

Miscellaneous

	Hrs	Rate	Total
Training	48	$ 150.00	$ 7,200.00
Contingencies			
Total Miscellaneous			**$ 7,200.00**

Energy demand analysis

Ligthing Kwh	413
HVAC Kwh	296
Ave. hourly energy Kwh	709
Average cost of per kWh, current cost	$ 0.19
Ave. monthly energy expenditure	$ 48,495.60
Ave. yearly energy expenditure projection-kWh	$ 581,947.20

Projected energy cost

Projected cost escalation		5 YEARS	10 YEARS
Energy cost escalation rate	6%	$ 34,916.83	$ 37,011.84
Inflation rate	3.0%	$ 17,458.42	$ 17,982.17
Total yearly cost escalation		$ 52,375.25	$ 54,994.01

Solar power energy generation

		5 YEARS	10 YEARS
Energy generation capacity kWh-AC		212	212
Average daily solar irradiance-hrs		5.5	5.5
Yearly energy generation kWh		425037	425037
5 Year Inflation Rate Rate per kWh	15.0%	$ 0.19	$ 0.22
Yearly cost savings by PV modules		$ 80,757.01	$ 92,870.56
Yearly aggregate savings		$ 133,132	$ 147,865

(c)

Figure 9.6 (*Continued*)

The costing estimate reflected in Figure 9.6 represents one of the main buildings referred to as the Water Education Museum and is a project funded by the Los Angeles Metropolitan Water District (MWD). The solar power generation of the entire campus is 540 W dc. Net ac power output including losses is estimated to be approximately 480 kW. At present the entire solar power generation system is paid for by the MWD; as a result the entire power generated by the system will be used by the Water and Life Museum, which represents approximately 70 to 75 percent of the overall electric demand load.

Feasibility Study Report

As mentioned in Chapters 3 and 4, the key to designing a viable solar power system begins with preparation of a feasibility report. A feasibility report is essentially a preliminary engineering design report that is intended to inform the end user about significant aspects of a project. The document therefore must include a thorough definition of the entire project from a material and financial perspective.

A well-prepared report must inform and educate the client and provide realistic engineering and financial projections to enable the user to weigh all aspects of a project from start to finish. The report must include a comprehensive technical and financial analysis of all aspects of the project, including particulars of local climatic conditions, solar power system installation alternatives, grid-integration requirements, electric power demand, and economic cost projection analysis. The report must also incorporate photographs, charts, and statistical graphs to illustrate and inform the client about the benefits of the solar power or sustainable energy system proposed. The following is a feasibility report prepared by the author for a public recreational facility that includes a community swimming pool, baseball and football fields, and several tennis courts.

Valley-Wide Recreation and Park District

901 West Esplanade Ave.
San Jacinto, CA 92581
Subject: Solar power preliminary study for Diamond Valley Lake Aquatic Facility, Community Building, Hemet, California
December 14, 2004

Dear Sirs,

The following solar power feasibility study reflects analysis of alternate approaches to grid-connected solar power cogeneration for the Diamond Valley Lake Aquatic Facility. The solar photovoltaic systems proposed are intended to provide a comprehensive electric energy cost saving solution for the duration of the guaranteed life of the equipment, which theoretically should span over 25 years and beyond.

Considering the 25,000-ft² extent of the recreation facility, which includes the community center, tennis courts, and baseball and soccer fields, we have proposed an alternate study that would allow the solar power cogeneration systems to expand in a modular fashion as the need arises.

In view of the fact that the present scope of the project includes only the pool, the bathrooms, and a small office building, the projected electric power demand is limited to that needed for pool filtration and the building's internal and external lights. Since the objective of a solar power cogeneration system is to save energy expenditure on the whole campus, we have based our analysis on the best possible power demand projection that would allow for tailoring of a solar power system that may meet the demand needs of the entire scope of the recreation facility. The following are the power demand projections, which are based on the existing civil engineering site plan:

- Present office, bathroom, and pool—185 kWh (as reflected in the electrical plans)
- Community center (future), 25,000 ft², 20 W/ft²—500 kWh (projected)
- Eight baseball fields, lighting load at 5000 W per field—40 kWh
- Eight soccer fields, lighting load at 5000 W per field—40 kWh
- Seven volleyball courts, lighting load at 1000 W per field—7 kWh
- Two basketball courts, lighting load at 3000 W per field—6 kWh
- Six tennis courts, lighting load at 3000 W per field—18 kWh
- Concessions building demand load—10 kWh
- Pathway lighting—10 kWh
- Projected total demand load—806 kWh
- Current demand load at 480 V, three-phase—2000-A service

As you may be aware, Southern California Edison (SCE) only provides a single service for each client, and in order to accommodate a 2000-A projected demand, the electrical service switchgear and equipment room must be designed in a fashion so as to provide sufficient expansion space and required underground conduits to meet the overall infrastructure needs. Likewise, integration of a solar power cogeneration system will mandate special provisions for incorporating grid-connection solar power transformers, inverters, and isolation transformers that should be housed within the main electrical room.

In general, the solar power cogeneration contribution, when intended to curb electric energy consumption, should be tailored in proportion to the overall power demand of the complex. The effective size of solar power cogenerators in a typical installation should be about 20 to 30 percent of the overall demand load. Solar power generation could also be designed to provide complete self-sufficiency for the entire complex during daytime operation, but also under a net metering agreement the system could be readily sized to produce surplus power that could be fed into the power grid if so desired.

A grid-connected solar power cogeneration system under a special service agreement with the primary electric power service provider could generate energy credits that will provide adequate compensation for all energy use during the absence of full insolation, such as during cloudy or rainy days and during nights. The power production capacity of a solar cogenerator system is directly proportional to the number of solar power panels installed, the efficiency of conversion devices, and the daily insolation, which is the effective daily solar aperture time for a geographic location. With reference to the local atmospheric conditions, the insolation time for Hemet is 5.5 to 6 hours of sunshine per day. The average insolation time (when sun rays impact the photovoltaic panel perpendicularly) is based on the mean value of sunny days throughout four seasons.

PROJECT SITE CLIMATIC CONDITIONS AND SURFACE METEOROLOGY

- The approximate latitude and longitude of the site are 33.50 N/116.20 W.
- The average annual temperature variation for the past 10 years has been recorded to be about 16.4°C.
- The average annual recorded wind speed is 3.58 m/s.
- The average daily solar radiation on horizontal surfaces [kWh/(m^2 · day)] is about 6 hours. On the average, recorded sunshine hours in summer exceed 8 hours and under worst-case conditions in winter, insolation will be about 3.2 hours.

With reference to the preceding site climatic data, yearly average solar temperature conditions, and daily solar irradiance, the site can be considered an ideal location for solar power cogeneration. Additionally, because of the specific characteristics of solar photovoltaic systems (both single and polycrystalline), it would be possible to harvest the optimum amount of electric energy.

PROPOSED SOLAR POWER COGENERATION CONFIGURATIONS

With the specific nature of the Diamond Valley Lake Aquatic Facility project in mind, solar power cogeneration would provide not only a significant amount of electric energy, but due to the aesthetic appearance of the polycrystalline PV panels, in addition to roof-mount arrays and parking lot carports, they could also be used in verandas, picnic areas, or park bench solar shades. Solar-powered lampposts with integrated battery packs could also be used for common area and parking lot illumination. In fact, with the availability of multicolor PV panels, it would be possible to integrate the landscape architecture and solar power to create a unique blend of technology and natural beauty. Figure 9.7 depicts a typical solar power farm installation.

Because of the availability of the vast surface area within the campus, it is also quite possible to set up a solar power park that could generate a significant amount of electric energy beyond the present and future needs of the project. In view of the limited roof area and the direction of the incline, the net available area that would be suitable for solar power panel installation will be limited to about 1500 to 2000 ft^2, which would effectively yield about 15 to 20 kWh of energy. At an average insolation time of 5.5 hours (conservative), energy contributions for the system will be as follows:

Figure 9.7 **Typical solar power farm installation.** *Graphics courtesy of UNIRAC.*

- Hourly electric power energy produced—15 to 20 kWh
- Daily total energy produced—82 to 110 kWh
- Total demand as per electric power calculations—182 kWh
- Total daily demand over 8 hours daily operation—1400 kWh
- Percent solar power contribution—6 to 8 percent
- Rebated energy—about 20,000 W (net ac power produced)
- Total installed cost—$160,000

Option I. Ground-mount solar power for existing aquatic facility

- Estimated CEC rebate amount—$72,000 (at $3.5/W) to $83,000 (at $4.0/W). (The rebate amount is subject to availability of funds at SCE or Southern California Gas.)
- Federal tax credit based on the balance of the net installed cost being 10 percent—$8600 to $9500.
- Net installed expenditure—$80,000 (at $3.5/W rebate) to $71,000 (at $4.00/W rebate).
- Savings over the next 25 years—about $600,000.
- Installed cost per ac watts—$7.6 to $8.0.
- Payback on investment—10 to 15 years.

Option II. Roof-mount solar power for community center building In
view of the limited roof area and the direction of the incline, the net available area

Figure 9.8   A typical tilted roof solar power system. *Photo courtesy of Atlantis Energy Systems.*

that would be suitable for solar power panel installation will be limited to about 12,500 ft², or 50 percent of the roof. If the roof design is made flat, then the effective area could be as high as 20,000 ft², which could yield substantially more power. Figure 9.8 is a photograph of a typical tilted roof solar power system. At an average insolation time of 5.5 hours (conservative), energy contributions for the system will be as follows:

- Hourly electric power energy produced—135 kWh.
- Daily total energy production—720 kWh.
- Total demand as per previous electric power projection—500 kWh.
- Total daily demand over 8 hours daily operation—4000 kWh.
- Solar power contribution—18 percent.
- Rebated energy—about 135,000 W (net ac power produced).
- Total installed cost—$1,087,000.
- Estimated CEC rebate amount—$478,000 (at $3.5/W) to $546,000 ($4.0/W). (The rebate amount is subject to availability of funds at SCE or Southern California Gas.)
- Federal tax credit based on the balance of the net installed cost being 10 percent—$54,000 to $610,000.
- Net installed expenditure—$519,000 (at $3.5/W rebate) to $460,000 (at $4.00/W rebate).
- Savings over the next 25 years—about $2,900,000.00.
- Installed cost per ac watts—$0.6 to $8.0.
- Payback on investment—8 to 12 years.

Figure 9.9   **Solar power canopy rendering.** *Courtesy of Integrated Solar Technologies.*

Option III. Parking stall solar power canopies   Another viable solar power installation option that is commonly used throughout the United States, Europe, and Australia is parking canopies. Solar power canopies are modular prefabricated units manufactured by a number of commercial entities that can be tailored to meet specific architectural aesthetics in a specific setting. Canopies when installed side by side could provide any desired amount of solar power. Figure 9.9 depicts graphics of a solar power canopy.

In general, the cost per wattage of fabricated solar power canopies is estimated to be about $1.50 to $2.00 depending on the structural design requirements dictated by the architecture. A typical parking stall when fully covered represents 100 ft^2 of usable area, which translates into about 1000 to 1500 W/h of solar power. As per the preceding estimate a group of 10 parking stalls can produce 15 kWh of energy and could provide as much power as the economics permit.

Depending on the cost of the canopy structure, the installed cost of a 1-kWh system could range from $9000 to $9500. The rebate amount would be about 40 percent, which would translate into $3000 to $3800 of CEC rebate.

POLLUTION ABATEMENT CONSIDERATION

According to a 1999 study report by the U.S. Department of Energy, 1 kW of energy produced by a coal-fired electric power–generating plant requires about 5 lb of coal. Likewise, generation of 1.5 kWh of electric energy per year requires about 7400 lb of coal that in turn produces 10,000 lb of carbon dioxide (CO_2).

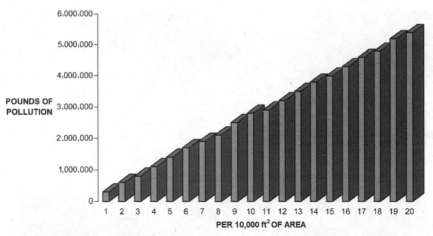

Figure 9.10   Commercial energy pollution graph: pounds of CO_2 emission per 10,000-ft^2 area.

Roughly speaking the calculated projection of the power demand for the project totals about 2500 to 3000 kWh. This will require between 12,000,000 and 15,000,000 lb of coal, thereby producing about 16,000,000 to 200,000,000 lb of carbon dioxide and contributing toward air pollution and global warming from greenhouse gases.

Solar power in turn if implemented as discussed here will substantially minimize the air pollution index. In fact, the EPA will soon be instituting an air pollution indexing system that will be factored in all future construction permits. At that time all major industrial projects will be required to meet and adhere to the air pollution standards and offset excess energy consumption by means of solar or renewable energy resources. Figure 9.10 is bar chart of commercial energy pollution graph.

The amount of power consumed in the United States is comparatively large, amounting to 25 percent of the total global energy. Thus, in order to avoid dwarfing of comparative values when comparing industrial nations, the graph in Figure 9.11 does not include energy consumption by the United States.

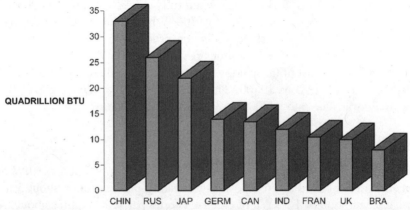

Figure 9.11   Top-10 energy consumers in the world by CO_2 emissions per area.

ENERGY ESCALATION COST PROJECTION

According to an Energy Information Administration data source published in 1999, California is among the top-10 energy consumers in the world, and this state alone consumes just as much energy as Brazil or the United Kingdom. Since the entire global crude oil reserves are estimated to last about 30 to 80 years and over 50 percent of the nation's energy is imported from abroad, it is inevitable that in the near future energy costs will undoubtedly surpass historical cost escalation averaging projections.

It is estimated that the cost of nonrenewable energy will, within the next decade, be increased by approximately 4 to 5 percent by producers. When compounded with a general inflation rate of 3 percent, the average energy cost over the next decade could be expected to rise at a rate of about 7 percent per year. This cost increase does not take into account other inflation factors such as regional conflicts, embargoes, and natural catastrophes.

In view of the fact that solar power cogeneration systems require nearly zero maintenance and are more reliable than any human-made power generation devices (the systems have an actual life span of 35 to 40 years and are guaranteed by the manufacturers for a period of 25 years), it is my opinion that in a near-perfect geographic setting such as Hemet, their integration into the mainstream of the architectural design will not only enhance the design aesthetics, but will generate considerable savings and mitigate adverse effects on the ecology and from global warming.

As indicated in the solar power cogeneration study, the average installed cost per watt of electric energy is approximately 40 percent of the total installed cost. The unit cost encompasses all turnkey cost components such as engineering design documentation, solar power components, PV support structures, electrical hardware, inverters, integration labor, and labor training.

PROJECT COST ANALYSIS AND STATE-SPONSORED REBATE FUND STATUS

Structures in the preceding costs include roof-mount support frames and simple carport canopies only. Special architectural monuments if required may necessitate some incremental cost adjustment. As per the California Energy Commission, all solar power cogeneration program rebate applications if applied for by December of 2004 will be subject to a 40 percent subsidy. Rebate allotments are strictly dependent on the amount of funding available at the time of application and are granted on a first come, first serve basis.

SYSTEM MAINTENANCE AND OPERATIONAL COSTS

As mentioned earlier, solar power systems have a near-zero maintenance requirement. However, to prevent marginal degradation in output performance from dust accumulation, solar arrays require a biyearly rinsing with a regular water hose. Since solar power arrays are completely modular, system expansion, module replacement, and troubleshooting are simple and require no special maintenance skills. All electronic dc-to-ac inverters are modular and can be replaced with a minimum of downtime.

An optional (and relatively inexpensive) computerized system monitoring console can provide a real-time performance status of the entire solar power cogeneration system. A software-based supervisory program featured in the monitoring system can also provide instantaneous indication of solar array performance and malfunction.

CONCLUSION

Even though a solar power cogeneration system requires a large initial capital investment, the long-term financial and ecological advantages are so significant that their deployment in the existing project should be given special consideration.

A solar power cogeneration system, if applied as per the recommendations reviewed here, will provide a considerable energy expenditure savings over the life span of the recreation facility and a hedge against unavoidable energy cost escalation.

10

PASSIVE SOLAR HEATING TECHNOLOGIES

Introduction

In this chapter we will review the basic principles of passive solar energy and applications. The term *passive* implies that solar power energy is harvested with direct exposure of fluids such as water or a fluid medium that absorbs the heat energy. Subsequently the harvested energy is converted to steam or vapor which in turn is used to drive turbines or provide evaporation energy in refrigerating and cooling equipment. Figure 10.1 depicts historical use of passive solar energy to power a printing press.

Solar power is the sun's energy without which life as we know it on our planet would cease to exist. Solar energy has been known and used by humankind throughout ages. As we all know, solar rays concentrated by a magnifying glass can provide intense heat energy that can burn wood or heat water to a boiling point. As discussed later, recent technological developments of this simple principle are currently being used to harness the solar energy and provide an abundance of electric power. Historically the principle of heating water to a boiling point was well-known by the French who in 1888 used solar power to drive printing machinery.

Please refer to Appendix D for a detailed solar power historical time line.

Passive Solar Water Heating

The simplest method of harvesting energy is exposing fluid-filled pipes to sun rays. Modern technology passive solar panels that heat water for pools and general household use are constructed from a combination of magnifying glasses and fluid-filled pipes. In some instances pipes carry special heat-absorbing fluids such as bromide which heats up quite rapidly. In other instances, water is heated and circulated by small pumps. In most instances pipes are painted black and are laid on a silver-colored

Figure 10.1   Historical use of passive solar energy to power a printing press.

reflective base that further concentrates the solar energy. Another purpose of silver back-boards is to prevent heat transmission to the roofs or support structures. Figure 10.2 depicts graphics of passive solar water heating panel and Figure 10.3 shows graphics of a residential passive solar water heater panel operation.

Pool Heating

Over the years, a wide variety of pool heating panel types has been introduced. Each has its intrinsic advantages and disadvantages. Figure 10.4 is a photograph of a passive solar power water heater in industrial use There are four primary types of solar pool collector design classifications:

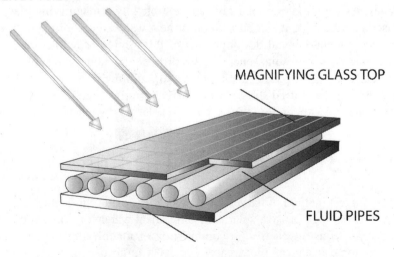

ENERGY FROM THE SUN

MAGNIFYING GLASS TOP

FLUID PIPES

SILVER REFLECTIVE SURFACE

Figure 10.2   Passive solar water heating panel.

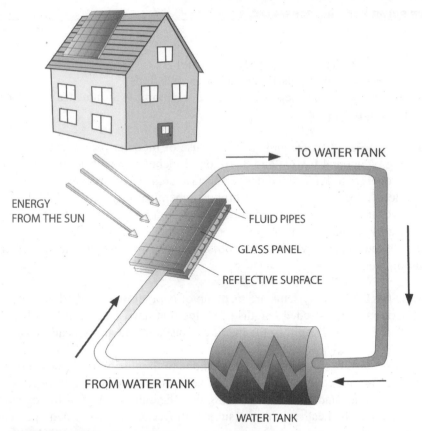

ENERGY FROM THE SUN

TO WATER TANK

FLUID PIPES

GLASS PANEL

REFLECTIVE SURFACE

FROM WATER TANK

WATER TANK

Figure 10.3 Household passive solar water heating panel.

Figure 10.4 Passive solar power water heater in industrial use.

Courtesy of Department of Energy.

- Rigid black plastic panels (polypropylene)
- Rubber mat or other plastic or rubber formulations
- Tube and fin metal panels with a copper or aluminum fin attached to copper tubing
- Plastic-type systems

Plastic panels   This technology makes use of modular panels with panel dimensions ranging about 4 ft wide and 8, 10, or 12 inches in length. Individual panels are coupled together to achieve the desired surface area. The principal advantages of this type of technology are lightness of product, chemical inertness, and high efficiency. The panels are also durable and can be mounted on racks. The products are available in a glazed version to accommodate for windy areas and colder climates. The disadvantage of the technology is its numerous system surface attachments, which can limit mounting locations.

Rubber mat   These systems are made up of parallel pipes, called headers, that are manufactured from extruded lengths of tubing that have stretching mats between the tubes. The length and width of the mat are adjustable and are typically custom-fit for each application, as seen in Figure 10.5.

The advantage of this technology is that due to the flexibility of the product, it can be installed around roof obstructions, like vent pipes. Installations require few, if any, roof penetrations, and are considered highly efficient. Because of the expandability of the product, the headers are less subject to freeze expansion damage. The main

Figure 10.5   **Rubber mat roof-mount solar water heater.**
Courtesy of UMA/Heliocol.

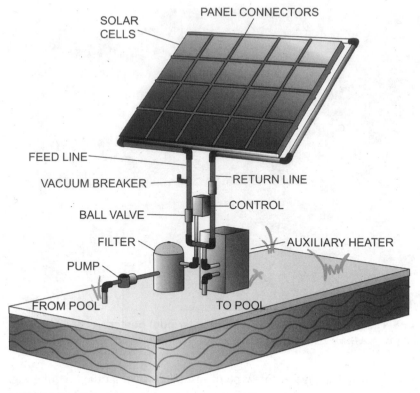

SOLAR CELLS
PANEL CONNECTORS
FEED LINE
VACUUM BREAKER
RETURN LINE
BALL VALVE
CONTROL
FILTER
AUXILIARY HEATER
PUMP
FROM POOL
TO POOL

Figure 10.6   Roof-mount solar water heating system diagram.

disadvantage of the system is that the mats are glued to the roof and can be difficult to remove without damaging either the roof or the solar panel. The installation also cannot be applied in rack-type installations. Figure 10.6 is a graphic presentation of a roof-mount solar water heating system diagram

Metal panels   These classes of products are constructed from copper waterways that are attached to either copper or aluminum fins. The fins collect the solar radiation and conduct it into the waterways. Advantages of these classes of product include their rigidity and durability of construction. Like rubber mats, glazed versions of these panels are also available for application in windy areas and cold climates. Figure 10.7 is a photograph of a residential roof-mount solar pool heater.

A significant disadvantage of this type of technology is that these panels require significantly more surface area, have low efficiency, and have no manufacturer's warranty.

Plastic pipe systems   In this technology, plastic pipes are connected in parallel or are configured in a circular pattern. The main advantage of the system is that installation can be done inexpensively and could easily be used as an overhead "trellis" for above-deck installations.

Figure 10.7   **Residential roof-mount solar pool heater.**
Courtesy of UMA/Heliocol.

The main disadvantage of this type of installation is that they require significantly larger surface area than other systems, and, like metal panels, they do not carry a manufacturer's warranty.

PANEL SELECTION

One of the most important considerations when selecting a pool heating system is the amount of panel surface area that is required to heat the pool. The relationship of the solar collector area to swimming pool surface area must be adequate in order to ensure that your pool achieves the temperatures you expect, generally in the high 70s to the low 80s at a minimum, during the swimming season. The percentage of solar panel surface area to pool surface area varies with geographic location and is affected by factors such as local microclimates, solar collector orientation, pool shading, and desired heating season.

It is very important to keep in mind that solar energy is a very dilute energy source. Only a limited amount of useful heat falls on each square foot of panel. Consequently, whatever type of solar system is used, a large panel area is needed to collect adequate amounts of energy.

In southern California, Texas, and Arizona, where there is abundant sunshine and warm temperatures, the swimming season stretches from April or May to September or October. To heat a pool during this period, it is necessary to install enough solar collectors to equal a minimum of 70 percent of the surface area of the swimming pool, when the solar panels are facing south.

Generally, it is desirable to mount the panels on a southerly exposure; however, an orientation within 45 degrees of south will not significantly decrease performance as

Figure 10.8   An industrial solar liquid heating panel installation.
Courtesy of Solargenix.

long as shading is avoided. A due-west exposure will work well if the square footage of the solar collector is increased to compensate. However, a due-east exposure is generally to be avoided, unless significantly more solar collectors are used. Figure 10.8 is an industrial solar liquid heating panel installation.

As the orientation moves away from the ideal, sizing should increase to 80 to 100 percent, or more for west or southeast orientations. If climatic conditions are less favorable, such as near the ocean, even more coverage may be required. In general, it is always recommended to exceed the minimums to offset changing weather patterns. However, there is a point of diminishing return, where more panels will not add significantly to the pool heating function. Table 10.1 shows the economics for a typical pool heating installation.

Heliocol solar collector sizing use   To determine the number of solar panels needed, divide the solar collector area needed by the total square feet of individual collectors. The following example demonstrates the use of the insolation chart.

- Calculate the solar panel requirement for a 14 ft × 28 ft pool located in Las Vegas, Nevada.
- Pool surface = 14 ft × 28 ft = 392 ft^2.
- Las Vegas is located in zone 5 which has a 0.52 multiplier.
- Collector area = 392 ft × 0.52 ft = 203.8 ft^2.
- Approximate number of panels required using Heliocol HC-40 panels = 5.1 or 5 panels.

TABLE 10.1   TYPICAL POOL HEATING INSTALLATION ECONOMICS

Pool size	7500 ft^2
Pool depth	5 ft
Total purchase price	$18,000
First year energy savings	$30,575 (fuel cost per therm = $1.00)
Ten-year energy savings	$38,505 (estimated)
Expected investment payback	5.3 years
Yearly average return (internal rate of return)	21%
Size of each panel	50 ft^2 (4~ – 12.5~)
Number of panels	144
Total required area	7200 ft^2
Panel tilt	2 degrees
Panel orientation	180 degrees
Total required area	7200 ft^2

Sizing is an art, as well as a science. There are so many factors that affect swimming pool heat losses that no one has yet come up with the "perfect" model or sizing calculation. Your company may already have a sizing guide or sizing method that works well for installations in your particular area, and you may not want to use the sizing method outlined in this chapter. Be sure to find out from your sales manager what sizing method or calculation you should use to determine the proper sizing for your systems.

This chapter outlines a sizing method that can be applied to any geographic area. If you follow the guidelines detailed in this chapter and have a thorough grasp of your geographic factors, you should be able to properly size all your solar system proposals with a reasonable degree of accuracy and confidence.

Average pool water temperatures ranging from 76°F to 80°F are usually considered comfortable. In northern states, however, 72°F is considered warm, and in the South, swimmers usually want the temperature to be 82°F.

Solar water heating system sizing guide   The following guideline addresses all factors which must be taken into account when designing a pool solar water heating system. It is assumed that the pool will be covered when the nighttime temperatures drop below 60°F. If you heat a pool, you should use a solar blanket. Not to do so is much like heating a house without a roof; the heat just goes right out the top! Use of a cover retains more than two-thirds of the collected heat needed to maintain a comfortable swimming temperature.

The key to properly sizing a system is taking into account all the environmental and physical factors that pertain to your area in general and the prospect's home in particular. There are 10 questions that you need to have answered to size a system. They are as follows:

1 How many months of the year do the owners swim in the pool?
2 How long can you reasonably extend their season taking into account their geographic location?
3 Will there be a backup heating system? What kind?
4 Does the pool have a screen enclosure?
5 Will they use a blanket?
6 Do they have a solar window?
7 Is the wind going to be a problem?
8 Is shading going to be a problem? How many hours a day?
9 What direction and at what angle will the collectors be mounted?
10 What is the surface area of the pool?

Some of these questions will be answered as part of your pool heating survey. The rest will be determined by measurement and inspection.

The following guideline will give a factor that represents how many square feet of solar collector area is needed in relation to the pool's surface area. Once determined, this factor is multiplied times the pool surface area. The resulting answer is divided by the selected collector area to determine the number of collectors required.

1 To begin, you will want to determine a sizing factor for optimum conditions in your geographic location. To obtain this information you need to consult with sales representative of a solar pool heating system. You can also contact the local weather bureau and ask for the mean daily solar radiation (Langley's) for the coldest month of the desired swimming season. Using the following table, determine the starting sizing factor by the corresponding Langley reading.

LANGLEY READING	SIZING FACTOR
200	1.05
250	0.96
300	0.85
350	0.75
400	0.67
450	0.60
500	0.55
550	0.51
600	0.48

2 For optimum efficiency, solar collectors should face south. If you are unable to face your system south, multiply the sizing factor by the applicable following figure:

East facing	1.25
West facing	1.15

Increase this figure if you have a roof with a pitch equal to or greater than 6/12. Decrease this figure if you have a roof with a pitch equal to or less than 4/12.

3 If the pool is shaded, you need to multiply the sizing factor by the following figure:

25% shaded	1
50% shaded	1.25
75% shaded	1.50
100% shaded	1.75

As a general rule of thumb, if there is a screen enclosure, multiply the factor by 1.25. If the pool is indoors, multiply the sizing factor by 2.00.

4 In the northern states, the best collector angle is the latitude minus 10°. This gradually changes as you move south until it reaches the latitude plus 10° in south Florida. For each 10° variance from the optimum angle multiply the sizing factor by the following figure:

DEVIATION	FIGURE
±10 degree	1.05
±20 degree	1.10
±30 degree	1.20

As a general rule of thumb, if collectors are laid flat, multiply the sizing factor by 1.10.

5 For the collector area use the following:

HELIOCOL COLLECTOR NO.	AREA (ft²)
HC-50	50
HC-40	40
HC-30	30

An Example The following example is used to illustrate sizing factor determination and the number of collectors needed on a partially shaded, flat roof for a 15 ft × 30 ft rectangular pool.

1 Langley's for coldest month of swimming is December.   0.55
2 Collectors are going to face south.
3 Pool shaded 25 percent (multiply by).   1.1
4 Collectors are going to go on a flat roof (multiply).   1.1
Sizing factor = 0.67.
5 Surface area of the pool.   450
6 Multiply by sizing factor.   0.67
Square feet of collector area needed.   302
7 HC-40s are going to be used on the project, so divide by 40 ft^2.
The number of HC-40 collectors needed = 8.

This guide is an approximation only. Wind speeds, humidity levels, desired pool temperature, and other factors can also affect proper solar pool system sizing. If the prospect does not want to use a cover, you may have to double or even triple the solar coverage to achieve the desired swimming season.

The choice of collector size and system configuration is dependent on the designer. What must be considered is the roof space as well as the associated cost. Installation of smaller collectors will be a great deal more difficult than larger ones. None of these rules are concrete, and a designer's best judgment should be followed.

SOME USEFUL SUGGESTED PRACTICES

The use of common sense when investing in solar pool heating is very important. First-time buyers should consider the following.

- Buy only from a licensed contractor, and check on their experience and reputation.
- Be aware that several factors should be considered when evaluating various system configurations. More solar panels, generally, mean your pool will be warmer.
- Use a pool cover, if possible.
- Make sure the system is sized properly. An inadequately sized system is guaranteed dissatisfaction.
- Beware of outrageous claims, such as "90-degree pool temperatures in December with no backup heater." No solar heating system can achieve such a performance.
- The contractor should produce evidence of adequate worker's compensation and liability insurance.
- Insurance certificates should be directly from the insurance company and not the contractor.
- Check the contractor's referrals before buying.
- Get a written description of the system, including the number of solar panels, size of panels, and the make and model numbers.
- Get a complete operation and maintenance manual and start-up demonstration.
- The price should not be the most important factor. But it should also not be dramatically different from prices of competing bidders for similar equipment.
- Be sure the contractor obtains a building permit if required.

Concentrator Solar Technologies

Concentrating solar power (CSP) technologies concentrate solar energy to produce high-temperature heat that is then converted into electricity. The three most advanced CSP technologies currently in use are parabolic troughs (PT), central receivers (CR), and dish engines (DE). CSP technologies are considered one of today's most efficient power plants; they can readily substitute solar heat for fossil fuels, fully or partially, to reduce emissions and provide additional power at peak times. Dish engines are better suited for distributed power, from 10 kW to 10 MW, while parabolic troughs and central receivers are suited for larger central power plants, 30 to 200 MW and higher. Figure 10.9 shows a graphic depiction of a passive parabolic concentrator used in solar electric power generating plants.

The solar resource for generating power from parabolic concentrating systems is very plentiful, which can provide sufficient electric power for the entire country if it could be arranged to cover only about 9 percent of the state of Nevada, which would amount to a plot of land 100 miles square.

The amount of power generated by a concentrating solar power plant depends on the amount of direct sunlight. Like photovoltaic concentrators, these technologies use only direct beams of sunlight to concentrate the thermal energy of the sun.

The southwestern United States potentially offers an excellent opportunity for developing concentrating solar power technologies. As is well-known, peak power demand generated as a result of air-conditioning systems can be offset by Solar Electric Generating System (SEGS) resource plants that operate for nearly 100 percent of the on-peak hours of southern California Edison.

Concentrating solar power systems can be sized from 2 to 10 kW or could be large enough to supply grid-connected power of up to 200 MW. Some existing systems

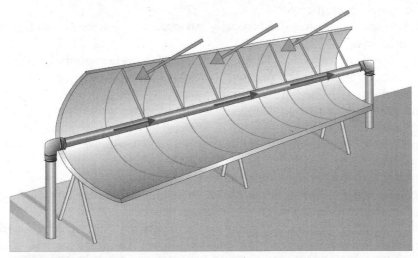

Figure 10.9   Passive parabolic concentrator used in solar power generating plants.

installed use thermal storage during cloudy periods and are combined with natural gas resulting in hybrid power plants that provide grid-connected dispatchable power. Solar power–driven electric generator conversion efficiencies make concentrating technologies a viable renewable energy resource in the Southwest. The United States Congress recently requested the Department of Energy to develop a plan for installing 1000 MW of concentrating solar power in the Southwest over the next 5 years. Concentrating solar power technologies are also considered as an excellent source for providing thermal energy for commercial and industrial processes.

BENEFITS

CSP technologies incorporating storage do not burn any fossil fuels and produce zero greenhouse gas, NO_x, and SO_x emissions. They are also proven and reliable. For the past decade the SEGS plants have operated successfully in the southern California desert, providing enough power for 100,000 homes. Plants with cost-effective storage or natural gas hybridization can deliver power to the utility grid whenever that power is needed, not just when the sun is shining. Existing CSP plants produce power now for around 11¢/kWh (including both capital and operating costs) with projected costs dropping below 4¢/kWh within the next 20 years as technology refinements and economies of scale are implemented. Because CSP uses relatively conventional technologies and materials (glass, concrete, steel, and standard utility-scale turbines), production capacity can be scaled up to several hundred megawatts per year rapidly.

EMISSIONS

Emissions benefits of CSP technologies depend on many factors including whether they have their own storage capacity or are hybridized with other electricity or heat production technologies. CSP technologies with storage produce zero emissions, and hybrid technologies can reduce emissions by 50 percent or more.

THROUGH PARABOLIC HEATING SYSTEM TECHNOLOGIES

In this technology a large field of parabolic systems that are secured on a single-axis solar tracking support are installed in a modular parallel-row configuration aligned in a north-south horizontal direction. Each of the solar parabolic collectors track the movement of the sun from east to west during daytime hours and focus the sun's rays to linear receiver tubing that circulates a heat transfer fluid (HTF). The heated fluid in turn passes through a series of heat exchanger chambers where the heat is transferred as superheated vapor that drives steam turbines. After propelling the turbine, the spent steam is condensed and returned to the heat exchanger via condensate pumps. Figure 10.10 is a photograph of a parabolic heater installation.

At present the technology has been successfully applied in thermal electric power generation. A 354-MW solar power–generated electric plant installed in 1984 in the California Mojave Desert has been in operation with remarkable success.

Figure 10.10   **Solar parabolic heater installation.** *Photo courtesy of Solargenix.*

SOLAR TOWER TECHNOLOGY

Another use of solar concentrator technology that generates electric power from the sun is a construction that focuses concentrated solar radiation on a tower-mounted heat exchanger. The system basically is configured from thousands of sun-tracking mirrors, commonly referred to as heliostats, that reflect the sun's rays onto the tower.

The receiver contains a fluid that once heated, by a similar method to that of the parabolic system, transfers the absorbed heat in the heat exchanger to produce steam that then drives a turbine to produce electricity. Figure 10.11 is a photograph of a parabolic solar heating system installation.

Power generated from this technology produces up to 400 MW of electricity. The heat transfer fluid, usually a molten liquid salt, can be raised to 550°F. The HTF is stored in an insulated storage tank and used in the absence of solar ray harvesting.

Recently a solar pilot plant located in southern California called Solar Two, which uses nitrate salt technology, has been producing 10 MW of grid-connected electricity with a sufficient thermal storage tank to maintain power production for 3 hours, which has rendered the technology as viable for commercial use.

Solar Cooling and Air Conditioning

Most of us associate cooling, refrigeration, and air conditioning as self-contained electromechanical devices connected to an electric power source that provide conditioned air for spaces in which we live as well as refrigerate our food stuff and groceries.

Figure 10.11   Parabolic solar heating system installation.
Photo courtesy of Solargenix.

Technically speaking technology that makes the refrigeration possible is based upon basic fundamental concepts of physics called heat transfer. Cold is essentially the absence of heat; likewise darkness is the absence of light.

The branch of physics that deals with the mechanics of heat transfer is called thermodynamics. There are two principal universal laws of thermodynamics. The first law concerns the conservation of energy, which states that energy neither can be created nor destroyed; however, it can be converted from one type to another. The second law of thermodynamics deals with the equalization and transfer of energy from a higher state to a lower one. Simply stated, energy is always transferred from a higher potential or state to a lower one, until two energy sources achieve exact equilibrium. Heat is essentially defined as a form of energy created as a result of transformation of another form of energy, a common example of which is when two solid bodies are rubbed together, which results in friction heat. In general, heat is energy in a transfer state, because it does not stay in any specific position and constantly moves from a warm object to a colder one, until such time, as per the second law of thermodynamics, both bodies reach heat equilibrium.

It should be noted that volume, size, or mass of objects are completely irrelevant in the heat transfer process; only the state of heat energy levels are factors in the energy balance equation. With this principle in mind, heat energy flows from a small object, such as a hot cup of coffee, to one with much larger mass, such as your hand. The rate of travel of heat is directly proportional to the difference in temperature between the two objects.

Heat travels in three forms, namely, radiation, conduction, and convection. As radiation, heat is transferred as a waveform similar to radio, microwaves, or light. For example, the sun transfers its energy to Earth by rays or radiation. In conduction, heat energy flows from one medium or substance to another by physical contact. Convection on the other hand is the flow of heat between air, gas, liquid, and a fluid medium. In refrigeration the basic principles are based on the second law of thermodynamics, that is, transfer or removal of heat from a higher energy medium to a lower one by means of convection. Figure 10.12 is a graphic depiction of refrigeration evaporation and condensation cycle.

TEMPERATURE

Temperature is a scale for measuring heat intensity with a directional flow of energy. Water freezes at 0°C (32°F) and boils at 100°C (212°F). Temperature scales are simply temperature differences between freezing and boiling water temperatures measured at

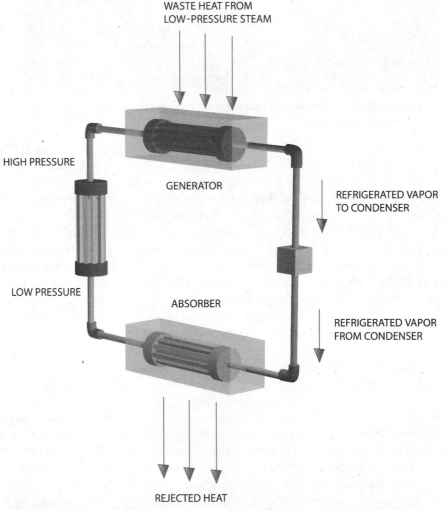

Figure 10.12   **Refrigeration Evaporation and Condensation cycle.**

Courtesy of Vector Delta Design Group, Inc.

sea level. As mentioned earlier, based on the second law of thermodynamics, heat transfer or measurement of temperature is not dependent on the quantity of heat.

MOLECULAR AGITATION

Depending on the state of heat energy, most substances in general can exist in vapor, liquid, and solid states. As an example depending on the heat energy level, water can exist as solid ice when it is frozen, a liquid at room temperature, and a vapor when it is heated above its boiling temperature of 212°F. In each of the states, water is within or without the two boundary temperatures of 32°F and 212°F.

Steam will condense back to the water state if heat energy is removed from it. Water will change into its solid state (ice) when sufficient heat energy is removed from it. The processes can be reversed when heat energy is introduced into the medium.

The state of change is related to the fact that in various substances, depending upon the presence or absence of heat energy, a phenomenon referred to as atomic thermal agitation causes expansion and contraction of molecules. A close contraction of molecules forms solids, and a larger separation transforms matter into liquid and gaseous states. In border state energy conditions, an excess lack (beyond the solid state) or excess surplus (beyond the gaseous state) of energy creates the states referred to as supercooled and superheated, respectively.

PRINCIPLES OF REFRIGERATION

Refrigeration is accomplished by two distinct processes. In one process, referred to as the compression cycle, a medium, such as Freon gas, is first given heat energy by compression, which turns the gas into a liquid. Then in a subsequent cycle, energy is removed from the liquid, in a form of evaporation or gas expansion, which disperses the gas molecules and turns the surrounding chamber into a cold environment.

A medium of energy-absorbing liquid such as water or air when circulated within the so-called evaporation chamber gives up its heat energy to the expanded gas. The cold water or air is in turn circulated by means of pumps into environments that have higher ambient heat energy levels. The circulated cold air in turn exchanges or passes the cold air into the ambient space through radiator tubes or fins, thus lowering the energy of the environment.

Temperature control is realized by the opening and closing of cold medium circulating tube valves or air duct control vanes, modulated by a local temperature-sensing device, such as a thermostat or a set point control mechanism.

COOLING TECHNOLOGIES

There are two types of refrigeration technologies currently in use, namely, electric vapor-compression (Freon gas) and heat-driven absorption cooling.

Absorption cooling chillers are operated by steam, hot water, or fossil fuel burners or combinations of these. There are two types of absorption chillers; one uses lithium bromide (LiBr) as an energy conversion medium and water as a refrigerant. In this type of technology the lowest temperature achieved is limited to 40°F. Another absorption chiller technology uses ammonia as the energy conversion medium and a mix of ammonia and water as the refrigerant. The maximum temperature limit for this technology is 20°F. Both of the technologies discussed have been around for about 100 years.

The basic principle of absorption chillers is gasification of LiBr or ammonia. Gasification takes place when either of the media is exposed to heat. Heat could be derived from fossil fuel gas burners, hot water obtained from geothermal energy, passive solar water heaters, or microturbine generators, which use landfill gases to produce electricity and heat energy.

Coefficient of Performance   The energy efficiency of an air-conditioning system is defined by a coefficient of performance (COP), which is defined as the ratio of cooling energy to the energy supplied to the unit. A ton of cooling energy is 12,000 British thermal units per hour (Btu/h), which as defined in olden days, is the energy required to remove heat from a space obtained through melting a ton of ice. One ton or 12,000 Btu is equal to 3413 W of electric power.

Based on the preceding definitions, the COP of an air-conditioning unit that requires 1500 W of electric power per ton is equal to 12,000 Btu/h (1 kW = 3413 BTU) therefore, 1.5 Kw × 3414 = 5121 BTU of supplied energy, therefore the COP = 12000 BTU cooling energy/ 5121 BTU supplied energy = 2.343, a lower electric energy requirement will increase the COP rating, which brings us to the conclusion that the lower the amount of energy input, the better the efficiency.

SOLAR POWER COOLING AND AIR CONDITIONING

A combination use of passive solar and natural gas-fired media evaporation has given rise to a generation of hybrid absorption chillers that can produce a large tonnage of cooling energy by use of solar- or geothermal-heated water. A class of absorption, which commonly uses LiBr, that has been commercially available for some time uses natural gas and solar power as the main sources of energy. A 1000-ton absorption chiller can reduce electric energy consumption by an average of 1 MW or 1 million watts, which will have a very significant impact on reducing the electric power consumption and resulting environmental pollutions as described in earlier chapters.

Desiccant evaporators   Another solar power cooling technology makes use of solar-desiccant-evaporator air conditioning, which reduces outside air humidity and passes it through an ultraefficient evaporative cooling system. This cooling process, which uses an indirect evaporative process, minimizes the air humidity, and this makes the use of this technology quite effective in coastal and humid areas. A typical building cooling capacity is shown in Table 10.2.

TABLE 10.2 TYPICAL BUILDING COOLING CAPACITIES		
SPACE	SIZE	COOLING TONS
Medium office	50,000	100–150
Hospital	150,000	400–600
Hotel	250,000	400–500
High school	50,000	100–400
Retail store	160,000	170–400

Direct Solar Power Generation

The following project undertaken by Solargenix Energy makes use of special parabolic reflectors that concentrate solar energy rays into circular pipes that are located at the focal center of the parabola. The concentrated reflection of energy elevates the temperature of the circulating mineral liquid oil within the pipes, raising the temperature to such levels that allow considerable steam generation via special heat exchangers that drive power turbines. The following abstract reflects a viable electric power generation in Arizona.

RED ROCK, ARIZ.—APS today broke ground on Arizona's first commercial solar trough power plant and the first such facility constructed in the United States since 1990.

Located at the company's Saguaro Power Plant in Red Rock, about 30 miles north of Tucson, the APS Saguaro Solar Trough Generating Station will have a one megawatt (MW) generating capacity, enough to provide for the energy needs of approximately 200 average-size homes. APS has contracted with Solargenix Energy to construct and provide the solar thermal technology for the plant, which is expected to come online in April 2005. Solargenix, formerly Duke Solar, is based out of Raleigh, North Carolina. Solargenix has partnered with Ormat who will provide the engine to convert the solar heat, collected by the Solargenix solar collectors, into electricity.

"The APS Saguaro Solar Trough Power Plant presents a unique opportunity to further expand our renewable energy portfolio," said Peter Johnston, manager of Technology Development for APS. "We are committed to developing clean renewable energy sources today that will fuel tomorrow's economy. We believe solar-trough technology can be part of a renewable solution.

The company's solar-trough technology uses parabolic shaped reflectors (or mirrors) to concentrate the sun's rays to heat a mineral oil between 250 and 570 degrees. The fluid then enters the Ormat engine passing first through a heat exchanger to vaporize a secondary working fluid. The vapor is used to spin a turbine, making electricity. It is then condensed back into a liquid before being vaporized once again.

Historically, solar-trough technology has required tens of megawatts of plant installation to produce steam from water to turn generation turbines. The significant first cost of multi-megawatt power plants had precluded their use in the APS solar portfolio. This solar trough system combines the relatively low cost of parabolic solar trough thermal technology with the commercially available, smaller turbines usually associated with low temperature geothermal generation plants, such as the Ormat unit being used for this project.

In addition to generating electricity for APS customers, the solar trough plant will help APS meet the goals of the Arizona Corporation Commission's Environmental Portfolio Standard, which requires APS to generate 1.1 percent of its energy through renewable sources—60 percent through solar—by 2007. APS owns and operates approximately 4.5 MW of photovoltaic solar generation around the state and has partnered on a 3-MW biomass plant in Eager, which came online in February, and a 15-megawatt wind farm to be constructed near St. Johns. APS, Arizona's largest and longest-serving electricity utility, serves about 902,000 customers in 11 of the state's counties.

Innovations in Passive Solar Power Technology

The following are a few innovations under development by Energy Innovations, a subsidiary of Idea Labs located in Pasadena, California.

SUNFLOWER 250

The experimental solar power tracking concentrator shown in Figure 10.13 consists of a 3-m square platform equipped with adjustable motorized reflective mirrors that concentrate solar rays onto focally located solar cells that produce electricity. The prototype units are currently undergoing testing in Pasadena, California, and throughout various geographic locations. Initial tests have shown very promising performance.

Manufactured units are expected to be available within the next couple of years. They will be ground mounted on stands and connected via internal wiring. The undercarriage will be shielded by a wind guard around the perimeter. Each unit is expected to produce a peak power rating of 200 W.

The continuous sun-tracking mechanism of the design enables the Sunflower to produce 30 percent more energy than a traditional flat PV panel of similar rating. At present, development continues to provide a wind mitigation solution.

STIRLING-ENGINE SUNFLOWER

The Stirling–engine Sunflower is a radical concept since it does not use a stationary photovoltaic cell technology; rather it is constructed from lightweight polished aluminized

Figure 10.13   **Sunflower 250 prototype.** *Photo courtesy of Energy Innovations, Pasadena, California.*

plastic reflector petals that are each adjusted by a microprocessor-based motor controller that enables the petals to track the sun in an independent fashion. This heat engine is essentially used to produce hot water by concentrating solar rays onto a low-profile water chamber.

At present the technology is being refined to produce higher-efficiency and more cost-effective production models, and the company is working on larger-scale models for use in large-scale solar water heating installations.

UNIT CONVERSION AND DESIGN
REFERENCE TABLES

Renewable Energy Tables and Important Solar Power Facts

1 Recent analysis by the Department of Energy (DOE) shows that by year 2025, one-half of new U.S. electricity generation could come from the sun.

2 In 2005 the United States generated only 4 GW (1 GW is 1000 MW) of solar power. By the year 2030, it is estimated to be 200 GW.

3 A typical nuclear power plant generates about 1 GW of electric power, which is equal to 5 GW of solar power (daily power generation is limited to an average of 5 to 6 hours per day).

4 Global sales of solar power systems have been growing at a rate of 35 percent in the past few years.

5 It is projected that by the year 2020, the United States will be producing about 7.2 GW of solar power per year.

6 The shipment of U.S. solar power systems has fallen by 10 percent annually, but has increased by 45 percent throughout Europe.

7 In the past 4 years the annual sales growth globally has been 35 percent.

8 Present cost of solar power modules on the average is $2.33/W. By 2030 it should be about $0.38/W.

9 World production of solar power is 1 GW/year.

10 Germany has a $0.50/W grid feed incentive that will be valid for the next 20 years. The incentive is to be decreased by 5 percent per year.

11 In the past few years, Germany installed 130 MW of solar power per year.

12 Japan has a 50 percent subsidy for solar power installations of 3- to 4-kW systems and has about 800 MW of grid-connected solar power systems. Solar power in Japan has been in effect since 1994.

13 California, in 1996, set aside $540 million for renewable energy, which has provided a $4.50/W to $3.00/W buyback as a rebate.

14 In the years 2015 through 2024, it is estimated that California could produce an estimated $40 billion of solar power sales.

15 In the United States, 20 states have a solar rebate program. Nevada and Arizona have set aside a state budget for solar programs.

16 Projected U.S. solar power statistics are shown in the following table:

	2004	2005
Base installed cost per watt	$6.50–$9.00	$1.93
Annual power production (MW)	120	31,000
Employment	20,000	350,000
Cell efficiency (%)	20	22–40
Module performance (%)	8–15	20–30
System performance (%)	6–12	18–25

17 Total U.S. production has been just about 18 percent of global production.

18 For each megawatt of solar power produced, we employ 32 people.

19 A solar power collector, sized 100 × 100 mi, in the southwest United States could produce sufficient electric power to satisfy the country's yearly energy needs.

20 For every kilowatt of power produced by nuclear or fossil fuel plants, $1/2$ gallon of water is used for scrubbing, cleaning, and cooling. Solar power practically does not require any water usage.

21 Solar power cogeneration has a significant impact:

■ Boosts economic development
■ Lowers cost of peak power
■ Provides greater grid stability
■ Lowers air pollution
■ Lowers greenhouse gas emissions
■ Lowers water consumption and contamination

22 A mere 6.7-mi/gal efficiency increase in cars driven in the United States could offset our share of imported Saudi oil.

23 Types of solar power technology at present:

■ Crystalline
■ Polycrystalline

- Amorphous
- Thin- and thick-film technologies

24 Types of solar power technology in the future:

- Plastic solar cells
- Nano-structured materials
- Dye-synthesized cells

Energy Conversion Table

ENERGY UNITS
1 J (joule) = 1 W · s = 4.1868 cal
1 GJ (gigajoule) = 10 E9 J
1 TJ (terajoule) = 10 E12 J
1 PJ (petajoule) = 10 E15 J
1 kWh (kilowatt-hour) = 3,600,000 J
1 toe (tonne oil equivalent) = 7.4 barrels of crude oil in primary energy
= 7.8 barrels in total final consumption
= 1270 m³ of natural gas
= 2.3 metric tonnes of coal
Mtoe (million tonne oil equivalent) = 41.868 PJ
POWER
Electric power is usually measured in watts (W), kilowatts (kW), megawatts (MW), and so forth. Power is energy transfer per unit of time. Power (e.g., in watts) may be measured at any point in time, whereas energy (e.g., in kilowatt-hours) has to be measured over a certain period, for example, a second, an hour, or a year.
1 kW = 1000 W
1 MW = 1,000,000 W
1 GW = 1000 MW
1 TW = 1,000,000 MW
UNIT ABBREVIATIONS
m = meter = 3.28 feet (ft)
s = second
h = hour
W = watt

(Continued)

hp = horsepower

J = joule

cal = calorie

toe = tonnes of oil equivalent

Hz = hertz (cycles per second)

10 E–12 = pico (p) = 1/1000,000,000,000

10 E–9 = nano (n) = 1/1,000,000,000

10 E–6 = micro (μ) = 1/1000,000

10 E–3 = milli (m) = 1/1000

10 E–3 = kilo (k) = 1000 = thousands

10 E–6 = mega (M) = 1,000,000 = millions

10 E–9 = giga (G) = 1,000,000,000

10 E–12 = tera (T) = 1,000,000,000,000

10 E–15 = peta (P) = 1,000,000,000,000,000

WIND SPEEDS

1 m/s = 3.6 km/h = 2.187 mi/h = 1.944 knots

1 knot = 1 nautical mile per hour = 0.5144 m/s = 1.852 km/h = 1.125 mi/h

Voltage Drop Formulas and DC Cable Charts

Single-phase VD = A ft × 2K/C.M.

Three-phase VD = A ft × 2K × 0.866/C.M.

Three-phase VD = A ft × 2K × 0.866 × 1.5/C.M.(for two pole systems)

where A = amperes

   L = distance from source of supply to load

   C.M. = cross-sectional area of conductor in circular mills:

      K = 12 for copper more than 50 percent loading

      K = 11 for copper less than 50 percent loading

      K = 18 for aluminum

| VOLTAGE DROP CALCULATION FOR COPPER WIRES | | | | |
WIRE	THHN AMPACITY	THWN AMPACITY	MCM	CONDUIT DIAMETER (IN)
2,000			2,016,252	
1,750			1,738,503	
1,500			1,490,944	
1,250			1,245,699	
1,000	615	545	999,424	
900	595	520	907,924	
800	565	490	792,756	
750	535	475	751,581	
700	520	460	698,389	4
600	475	420	597,861	4
500	430	380	497,872	4
400	380	335	400,192	4
350	350	310	348,133	3
300	320	285	299,700	3
250	290	255	248,788	3
4/0	260	230	211,600	$2\frac{1}{2}$
3/0	225	200	167,000	2
2/0	195	175	133,100	2
1/0	170	150	105,600	2
1	150	130	83,690	$1\frac{1}{2}$
2	130	115	66,360	$1\frac{1}{4}$
3	110	100	52,620	$1\frac{1}{2}$
4	95	85	41,740	$1\frac{1}{2}$
6	75	65	26,240	1
8	55	50	15,510	$\frac{3}{4}$
10	30	30	10,380	$\frac{1}{2}$
12	20	20	6,530	$\frac{1}{2}$

24 NEC ALLOWED CABLE DISTANCES FOR 0240 VOLT AC OR DC CABLE – CHART FOR VOLTAGE DROP OF 2% VOLT AC OR DC CABLE CHART = VOLTAGE DROP OF 2% NEC CODE ALLOWED CABLE DISTANCES

AMPS	WATTS	AWG #14	AWG #12	AWG #10	AWG #8	AWG #6	AWG #4	AWG #2	AWG #1/0	AWG #2/0	AWG #3/0
2	480	338	525								
4	960	150	262	413							
6	1,440	113	180	262							
8	1,920	82	180	218	338	266					
10	2,400	67	105	173	270	427					
15	3,600	45	67	105	180	285	450				
20	4,800		52	82	144	218	338	540			
25	6,000			67	105	173	270	434			
30	7,200			53	90	142	225	360	578		
40	9,600				67	250	173	270	434	540	
50	12,000				54	82	137	218	345	434	547

120 NEC ALLOWED CABLE DISTANCES FOR 120 VOLT AC OR DC CABLE – CHART FOR VOLTAGE DROP OF 2% VOLT VOLT AC OR DC CABLE CHART = VOLTAGE DROP OF 2% NEC CODE ALLOWED CABLE DISTANCES

AMPS	WATTS	AWG #14	AWG #12	AWG #10	AWG #8	AWG #6	AWG #4	AWG #2	AWG #1/0	AWG #2/0	AWG #3/0
2	240	169	262								
4	480	75	131	206							
6	720	56	90	131	225						
8	960	41	90	109	169	266					
10	1,200	34	52	86	135	214					
15	1,800	22	34	52	90	142	225				
20	2,400		26	41	72	109	169	270			
25	3,000			34	52	86	135	217			
30	3,600			26	45	71	112	180	289		
40	4,800				34	125	86	135	217	270	
50	6,000				27	41	68	109	172	217	274

295

4<8NEC ALLOWED CABLE DISTANCES FOR 48 VOLT DC — CHART FOR VOLTAGE DROP OF 2% VOLT VOLT DC CABLE CHART = VOLTAGE DROP OF 2% NEC CODE ALLOWED CABLE DISTANCES

AMPS	WATTS	AWG # 14	AWG # 12	AWG # 10	AWG # 8	AWG # 6	AWG # 4	AWG # 2	AWG # 1/0	AWG # 2/0	AWG # 3/0
1	48	135	210	330	540						
2	96	67	105	166	270	426					
4	192	30	53	82	135	214					
6	288	22	36	53	90	142	226				
8	384	17	26	43	67	106	173				
10	480	14	21	34	54	86	135	216			
15	720	9	14	21	36	57	90	144	231		
20	960		10	17	30	43	67	108	174	216	274
35	1,200			14	21	34	214	86	138	174	219
30	1,440			10	18	29	45	72	115	138	182
40	1,920				14	21	34	54	86	115	137
50	2,400				9	17	27	43	69	86	138

24 NEC ALLOWED CABLE DISTANCES FOR 24 VOLT DC–CHART FOR VOLTAGE DROP OF 2% VOLT VOLT DC CABLE CHART = VOLTAGE DROP OF 2% NEC CODE ALLOWED CABLE DISTANCES

AMPS	WATTS	AWG #14	AWG #12	AWG #10	AWG #8	AWG #6	AWG #4	AWG #2	AWG #1/0	AWG #2/0	AWG #3/0
1	24	68	105	165	270						
2	48	34	52	83	135	213					
4	96	15	26	41	68	107					
6	144	11	18	26	45	71	113				
8	192	8	13	22	34	53	86				
10	240	7	10	17	27	43	68	108			
15	360	4	7	10	18	28	45	72	116		
20	480		5	8	15	22	34	54	87	108	137
25	600			7	10	17	27	43	69	87	110
30	720			5	9	14	22	36	58	69	91
40	960				7	10	17	27	43	58	68
50	1,200				4	8	14	22	34	43	89

12 NEC ALLOWED CABLE DISTANCES FOR 12 VOLT DC – CHART FOR VOLTAGE DROP OF 2% VOLT OLT DC CABLE CHART = VOLTAGE DROP OF 2% NEC CODE ALLOWED CABLE DISTANCES

AMPS	WATTS	AWG #14	AWG #12	AWG #10	AWG #8	AWG #6	AWG #4	AWG #2	AWG #1/0	AWG #2/0	AWG #3/0
1	12	84	131	206	337	532					
2	24	42	66	103	168	266	432	675			
4	48	18	33	52	84	133	216	337	543	672	
6	72	14	22	33	56	89	141	225	360	450	570
8	96	10	16	27	42	66	108	168	272	338	427
10	120	9	13	22	33	53	84	135	218	270	342
15	180	6	9	13	22	35	56	90	144	180	228
20	240		7	10	16	27	42	67	108	135	171
25	300			8	13	22	33	54	86	108	137
30	360			7	11	18	28	45	72	90	114
50	480				8	13	21	33	54	67	85

CROSS REFERENCE OF AMERICAN WIRE GAUGE (AWG) AND METRIC SYSTEM (mm)			
AWG	mm²	AWG	mm²
30	0.05	6	16
28	0.08	4	25
26	0.14	2	35
24	0.25	1	50
22	0.34	1/0	55
21	0.38	2/0	70
20	0.5	3/0	95
18	0.75	4/0	120
17	1	300 MCM	150
16	1.5	350 MCM	185
14	2.5	500 MCM	240
12	4	600 MCM	300
10	6	750 MCM	400
8	10	1,000 MCM	500

Solar Photovoltaic Module Tilt Angle Correction Table

The following table represent multiplier which must be used to correct losses associated with tilt angles.

SOLAR PANEL ORIENTATION TILT CORRECTION FACTOR						
	COLLECTOR TILT ANGLE FROM HORIZONTAL (DEGREES)					
	0	15	30	45	60	90
South	0.89	0.97	1.00	0.97	0.88	0.56
SSE or SSW	0.89	0.97	0.99	0.96	0.87	0.57
SE or SW	0.89	0.95	0.96	0.93	0.85	0.59
ESE or WSW	0.89	0.92	0.91	0.87	0.79	0.57
East, west	0.89	0.88	0.84	0.78	0.7	0.51

TITL ANGLE EFFICIENCY MULTIPLIER TABLE

	COLLECTOR TILT ANGLE FROM HORIZONTAL (DEGREES)					
	0	15	30	45	60	90
FRESNO						
South	0.90	0.98	1.00	0.96	0.87	0.55
SSE, SSW	0.90	0.97	0.99	0.96	0.87	0.56
SE, SW	0.90	0.95	0.96	0.92	0.84	0.68
ESE, WSW	0.90	0.92	0.91	0.87	0.79	0.57
East, west	0.90	0.88	0.86	0.78	0.70	0.51
DAGGETT						
South	0.88	0.97	1.00	0.97	0.88	0.56
SSE, SSW	0.88	0.96	0.99	0.96	0.87	0.58
SE, SW	0.88	0.94	0.96	0.93	0.85	0.59
ESE, WSW	0.88	0.91	0.91	0.86	0.78	0.57
East, west	0.88	0.87	0.83	0.77	0.69	0.51
SANTA MARIA						
South	0.89	0.97	1.00	0.97	0.88	0.57
SSE, SSW	0.89	0.97	0.99	0.96	0.87	0.58
SE, SW	0.89	0.95	0.96	0.93	0.86	0.59
ESE, WSW	0.89	0.92	0.91	0.87	0.79	0.67
East, west	0.89	0.88	0.84	0.78	0.70	0.52
LOS ANGELES						
South	0.89	0.97	1.00	0.97	0.88	0.57
SSE, SSW	0.89	0.97	0.99	0.96	0.87	0.58
SE, SW	0.89	0.95	0.96	0.93	0.85	0.69
ESE, WSW	0.89	0.92	0.91	0.87	0.79	0.57
East, west	0.89	0.88	0.85	0.78	0.70	0.51
SAN DIEGO						
South	0.89	0.98	1.00	0.97	0.88	0.57
SSE, SSW	0.89	0.97	0.99	0.96	0.87	0.58
SE, SW	0.89	0.95	0.96	0.92	0.54	0.59
ESE, WSW	0.89	0.92	0.91	0.87	0.79	0.57
East, west	0.89	0.88	0.85	0.78	0.70	0.51

Solar Insolation Table for Major Cities in the United States*

STATE	CITY	HIGH	LOW	AVG.	STATE	CITY	HIGH	LOW	AVG.
AK	Fairbanks	5.87	2.12	3.99	GA	Griffin	5.41	4.26	4.99
AK	Matanuska	5.24	1.74	3.55	HI	Honolulu	6.71	5.59	6.02
AL	Montgomery	4.69	3.37	4.23	IA	Ames	4.80	3.73	4.40
AR	Bethel	6.29	2.37	3.81	IL	Boise	5.83	3.33	4.92
AR	Little Rock	5.29	3.88	4.69	IL	Twin Falls	5.42	3.42	4.70
AZ	Tucson	7.42	6.01	6.57	IL	Chicago	4.08	1.47	3.14
AZ	Page	7.30	5.65	6.36	IN	Indianapolis	5.02	2.55	4.21
AZ	Phoenix	7.13	5.78	6.58	KS	Manhattan	5.08	3.62	4.57
CA	Santa Maria	6.52	5.42	5.94	KS	Dodge City	4.14	5.28	5.79
CA	Riverside	6.35	5.35	5.87	KY	Lexington	5.97	3.60	4.94
CA	Davis	6.09	3.31	5.10	LA	Lake Charles	5.73	4.29	4.93
CA	Fresno	6.19	3.42	5.38	LA	New Orleans	5.71	3.63	4.92
CA	Los Angeles	6.14	5.03	5.62	LA	Shreveport	4.99	3.87	4.63
CA	Soda Springs	6.47	4.40	5.60	MA	E. Wareham	4.48	3.06	3.99
CA	La Jolla	5.24	4.29	4.77	MA	Boston	4.27	2.99	3.84
CA	Inyokern	8.70	6.87	7.66	MA	Blue Hill	4.38	3.33	4.05
CO	Grandbaby	7.47	5.15	5.69	MA	Natick	4.62	3.09	4.10
CO	Grand Lake	5.86	3.56	5.08	MA	Lynn	4.60	2.33	3.79
CO	Grand Junction	6.34	5.23	5.85	MD	Silver Hill	4.71	3.84	4.47
CO	Boulder	5.72	4.44	4.87	ME	Caribou	5.62	2.57	4.19
DC	Washington	4.69	3.37	4.23	ME	Portland	5.23	3.56	4.51
FL	Apalachicola	5.98	4.92	5.49	MI	Sault Ste. Marie	4.83	2.33	4.20
FL	Belie Is.	5.31	4.58	4.99	MI	E. Lansing	4.71	2.70	4.00
FL	Miami	6.26	5.05	5.62	MN	St. Cloud	5.43	3.53	4.53
FL	Gainesville	5.81	4.71	5.27	MO	Columbia	5.50	3.97	4.73
FL	Tampa	6.16	5.26	5.67	MO	St. Louis	4.87	3.24	4.38
GA	Atlanta	5.16	4.09	4.74	MS	Meridian	4.86	3.64	4.43

(Continued)

STATE	CITY	HIGH	LOW	AVG.	STATE	CITY	HIGH	LOW	AVG.
MT	Glasgow	5.97	4.09	5.15	PA	Pittsburg	4.19	1.45	3.28
MT	Great Falls	5.70	3.66	4.93	PA	State College	4.44	2.79	3.91
MT	Summit	5.17	2.36	3.99	RI	Newport	4.69	3.58	4.23
NM	Albuquerque	7.16	6.21	6.77	SC	Charleston	5.72	4.23	5.06
NB	Lincoln	5.40	4.38	4.79	SD	Rapid City	5.91	4.56	5.23
NB	N. Omaha	5.28	4.26	4.90	TN	Nashville	5.2	3.14	4.45
NC	Cape Hatteras	5.81	4.69	5.31	TN	Oak Ridge	5.06	3.22	4.37
NC	Greensboro	5.05	4.00	4.71	TX	San Antonio	5.88	4.65	5.3
ND	Bismarck	5.48	3.97	5.01	TX	Brownsville	5.49	4.42	4.92
NJ	Sea Brook	4.76	3.20	4.21	TX	El Paso	7.42	5.87	6.72
NV	Las Vegas	7.13	5.84	6.41	TX	Midland	6.33	5.23	5.83
NV	Ely	6.48	5.49	5.98	TX	Fort Worth	6.00	4.80	5.43
NY	Binghamton	3.93	1.62	3.16	UT	Salt Lake City	6.09	3.78	5.26
NY	Ithaca	4.57	2.29	3.79	UT	Flaming Gorge	6.63	5.48	5.83
NY	Schenectady	3.92	2.53	3.55	VA	Richmond	4.50	3.37	4.13
NY	Rochester	4.22	1.58	3.31	WA	Seattle	4.83	1.60	3.57
NY	New York City	4.97	3.03	4.08	WA	Richland	6.13	2.01	4.44
OH	Columbus	5.26	2.66	4.15	WA	Pullman	6.07	2.90	4.73
OH	Cleveland	4.79	2.69	3.94	WA	Spokane	5.53	1.16	4.48
OK	Stillwater	5.52	4.22	4.99	WA	Prosser	6.21	3.06	5.03
OK	Oklahoma City	6.26	4.98	5.59	WI	Madison	4.85	3.28	4.29
OR	Astoria	4.76	1.99	3.72	WV	Charleston	4.12	2.47	3.65
OR	Corvallis	5.71	1.90	4.03	WY	Lander	6.81	5.50	6.06
OR	Medford	5.84	2.02	4.51					

*Values are given in kilowatt-hours per square meter per day.

Longitude and Latitude Tables

	LONGITUDE	LATITUDE
ALABAMA		
Alexander City	32° 57'N	85° 57'W
Anniston AP	33° 35'N	85° 51'W
Auburn	32° 36'N	85° 30'W
Birmingham AP	33° 34'N	86° 45'W
Decatur	34° 37'N	86° 59'W
Dothan AP	31° 19'N	85° 27'W
Florence AP	34° 48'N	87° 40'W
Gadsden	34° 1'N	86° 0'W
Huntsville AP	34° 42'N	86° 35'W
Mobile AP	30° 41'N	88° 15'W
Mobile Co	30° 40'N	88° 15'W
Montgomery AP	32° 23'N	86° 22'W
Selma-Craig AFB	32° 20'N	87° 59'W
Talladega	33° 27'N	86° 6'W
Tuscaloosa AP	33° 13'N	87° 37'W
ALASKA		
Anchorage AP	61° 10'N	150° 1'W
Barrow	71° 18'N	156° 47'W
Fairbanks AP	64° 49'N	147° 52'W
Juneau AP	58° 22'N	134° 35'W
Kodiak	57° 45'N	152° 29'W
Nome AP	64° 30'N	165° 26'W
ARIZONA		
Douglas AP	31° 27'N	109° 36'W

	LONGITUDE	LATITUDE
Burbank AP	34° 12'N	118° 21'W
Chico	39° 48'N	121° 51'W
Concord	37° 58'N	121° 59'W
Covina	34° 5'N	117° 52'W
Crescent City AP	41° 46'N	124° 12'W
Downey	33° 56'N	118° 8'W
El Cajon	32° 49'N	116° 58'W
El Cerrito AP	32° 49'N	115° 40'W
Escondido	33° 7'N	117° 5'W
Eureka/Arcata AP	40° 59'N	124° 6'W
Fairfield-Trafis AFB	38° 16'N	121° 56'W
Fresno AP	36° 46'N	119° 43'W
Hamilton AFB	38° 4'N	122° 30'W
Laguna Beach	33° 33'N	117° 47'W
Livermore	37° 42'N	121° 57'W
Lompoc, Vandenberg AFB	34° 43'N	120° 34'W
Long Beach AP	33° 49'N	118° 9'W
Los Angeles AP	33° 56'N	118° 24'W
Los Angeles Co	34° 3'N	118° 14'W
Merced-Castle AFB	37° 23'N	120° 34'W
Modesto	37° 39'N	121° 0'W
Monterey	36° 36'N	121° 54'W
Napa	38° 13'N	122° 17'W
Needles AP	34° 36'N	114° 37'W
Oakland AP	37° 49'N	122° 19'W

City	Lat.	Long.
Flagstaff AP	35° 8'N	111° 40'W
Fort Huachuca AP	31° 35'N	110° 20'W
Kingman AP	35° 12'N	114° 1'W
Nogales	31° 21'N	110° 55'W
Phoenix AP	33° 26'N	112° 1'W
Prescott AP	34° 39'N	112° 26'W
Tucson AP	32° 7'N	110° 56'W
Winslow AP	35° 1'N	110° 44'W
Yuma AP	32° 39'N	114° 37'W
ARKANSAS		
Blytheville AFB	35° 57'N	89° 57'W
Camden	33° 36'N	92° 49'W
El Dorado AP	33° 13'N	92° 49'W
Fayetteville AP	36° 0'N	94° 10'W
Fort Smith AP	35° 20'N	94° 22'W
Hot Springs	34° 29'N	93° 6'W
Jonesboro	35° 50'N	90° 42' W
Little Rock AP	34° 44'N	92° 14'W
Pine Bluff AP	34° 18'N	92° 5'W
Texarkana AP	33° 27'N	93° 59' W
CALIFORNIA		
Bakersfield AP	35° 25'N	119° 3'W
Barstow AP	34° 51'N	116° 47'W
Blythe AP	33° 37'N	114° 43'W
Oceanside	33° 14'N	117° 25'W
Ontario	34° 3'N	117° 36'W
Oxnard	34° 12'N	119° 11'W
Palmdale AP	34° 38'N	118° 6'W
Palm Springs	33° 49'N	116° 32'W
Pasadena	34° 9'N	118° 9'W
Petaluma	38° 14'N	122° 38'W
Pomona Co	34° 3'N	117° 45'W
Redding AP	40° 31'N	122° 18'W
Redlands	34° 3'N	117° 11'W
Richmond	37° 56'N	122° 21'W
Riverside-March AFB	33° 54'N	117° 15'W
Sacramento AP	38° 31'N	121° 30'W
Salinas AP	36° 40'N	121° 36'W
San Bernardino, Norton AFB	34° 8'N	117° 16'W
San Diego AP	32° 44'N	117° 10'W
San Fernando	34° 17'N	118° 28'W
San Francisco AP	37° 37'N	122° 23'W
San Francisco Co	37° 46'N	122° 26'W
San Jose AP	37° 22'N	121° 56'W
San Luis Obispo	35° 20'N	120° 43'W
Santa Ana AP	33° 45'N	117° 52'W
Santa Barbara MAP	34° 26'N	119° 50'W

(Continued)

	LONGITUDE	LATITUDE
CALIFORNIA (Continued)		
Santa Cruz	36° 59'N	122° 1'W
Santa Maria AP	34° 54'N	120° 27'W
Santa Monica CIC	34° 1'N	118° 29'W
Santa Paula	34° 21'N	119° 5'W
Santa Rosa	38° 31'N	122° 49'W
Stockton AP	37° 54'N	121° 15'W
Ukiah	39° 9'N	123° 12'W
Visalia	36° 20'N	119° 18'W
Yreka	41° 43'N	122° 38'W
Yuba City	39° 8'N	121° 36'W
COLORADO		
Alamosa AP	37° 27'N	105° 52'W
Boulder	40° 0'N	105° 16'W
Colorado Springs AP	38° 49'N	104° 43'W
Denver AP	39° 45'N	104° 52'W
Durango	37° 17'N	107° 53'W
Fort Collins	40° 45'N	105° 5'W
Grand Junction AP	39° 7'N	108° 32'W
Greeley	40° 26'N	104° 38'W
La Junta AP	38° 3'N	103° 30'W
Leadville	39° 15'N	106° 18'W
Pueblo AP	38° 18'N	104° 29'W
Sterling	40° 37'N	103° 12'W

	LONGITUDE	LATITUDE
Key West AP	24° 33'N	81° 45'W
Lakeland Co	28° 2'N	81° 57'W
Miami AP	25° 48'N	80° 16'W
Miami Beach Co	25° 47'N	80° 17'W
Ocala	29° 11'N	82° 8'W
Orlando AP	28° 33'N	81° 23'W
Panama City, Tyndall AFB	30° 4'N	85° 35'W
Pensacola Co	30° 25'N	87° 13'W
St. Augustine	29° 58'N	81° 20'W
St. Petersburg	27° 46'N	82° 80'W
Sarasota	27° 23'N	82° 33'W
Stanford	28° 46'N	81° 17'W
Tallahassee AP	30° 23'N	84° 22'W
Tampa AP	27° 58'N	82° 32'W
West Palm Beach AP	26° 41'N	80° 6'W
GEORGIA		
Albany, Turner AFB	31° 36'N	84° 5'W
Americus	32° 3'N	84° 14'W
Athens	33° 57'N	83° 19'W
Atlanta AP	33° 39'N	84° 26'W
Augusta AP	33° 22'N	81° 58'W
Brunswick	31° 15'N	81° 29'W
Columbus, Lawson AFB	32° 31'N	84° 56'W

City	Lat	Long	City	Lat	Long
Trinidad	37° 15'N	104° 20'W	Dalton	34° 34'N	84° 57'W
CONNECTICUT			Dublin	32° 20'N	82° 54'W
Bridgeport AP	41° 11'N	73° 11'W	Gainesville	34° 11'N	83° 41'W
Hartford, Brainard Field	41° 44'N	72° 39'W	Griffin	33° 13'N	84° 16'W
New Haven AP	41° 19'N	73° 55'W	LaGrange	33° 1'N	85° 4'W
New London	41° 21'N	72° 6'W	Macon AP	32° 42'N	83° 39'W
Norwalk	41° 7'N	73° 25'W	Marietta, Dobbins AFB	33° 55'N	84° 31'W
Norwich	41° 32'N	72° 4'W	Savannah	32° 8'N	81° 12'W
Waterbury	41° 35'N	73° 4'W	Valdosta-Moody AFB	30° 58'N	83° 12'W
Windsor Locks, Bradley Fld	41° 56'N	72° 41'W	Waycross	31° 15'N	82° 24'W
DELAWARE			**HAWAII**		
Dover AFB	39° 8'N	75° 28'W	Hilo AP	19° 43'N	155° 5'W
Wilmington AP	39° 40'N	75° 36'W	Honolulu AP	21° 20'N	157° 55'W
DISTRICT OF COLUMBIA			Kaneohe Bay MCAS	21° 27'N	157° 46'W
Andrews AFB	38° 5'N	76° 5'W	Wahiawa	21° 3'N	158° 2'W
Washington, National AP	38° 51'N	77° 2'W	**IDAHO**		
FLORIDA			Boise AP	43° 34'N	116° 13'W
Belle Glade	26° 39'N	80° 39'W	Burley	42° 32'N	113° 46'W
Cape Kennedy AP	28° 29'N	80° 34'W	Coeur D'Alene AP	47° 46'N	116° 49'W
Daytona Beach AP	29° 11'N	81° 3'W	Idaho Falls AP	43° 31'N	112° 4'W
E Fort Lauderdale	26° 4'N	80° 9'W	Lewiston AP	46° 23'N	117° 1'W
Fort Myers AP	26° 35'N	81° 52'W	Moscow	46° 44'N	116° 58'W
Fort Pierce	27° 28'N	80° 21'W	Mountain Home AFB	43° 2'N	115° 54'W
Gainesville AP	29° 41'N	82° 16'W	Pocatello AP	42° 55'N	112° 36'W
Jacksonville AP	30° 30'N	81° 42'W	Twin Falls AP	42° 29'N	114° 29'W

(Continued)

	LONGITUDE	LATITUDE		LONGITUDE	LATITUDE
ILLINOIS			Richmond AP	39° 46'N	84° 50'W
Aurora	41° 45'N	88° 20'W	Shelbyville	39° 31'N	85° 47'W
Belleville, Scott AFB	38° 33'N	89° 51'W	South Bend AP	41° 42'N	86° 19'W
Bloomington	40° 29'N	88° 57'W	Terre Haute AP	39° 27'N	87° 18'W
Carbondale	37° 47'N	89° 15'W	Valparaiso	41° 31'N	87° 2'W
Champaign/Urbana	40° 2'N	88° 17'W	Vincennes	38° 41'N	87° 32'W
Chicago, Midway AP	41° 47'N	87° 45'W	**IOWA**		
Chicago, O'Hare AP	41° 59'N	87° 54'W	Ames	42° 2'N	93° 48'W
Chicago Co	41° 53'N	87° 38'W	Burlington AP	40° 47'N	91° 7'W
Danville	40° 12'N	87° 36'W	Cedar Rapids AP	41° 53'N	91° 42'W
Decatur	39° 50'N	88° 52'W	Clinton	41° 50'N	90° 13'W
Dixon	41° 50'N	89° 29'W	Council Bluffs	41° 20'N	95° 49'W
Elgin	42° 2'N	88° 16'W	Des Moines AP	41° 32'N	93° 39'W
Freeport	42° 18'N	89° 37'W	Dubuque	42° 24'N	90° 42'W
Galesburg	40° 56'N	90° 26'W	Fort Dodge	42° 33'N	94° 11'W
Greenville	38° 53'N	89° 24'W	Iowa City	41° 38'N	91° 33'W
Joliet	41° 31'N	88° 10'W	Keokuk	40° 24'N	91° 24'W
Kankakee	41° 5'N	87° 55'W	Marshalltown	42° 4'N	92° 56'W
La Salle/Peru	41° 19'N	89° 6'W	Mason City AP	43° 9'N	93° 20'W
Macomb	40° 28'N	90° 40'W	Newton	41° 41'N	93° 2'W
Moline AP	41° 27'N	90° 31'W	Ottumwa AP	41° 6'N	92° 27'W
Mt Vernon	38° 19'N	88° 52'W	Sioux City AP	42° 24'N	96° 23'W
Peoria AP	40° 40'N	89° 41'W	Waterloo	42° 33'N	92° 24'W

	Lat.	Long.
Quincy AP	39° 57'N	91° 12'W
Rantoul, Chanute AFB	40° 18'N	88° 8'W
Rockford	42° 21'N	89° 3'W
Springfield AP	39° 50'N	89° 40'W
Waukegan	42° 21'N	87° 53'W
INDIANA		
Anderson	40° 6'N	85° 37'W
Bedford	38° 51'N	86° 30'W
Bloomington	39° 8'N	86° 37'W
Columbus, Bakalar AFB	39° 16'N	85° 54'W
Crawfordsville	40° 3'N	86° 54'W
Evansville AP	38° 3'N	87° 32'W
Fort Wayne AP	41° 0'N	85° 12'W
Goshen AP	41° 32'N	85° 48'W
Hobart	41° 32'N	87° 15'W
Huntington	40° 53'N	85° 30'W
Indianapolis AP	39° 44'N	86° 17'W
Jeffersonville	38° 17'N	85° 45'W
Kokomo	40° 25'N	86° 3'W
Lafayette	40° 2'N	86° 5'W
La Porte	41° 36'N	86° 43'W
Marion	40° 29'N	85° 41'W
Muncie	40° 11'N	85° 21'W
Peru, Grissom AFB	40° 39'N	86° 9'W
KANSAS		
Atchison	39° 34'N	95° 7'W
Chanute AP	37° 40'N	95° 29'W
Dodge City AP	37° 46'N	99° 58'W
El Dorado	37° 49'N	96° 50'W
Emporia	38° 20'N	96° 12'W
Garden City AP	37° 56'N	100° 44'W
Goodland AP	39° 22'N	101° 42'W
Great Bend	38° 21'N	98° 52'W
Hutchinson AP	38° 4'N	97° 52'W
Liberal	37° 3'N	100° 58'W
Manhattan, Ft Riley	39° 3'N	96° 46'W
Parsons	37° 20'N	95° 31'W
Russell AP	38° 52'N	98° 49'W
Salina	38° 48'N	97° 39'W
Topeka AP	39° 4'N	95° 38'W
Wichita AP	37° 39'N	97° 25'W
KENTUCKY		
Ashland	38° 33'N	82° 44'W
Bowling Green AP	35° 58'N	86° 28'W
Corbin AP	36° 57'N	84° 6'W
Covington AP	39° 3'N	84° 40'W
Hopkinsville, Ft Campbell	36° 40'N	87° 29'W
Lexington AP	38° 2'N	84° 36'W

(Continued)

	LONGITUDE	LATITUDE		LONGITUDE	LATITUDE
KENTUCKY (Continued)			Battle Creek AP	42° 19'N	85° 15'W
Louisville AP	38° 11'N	85° 44'W	Benton Harbor AP	42° 8'N	86° 26'W
Madisonville	37° 19'N	87° 29'W	Detroit	42° 25'N	83° 1'W
Owensboro	37° 45'N	87° 10'W	Escanaba	45° 44'N	87° 5'W
Paducah AP	37° 4'N	88° 46'W	Flint AP	42° 58'N	83° 44'W
LOUISIANA			Grand Rapids AP	42° 53'N	85° 31'W
Alexandria AP	31° 24'N	92° 18'W	Holland	42° 42'N	86° 6'W
Baton Rouge AP	30° 32'N	91° 9'W	Jackson AP	42° 16'N	84° 28'W
Bogalusa	30° 47'N	89° 52'W	Kalamazoo	42° 17'N	85° 36'W
Houma	29° 31'N	90° 40'W	Lansing AP	42° 47'N	84° 36'W
Lafayette AP	30° 12'N	92° 0'W	Marquette Co	46° 34'N	87° 24'W
Lake Charles AP	30° 7'N	93° 13'W	Mt Pleasant	43° 35'N	84° 46'W
Minden	32° 36'N	93° 18'W	Muskegon AP	43° 10'N	86° 14'W
Monroe AP	32° 31'N	92° 2'W	Pontiac	42° 40'N	83° 25'W
Natchitoches	31° 46'N	93° 5'W	Port Huron	42° 59'N	82° 25'W
New Orleans AP	29° 59'N	90° 15'W	Saginaw AP	43° 32'N	84° 5'W
Shreveport AP	32° 28'N	93° 49'W	Sault Ste. Marie AP	46° 28'N	84° 22'W
MAINE			Traverse City AP	44° 45'N	85° 35'W
Augusta AP	44° 19'N	69° 48'W	Ypsilanti	42° 14'N	83° 32'W
Bangor, Dow AFB	44° 48'N	68° 50'W	**MINNESOTA**		
Caribou AP	46° 52'N	68° 1'W	Albert Lea	43° 39'N	93° 21'W
Lewiston	44° 2'N	70° 15'W	Alexandria AP	45° 52'N	95° 23'W
Millinocket AP	45° 39'N	68° 42'W	Bemidji AP	47° 31'N	94° 56'W
Portland	43° 39'N	70° 19'W	Brainerd	46° 24'N	94° 8'W

Location	Latitude	Longitude
Waterville	44° 32'N	69° 40'W
MARYLAND		
Baltimore AP	39° 11'N	76° 40'W
Baltimore Co	39° 20'N	76° 25'W
Cumberland	39° 37'N	78° 46'W
Frederick AP	39° 27'N	77° 25'W
Hagerstown	39° 42'N	77° 44'W
Salisbury	38° 20'N	75° 30'W
MASSACHUSETTS		
Boston AP	42° 22'N	71° 2'W
Clinton	42° 24'N	71° 41'W
Fall River	41° 43'N	71° 8'W
Framingham	42° 17'N	71° 25'W
Gloucester	42° 35'N	70° 41'W
Greenfield	42° 3'N	72° 4'W
Lawrence	42° 42'N	71° 10'W
Lowell	42° 39'N	71° 19'W
New Bedford	41° 41'N	70° 58'W
Pittsfield AP	42° 26'N	73° 18'W
Springfield, Westover AFB	42° 12'N	72° 32'W
Taunton	41° 54'N	71° 4'W
Worcester AP	42° 16'N	71° 52'W
MICHIGAN		
Adrian	41° 55'N	84° 1'W
Alpena AP	45° 4'N	83° 26'W
Duluth AP	46° 50'N	92° 11'W
Faribault	44° 18'N	93° 16'W
Fergus Falls	46° 16'N	96° 4'W
International Falls AP	48° 34'N	93° 23'W
Mankato	44° 9'N	93° 59'W
Minneapolis/St. Paul AP	44° 53'N	93° 13'W
Rochester AP	43° 55'N	92° 30'W
St. Cloud AP	45° 35'N	94° 11'W
Virginia	47° 30'N	92° 33'W
Willmar	45° 7'N	95° 5'W
Winona	44° 3'N	91° 38'W
MISSISSIPPI		
Biloxi—Keesler AFB	30° 25'N	88° 55'W
Clarksdale	34° 12'N	90° 34'W
Columbus AFB	33° 39'N	88° 27'W
Greenville AFB	33° 29'N	90° 59'W
Greenwood	33° 30'N	90° 5'W
Hattiesburg	31° 16'N	89° 15'W
Jackson AP	32° 19'N	90° 5'W
Laurel	31° 40'N	89° 10'W
McComb AP	31° 15'N	90° 28'W
Meridian AP	32° 20'N	88° 45'W
Natchez	31° 33'N	91° 23'W
Tupelo	34° 16'N	88° 46'W
Vicksburg Co	32° 24'N	90° 47'W

(Continued)

	LONGITUDE	LATITUDE
MISSOURI		
Cape Girardeau	37° 14'N	89° 35'W
Columbia AP	38° 58'N	92° 22'W
Farmington AP	37° 46'N	90° 24'W
Hannibal	39° 42'N	91° 21'W
Jefferson City	38° 34'N	92° 11'W
Joplin AP	37° 9'N	94° 30'W
Kansas City AP	39° 7'N	94° 35'W
Kirksville AP	40° 6'N	92° 33'W
Mexico	39° 11'N	91° 54'W
Moberly	39° 24'N	92° 26'W
Poplar Bluff	36° 46'N	90° 25'W
Rolla	37° 59'N	91° 43'W
St. Joseph AP	39° 46'N	94° 55'W
St. Louis AP	38° 45'N	90° 23'W
St. Louis Co	38° 39'N	90° 38'W
Sedalia—Whiteman AFB	38° 43'N	93° 33'W
Sikeston	36° 53'N	89° 36'W
Springfield AP	37° 14'N	93° 23'W
MONTANA		
Billings AP	45° 48'N	108° 32'W
Bozeman	45° 47'N	111° 9'W
Butte AP	45° 57'N	112° 30'W
Cut Bank AP	48° 37'N	112° 22'W

	LONGITUDE	LATITUDE
Sidney AP	41° 13'N	103° 6'W
NEVADA		
Carson City	39° 10'N	119° 46'W
Elko AP	40° 50'N	115° 47'W
Ely AP	39° 17'N	114° 51'W
Las Vegas AP	36° 5'N	115° 10'W
Lovelock AP	40° 4'N	118° 33'W
Reno AP	39° 30'N	119° 47'W
Reno Co	39° 30'N	119° 47'W
Tonopah AP	38° 4'N	117° 5'W
Winnemucca AP	40° 54'N	117° 48'W
NEW HAMPSHIRE		
Berlin	44° 3'N	71° 1'W
Claremont	43° 2'N	72° 2'W
Concord AP	43° 12'N	71° 30'W
Keene	42° 55'N	72° 17'W
Laconia	43° 3'N	71° 3'W
Manchester, Grenier AFB	42° 56'N	71° 26'W
Portsmouth, Pease AFB	43° 4'N	70° 49'W
NEW JERSEY		
Atlantic City Co	39° 23'N	74° 26'W
Long Branch	40° 19'N	74° 1'W
Newark AP	40° 42'N	74° 10'W

Location	Lat	Long	Location	Lat	Long
Glasgow AP	48° 25'N	106° 32'W	New Brunswick	40° 29'N	74° 26'W
Glendive	47° 8'N	104° 48'W	Paterson	40° 54'N	74° 9'W
Great Falls AP	47° 29'N	111° 22'W	Phillipsburg	40° 41'N	75° 11'W
Havre	48° 34'N	109° 40'W	Trenton Co	40° 13'N	74° 46'W
Helena AP	46° 36'N	112° 0'W	Vineland	39° 29'N	75° 0'W
Kalispell AP	48° 18'N	114° 16'W	**NEW MEXICO**		
Lewiston AP	47° 4'N	109° 27'W	Alamogordo, Holloman AFB	32° 51'N	106° 6'W
Livingstown AP	45° 42'N	110° 26'W	Albuquerque AP	35° 3'N	106° 37'W
Miles City AP	46° 26'N	105° 52'W	Artesia	32° 46'N	104° 23'W
Missoula AP	46° 55'N	114° 5'W	Carlsbad AP	32° 20'N	104° 16'W
NEBRASKA			Clovis AP	34° 23'N	103° 19'W
Beatrice	40° 16'N	96° 45'W	Farmington AP	36° 44'N	108° 14'W
Chadron AP	42° 50'N	103° 5'W	Gallup	35° 31'N	108° 47'W
Columbus	41° 28'N	97° 20'W	Grants	35° 10'N	107° 54'W
Fremont	41° 26'N	96° 29'W	Hobbs AP	32° 45'N	103° 13'W
Grand Island AP	40° 59'N	98° 19'W	Las Cruces	32° 18'N	106° 55'W
Hastings	40° 36'N	98° 26'W	Los Alamos	35° 52'N	106° 19'W
Kearney	40° 44'N	99° 1'W	Raton AP	36° 45'N	104° 30'W
Lincoln Co	40° 51'N	96° 45'W	Roswell, Walker AFB	33° 18'N	104° 32'W
McCook	40° 12'N	100° 38'W	Santa Fe Co	35° 37'N	106° 5'W
Norfolk	41° 59'N	97° 26'W	Silver City AP	32° 38'N	108° 10'W
North Platte AP	41° 8'N	100° 41'W	Socorro AP	34° 3'N	106° 53'W
Omaha AP	41° 18'N	95° 54'W	Tucumcari AP	35° 11'N	103° 36'W
Scottsbluff AP	41° 52'N	103° 36'W			

(Continued)

	LONGITUDE	LATITUDE
NEW YORK		
Albany AP	42° 45'N	73° 48'W
Albany Co	42° 39'N	73° 45'W
Auburn	42° 54'N	76° 32'W
Batavia	43° 0'N	78° 11'W
Binghamton AP	42° 13'N	75° 59'W
Buffalo AP	42° 56'N	78° 44'W
Cortland	42° 36'N	76° 11'W
Dunkirk	42° 29'N	79° 16'W
Elmira AP	42° 10'N	76° 54'W
Geneva	42° 45'N	76° 54'W
Glens Falls	43° 20'N	73° 37'W
Gloversville	43° 2'N	74° 21'W
Hornell	42° 21'N	77° 42'W
Ithaca	42° 27'N	76° 29'W
Jamestown	42° 7'N	79° 14'W
Kingston	41° 56'N	74° 0'W
Lockport	43° 9'N	79° 15'W
Massena AP	44° 56'N	74° 51'W
Newburgh, Stewart AFB	41° 30'N	74° 6'W
NYC-Central Park	40° 47'N	73° 58'W
NYC-Kennedy AP	40° 39'N	73° 47'W
NYC-La Guardia AP	40° 46'N	73° 54'W

	LONGITUDE	LATITUDE
Jacksonville	34° 50'N	77° 37'W
Lumberton	34° 37'N	79° 4'W
New Bern AP	35° 5'N	77° 3'W
Raleigh/Durham AP	35° 52'N	78° 47'W
Rocky Mount	35° 58'N	77° 48'W
Wilmington AP	34° 16'N	77° 55'W
Winston-Salem AP	36° 8'N	80° 13'W
NORTH DAKOTA		
Bismarck AP	46° 46'N	100° 45'W
Devils Lake	48° 7'N	98° 54'W
Dickinson AP	46° 48'N	102° 48'W
Fargo AP	46° 54'N	96° 48'W
Grand Forks AP	47° 57'N	97° 24'W
Jamestown AP	46° 55'N	98° 41'W
Minot AP	48° 25'N	101° 21'W
Williston	48° 9'N	103° 35'W
OHIO		
Akron-Canton AP	40° 55'N	81° 26'W
Ashtabula	41° 51'N	80° 48'W
Athens	39° 20'N	82° 6'W
Bowling Green	41° 23'N	83° 38'W
Cambridge	40° 4'N	81° 35'W
Chillicothe	39° 21'N	83° 0'W

Location	Lat	Long	Location	Lat	Long
Niagara Falls AP	43° 6'N	79° 57'W	Cincinnati Co	39° 9'N	84° 31'W
Olean	42° 14'N	78° 22'W	Cleveland AP	41° 24'N	81° 51'W
Oneonta	42° 31'N	75° 4'W	Columbus AP	40° 0'N	82° 53'W
Oswego Co	43° 28'N	76° 33'W	Dayton AP	39° 54'N	84° 13'W
Plattsburg AFB	44° 39'N	73° 28'W	Defiance	41° 17'N	84° 23'W
Poughkeepsie	41° 38'N	73° 55'W	Findlay AP	41° 1'N	83° 40'W
Rochester AP	43° 7'N	77° 40'W	Fremont	41° 20'N	83° 7'W
Rome, Griffiss AFB	43° 14'N	75° 25'W	Hamilton	39° 24'N	84° 35'W
Schenectady	42° 51'N	73° 57'W	Lancaster	39° 44'N	82° 38'W
Suffolk County AFB	40° 51'N	72° 38'W	Lima	40° 42'N	84° 2'W
Syracuse AP	43° 7'N	76° 7'W	Mansfield AP	40° 49'N	82° 31'W
Utica	43° 9'N	75° 23'W	Marion	40° 36'N	83° 10'W
Watertown	43° 59'N	76° 1'W	Middletown	39° 31'N	84° 25'W
NORTH CAROLINA			Newark	40° 1'N	82° 28'W
Asheville AP	35° 26'N	82° 32'W	Norwalk	41° 16'N	82° 37'W
Charlotte AP	35° 13'N	80° 56'W	Portsmouth	38° 45'N	82° 55'W
Durham	35° 52'N	78° 47'W	Sandusky Co	41° 27'N	82° 43'W
Elizabeth City AP	36° 16'N	76° 11'W	Springfield	39° 50'N	83° 50'W
Fayetteville, Pope AFB	35° 10'N	79° 1'W	Steubenville	40° 23'N	80° 38'W
Goldsboro, Seymour-Johnson	35° 20'N	77° 58'W	Toledo AP	41° 36'N	83° 48'W
Greensboro AP	36° 5'N	79° 57'W	Warren	41° 20'N	80° 51'W
Greenville	35° 37'N	77° 25'W	Wooster	40° 47'N	81° 55'W
Henderson	36° 22'N	78° 25'W	Youngstown AP	41° 16'N	80° 40'W
Hickory	35° 45'N	81° 23'W	Zanesville AP	39° 57'N	81° 54'W

(Continued)

	LONGITUDE	LATITUDE		LONGITUDE	LATITUDE
OKLAHOMA			Scranton/Wilkes-Barre	41° 20'N	75° 44'W
Ada	34° 47'N	96° 41'W	State College	40° 48'N	77° 52'W
Altus AFB	34° 39'N	99° 16'W	Sunbury	40° 53'N	76° 46'W
Ardmore	34° 18'N	97° 1'W	Uniontown	39° 55'N	79° 43'W
Bartlesville	36° 45'N	96° 0'W	Warren	41° 51'N	79° 8'W
Chickasha	35° 3'N	97° 55'W	West Chester	39° 58'N	75° 38'W
Enid, Vance AFB	36° 21'N	97° 55'W	Williamsport AP	41° 15'N	76° 55'W
Lawton AP	34° 34'N	98° 25'W	York	39° 55'N	76° 45'W
McAlester	34° 50'N	95° 55'W	**RHODE ISLAND**		
Muskogee AP	35° 40'N	95° 22'W	Newport	41° 30'N	71° 20'W
Norman	35° 15'N	97° 29'W	Providence AP	41° 44'N	71° 26'W
Oklahoma City AP	35° 24'N	97° 36'W	**SOUTH CAROLINA**		
Ponca City	36° 44'N	97° 6'W	Anderson	34° 30'N	82° 43'W
Seminole	35° 14'N	96° 40'W	Charleston AFB	32° 54'N	80° 2'W
Stillwater	36° 10'N	97° 5'W	Charleston Co	32° 54'N	79° 58'W
Tulsa AP	36° 12'N	95° 54'W	Columbia AP	33° 57'N	81° 7'W
Woodward	36° 36'N	99° 31'W	Florence AP	34° 11'N	79° 43'W
OREGON			Georgetown	33° 23'N	79° 17'W
Albany	44° 38'N	123° 7'W	Greenville AP	34° 54'N	82° 13'W
Astoria AP	46° 9'N	123° 53'W	Greenwood	34° 10'N	82° 7'W
Baker AP	44° 50'N	117° 49'W	Orangeburg	33° 30'N	80° 52'W
Bend	44° 4'N	121° 19'W	Rock Hill	34° 59'N	80° 58'W
Corvallis	44° 30'N	123° 17'W	Spartanburg AP	34° 58'N	82° 0'W
Eugene AP	44° 7'N	123° 13'W	Sumter, Shaw AFB	33° 54'N	80° 22'W

	Lat.	Long.		Lat.	Long.
Grants Pass	42° 26'N	123° 19'W	**SOUTH DAKOTA**		
Klamath Falls AP	42° 9'N	121° 44'W	Aberdeen AP	45° 27'N	98° 26'W
Medford AP	42° 22'N	122° 52'W	Brookings	44° 18'N	96° 48'W
Pendleton AP	45° 41'N	118° 51'W	Huron AP	44° 23'N	98° 13'W
Portland AP	45° 36'N	122° 36'W	Mitchell	43° 41'N	98° 1'W
Portland Co	45° 32'N	122° 40'W	Pierre AP	44° 23'N	100° 17'W
Roseburg AP	43° 14'N	123° 22'W	Rapid City AP	44° 3'N	103° 4'W
Salem AP	44° 55'N	123° 1'W	Sioux Falls AP	43° 34'N	96° 44'W
The Dalles	45° 36'N	121° 12'W	Watertown AP	44° 55'N	97° 9'W
PENNSYLVANIA			Yankton	42° 55'N	97° 23'W
Allentown AP	40° 39'N	75° 26'W	**TENNESSEE**		
Altoona Co	40° 18'N	78° 19'W	Athens	35° 26'N	84° 35'W
Butler	40° 52'N	79° 54'W	Bristol-Tri City AP	36° 29'N	82° 24'W
Chambersburg	39° 56'N	77° 38'W	Chattanooga AP	35° 2'N	85° 12'W
Erie AP	42° 5'N	80° 11'W	Clarksville	36° 33'N	87° 22'W
Harrisburg AP	40° 12'N	76° 46'W	Columbia	35° 38'N	87° 2'W
Johnstown	40° 19'N	78° 50'W	Dyersburg	36° 1'N	89° 24'W
Lancaster	40° 7'N	76° 18'W	Greenville	36° 4'N	82° 50'W
Meadville	41° 38'N	80° 10'W	Jackson AP	35° 36'N	88° 55'W
New Castle	41° 1'N	80° 22'W	Knoxville AP	35° 49'N	83° 59'W
Philadelphia AP	39° 53'N	75° 15'W	Memphis AP	35° 3'N	90° 0'W
Pittsburgh AP	40° 30'N	80° 13'W	Murfreesboro	34° 55'N	86° 28'W
Pittsburgh Co	40° 27'N	80° 0'W	Nashville AP	36° 7'N	86° 41'W
Reading Co	40° 20'N	75° 38'W	Tullahoma	35° 23'N	86° 5'W

(Continued)

	LONGITUDE	LATITUDE		LONGITUDE	LATITUDE
TEXAS			Waco AP	31° 37'N	97° 13'W
Abilene AP	32° 25'N	99° 41'W	Wichita Falls AP	33° 58'N	98° 29'W
Alice AP	27° 44'N	98° 2'W	**UTAH**		
Amarillo AP	35° 14'N	100° 42'W	Cedar City AP	37° 42'N	113° 6'W
Austin AP	30° 18'N	97° 42'W	Logan	41° 45'N	111° 49'W
Bay City	29° 0'N	95° 58'W	Moab	38° 36'N	109° 36'W
Beaumont	29° 57'N	94° 1'W	Ogden AP	41° 12'N	112° 1'W
Beeville	28° 22'N	97° 40'W	Price	39° 37'N	110° 50'W
Big Spring AP	32° 18'N	101° 27'W	Provo	40° 13'N	111° 43'W
Brownsville AP	25° 54'N	97° 26'W	Richfield	38° 46'N	112° 5'W
Brownwood	31° 48'N	98° 57'W	St George Co	37° 2'N	113° 31'W
Bryan AP	30° 40'N	96° 33'W	Salt Lake City AP	40° 46'N	111° 58'W
Corpus Christi AP	27° 46'N	97° 30'W	Vernal AP	40° 27'N	109° 31'W
Corsicana	32° 5'N	96° 28'W	**VERMONT**		
Dallas AP	32° 51'N	96° 51'W	Barre	44° 12'N	72° 31'W
Del Rio, Laughlin AFB	29° 22'N	100° 47'W	Burlington AP	44° 28'N	73° 9'W
Denton	33° 12'N	97° 6'W	Rutland	43° 36'N	72° 58'W
Eagle Pass	28° 52'N	100° 32'W	**VIRGINIA**		
El Paso AP	31° 48'N	106° 24'W	Charlottesville	38° 2'N	78° 31'W
Fort Worth AP	32° 50'N	97° 3'W	Danville AP	36° 34'N	79° 20'W
Galveston AP	29° 18'N	94° 48'W	Fredericksburg	38° 18'N	77° 28'W
Greenville	33° 4'N	96° 3'W	Harrisonburg	38° 27'N	78° 54'W
Harlingen	26° 14'N	97° 39'W	Lynchburg AP	37° 20'N	79° 12'W
Houston AP	29° 58'N	95° 21'W	Norfolk AP	36° 54'N	76° 12'W

Location	Latitude	Longitude
Houston Co	29° 59'N	95° 22'W
Huntsville	30° 43'N	95° 33'W
Killeen, Robert Gray AAF	31° 5'N	97° 41'W
Lamesa	32° 42'N	101° 56'W
Laredo AFB	27° 32'N	99° 27'W
Longview	32° 28'N	94° 44'W
Lubbock AP	33° 39'N	101° 49'W
Lufkin AP	31° 25'N	94° 48'W
McAllen	26° 12'N	98° 13'W
Midland AP	31° 57'N	102° 11'W
Mineral Wells AP	32° 47'N	98° 4'W
Palestine Co	31° 47'N	95° 38'W
Pampa	35° 32'N	100° 59'W
Pecos	31° 25'N	103° 30'W
Plainview	34° 11'N	101° 42'W
Port Arthur AP	29° 57'N	94° 1'W
San Angelo,Goodfellow AFB	31° 26'N	100° 24'W
San Antonio AP	29° 32'N	98° 28'W
Sherman, Perrin AFB	33° 43'N	96° 40'W
Snyder	32° 43'N	100° 55'W
Temple	31° 6'N	97° 21'W
Tyler AP	32° 21'N	95° 16'W
Vernon	34° 10'N	99° 18'W
Victoria AP	28° 51'N	96° 55'W
Petersburg	37° 11'N	77° 31'W
Richmond AP	37° 30'N	77° 20'W
Roanoke AP	37° 19'N	79° 58'W
Staunton	38° 16'N	78° 54'W
Winchester	39° 12'N	78° 10'W
WASHINGTON		
Aberdeen	46° 59'N	123° 49'W
Bellingham AP	48° 48'N	122° 32'W
Bremerton	47° 34'N	122° 40'W
Ellensburg AP	47° 2'N	120° 31'W
Everett, Paine AFB	47° 55'N	122° 17'W
Kennewick	46° 13'N	119° 8'W
Longview	46° 10'N	122° 56'W
Moses Lake, Larson AFB	47° 12'N	119° 19'W
Olympia AP	46° 58'N	122° 54'W
Port Angeles	48° 7'N	123° 26'W
Seattle-Boeing Field	47° 32'N	122° 18'W
Seattle Co	47° 39'N	122° 18'W
Seattle-Tacoma AP	47° 27'N	122° 18'W
Spokane AP	47° 38'N	117° 31'W
Tacoma, McChord AFB	47° 15'N	122° 30'W
Walla Walla AP	46° 6'N	118° 17'W
Wenatchee	47° 25'N	120° 19'W
Yakima AP	46° 34'N	120° 32'W

(Continued)

	LONGITUDE	LATITUDE
WEST VIRGINIA		
Beckley	37° 47'N	81° 7'W
Bluefield AP	37° 18'N	81° 13'W
Charleston AP	38° 22'N	81° 36'W
Clarksburg	39° 16'N	80° 21'W
Elkins AP	38° 53'N	79° 51'W
Huntington Co	38° 25'N	82° 30'W
Martinsburg AP	39° 24'N	77° 59'W
Morgantown AP	39° 39'N	79° 55'W
Parkersburg Co	39° 16'N	81° 34'W
Wheeling	40° 7'N	80° 42'W
WISCONSIN		
Appleton	44° 15'N	88° 23'W
Ashland	46° 34'N	90° 58'W
Beloit	42° 30'N	89° 2'W
Eau Claire AP	44° 52'N	91° 29'W
Fond Du Lac	43° 48'N	88° 27'W
Green Bay AP	44° 29'N	88° 8'W
La Crosse AP	43° 52'N	91° 15'W
Madison AP	43° 8'N	89° 20'W

	LONGITUDE	LATITUDE
Manitowoc	44° 6'N	87° 41'W
Marinette	45° 6'N	87° 38'W
Milwaukee AP	42° 57'N	87° 54'W
Racine	42° 43'N	87° 51'W
Sheboygan	43° 45'N	87° 43'W
Stevens Point	44° 30'N	89° 34'W
Waukesha	43° 1'N	88° 14'W
Wausau AP	44° 55'N	89° 37'W
WYOMING		
Casper AP	42° 55'N	106° 28'W
Cheyenne	41° 9'N	104° 49'W
Cody AP	44° 33'N	109° 4'W
Evanston	41° 16'N	110° 57'W
Lander AP	42° 49'N	108° 44'W
Laramie AP	41° 19'N	105° 41'W
Newcastle	43° 51'N	104° 13'W
Rawlins	41° 48'N	107° 12'W
Rock Springs AP	41° 36'N	109° 0'W
Sheridan AP	44° 46'N	106° 58'W
Torrington	42° 5'N	104° 13'W

AP = airport, AFB = air force base.

CANADA LONGITUDES AND LATITUDES

	LONGITUDE	LATITUDE		LONGITUDE	LATITUDE
ALBERTA			Trail	49° 8~ N	117° 44~ W
Calgary AP	51° 6~ N	114° 1~ W	Vancouver AP	49° 11~ N	123° 10~ W
Edmonton AP	53° 34~ N	113° 31~ W	Victoria Co	48° 25~ N	123° 19~ W
Grande Prairie AP	55° 11~ N	118° 53~ W	**MANITOBA**		
Jasper	52° 53~ N	118° 4~ W	Brandon	49° 52~ N	99° 59~ W
Lethbridge AP	49° 38~ N	112° 48~ W	Churchill AP	58° 45~ N	94° 4~ W
McMurray AP	56° 39~ N	111° 13~ W	Dauphin AP	51° 6~ N	100° 3~ W
Medicine Hat AP	50° 1~ N	110° 43~ W	Flin Flon	54° 46~ N	101° 51~ W
Red Deer AP	52° 11~ N	113° 54~ W	Portage La Prairie AP	49° 54~ N	98° 16~ W
BRITISH COLUMBIA			The Pas AP	53° 58~ N	101° 6~ W
Dawson Creek	55° 44~ N	120° 11~ W	Winnipeg AP	49° 54~ N	97° 14~ W
Fort Nelson AP	58° 50~ N	122° 35~ W	**NEW BRUNSWICK**		
Kamloops Co	50° 43~ N	120° 25~ W	Campbellton Co	48° 0~ N	66° 40~ W
Nanaimo	49° 11~ N	123° 58~ W	Chatham AP	47° 1~ N	65° 27~ W
New Westminster	49° 13~ N	122° 54~ W	Edmundston Co	47° 22~ N	68° 20~ W
Penticton AP	49° 28~ N	119° 36~ W	Fredericton AP	45° 52~ N	66° 32~ W
Prince George AP	53° 53~ N	122° 41~ W	Moncton AP	46° 7~ N	64° 41~ W
Prince Rupert Co	54° 17~ N	130° 23~ W	Saint John AP	45° 19~ N	65° 53~ W

(Continued)

	LONGITUDE	LATITUDE		LONGITUDE	LATITUDE
NEWFOUNDLAND			Sudbury AP	46° 37~ N	80° 48~ W
Corner Brook	48° 58~ N	57° 57~ W	Thunder Bay AP	48° 22~ N	89° 19~ W
Gander AP	48° 57~ N	54° 34~ W	Timmins AP	48° 34~ N	81° 22~ W
Goose Bay AP	53° 19~ N	60° 25~ W	Toronto AP	43° 41~ N	79° 38~ W
St John's AP	47° 37~ N	52° 45~ W	Windsor AP	42° 16~ N	82° 58~ W
Stephenville AP	48° 32~ N	58° 33~ W	**PRINCE EDWARD ISLAND**		
NORTHWEST TERRITORIES			Charlottetown AP	46° 17~ N	63° 8~ W
Fort Smith AP	60° 1~ N	111° 58~ W	Summerside AP	46° 26~ N	63° 50~ W
Frobisher AP	63° 45~ N	68° 33~ W	**QUEBEC**		
Inuvik	68° 18~ N	133° 29~ W	Bagotville AP	48° 20~ N	71° 0~ W
Resolute AP	74° 43~ N	94° 59~ W	Chicoutimi	48° 25~ N	71° 5~ W
Yellowknife AP	62° 28~ N	114° 27~ W	Drummondville	45° 53~ N	72° 29~ W
NOVA SCOTIA			Granby	45° 23~ N	72° 42~ W
Amherst	45° 49~ N	64° 13~ W	Hull	45° 26~ N	75° 44~ W
Halifax AP	44° 39~ N	63° 34~ W	Megantic AP	45° 35~ N	70° 52~ W
Kentville	45° 3~ N	64° 36~ W	Montreal AP	45° 28~ N	73° 45~ W
New Glasgow	45° 37~ N	62° 37~ W	Quebec AP	46° 48~ N	71° 23~ W
Sydney AP	46° 10~ N	60° 3~ W	Rimouski	48° 27~ N	68° 32~ W
Truro Co	45° 22~ N	63° 16~ W	St Jean	45° 18~ N	73° 16~ W
Yarmouth AP	43° 50~ N	66° 5~ W	St Jerome	45° 48~ N	74° 1~ W

ONTARIO

Belleville	44° 9~ N	77° 24~ W
Chatham	42° 24~ N	82° 12~ W
Cornwall	45° 1~ N	74° 45~ W
Hamilton	43° 16~ N	79° 54~ W
Kapuskasing AP	49° 25~ N	82° 28~ W
Kenora AP	49° 48~ N	94° 22~ W
Kingston	44° 16~ N	76° 30~ W
Kitchener	43° 26~ N	80° 30~ W
London AP	43° 2~ N	81° 9~ W
North Bay AP	46° 22~ N	79° 25~ W
Oshawa	43° 54~ N	78° 52~ W
Ottawa AP	45° 19~ N	75° 40~ W
Owen Sound	44° 34~ N	80° 55~ W
Peterborough	44° 17~ N	78° 19~ W
St Catharines	43° 11~ N	79° 14~ W
Sarnia	42° 58~ N	82° 22~ W
Sault Ste Marie AP	46° 32~ N	84° 30~ W
Sept. Iles AP	50° 13~ N	66° 16~ W
Shawinigan	46° 34~ N	72° 43~ W
Sherbrooke Co	45° 24~ N	71° 54~ W
Thetford Mines	46° 4~ N	71° 19~ W
Trois Rivieres	46° 21~ N	72° 35~ W
Val D'or AP	48° 3~ N	77° 47~ W
Valleyfield	45° 16~ N	74° 6~ W

SASKATCHEWAN

Estevan AP	49° 4~ N	103° 0~ W
Moose Jaw AP	50° 20~ N	105° 33~ W
North Battleford AP	52° 46~ N	108° 15~ W
Prince Albert AP	53° 13~ N	105° 41~ W
Regina AP	50° 26~ N	104° 40~ W
Saskatoon AP	52° 10~ N	106° 41~ W
Swift Current AP	50° 17~ N	107° 41~ W
Yorkton AP	51° 16~ N	102° 28~ W

YUKON TERRITORY

Whitehorse AP	60° 43~ N	135° 4~ W

INTERNATIONAL LONGITUDES AND LATITUDES

	LONGITUDE	LATITUDE
AFGHANISTAN		
Kabul	34° 35~ N	69° 12~ E
ALGERIA		
Algiers	36° 46~ N	30° 3~ E
ARGENTINA		
Buenos Aires	34° 35~ S	58° 29~ W
Cordoba	31° 22~ S	64° 15~ W
Tucuman	26° 50~ S	65° 10~ W
AUSTRALIA		
Adelaide	34° 56~ S	138° 35~ E
Alice Springs	23° 48~ S	133° 53~ E
Brisbane	27° 28~ S	153° 2~ E
Darwin	12° 28~ S	130° 51~ E
Melbourne	37° 49~ S	144° 58~ E
Perth	31° 57~ S	115° 51~ E
Sydney	33° 52~ S	151° 12~ E
AUSTRIA		
Vienna	48° 15~ N	16° 22~ E
AZORES		
Lajes (Terceira)	38° 45~ N	27° 5~ W
BAHAMAS		
Nassau	25° 5~ N	77° 21~ W

	LONGITUDE	LATITUDE
BURMA		
Mandalay	21° 59~ N	96° 6~ E
Rangoon	16° 47~ N	96° 9~ E
CAMBODIA		
Phnom Penh	11° 33~ N	104° 51~ E
CHILE		
Punta Arenas	53° 10~ S	70° 54~ W
Santiago	33° 27~ S	70° 42~ W
Valparaiso	33° 1~ S	71° 38~ W
CHINA		
Chongquing	29° 33~ N	106° 33~ E
Shanghai	31° 12~ N	121° 26~ E
COLOMBIA		
Baranquilla	10° 59~ N	74° 48~ W
Bogota	4° 36~ N	74° 5~ W
Cali	3° 25~ N	76° 30~ W
Medellin	6° 13~ N	75° 36~ W
CONGO		
Brazzaville	4° 15~ S	15° 15~ E
CUBA		
Guantanamo Bay	19° 54~ N	75° 9~ W
Havana	23° 8~ N	82° 21~ W

Location	Latitude	Longitude
BANGLADESH		
Chittagong	22° 21~ N	91° 50~ E
BELGIUM		
Brussels	50° 48~ N	4°21~ E
BELIZE		
Belize	17° 31~ N	88° 11~ W
BERMUDA		
Kindley AFB	33° 22~ N	64° 41~ W
BOLIVIA		
La Paz	16° 30~ S	68° 9~ W
BRAZIL		
Belem	1° 27~ S	48° 29~ W
Belo Horizonte	19° 56~ S	43° 57~ W
Brasilia	15° 52~ S	47° 55~ W
Curitiba	25° 25~ S	49° 17~ W
Fortaleza	3° 46~ S	38° 33~ W
Porto Alegre	30° 2~ S	51° 13~ W
Recife	8° 4~ S	34° 53~ W
Rio de Janeiro	22° 55~ S	43° 12~ W
Salvador	13° 0~ S	38° 30~ W
Sao Paulo	23° 33~ S	46° 38~ W
BULGARIA		
Sofia	42° 42~ N	23° 20~ E
CZECHOSLOVAKIA		
Prague	50° 5~ N	14° 25~ E
DENMARK		
Copenhagen	55° 41~ N	12° 33~ E
DOMINICAN REPUBLIC		
Santo Domingo	18° 29~ N	69° 54~ W
EGYPT		
Cairo	29° 52~ N	31° 20~ E
EL SALVADOR		
San Salvador	13° 42~ N	89° 13~ W
EQUADOR		
Guayaquil	2° 0~ S	79° 53~ W
Quito	0° 13~ S	78° 32~ W
ETHIOPIA		
Addis Ababa	90° 2~ N	38° 45~ E
Asmara	15° 17~ N	38° 55~ E
FINLAND		
Helsinki	60° 10~ N	24° 57~ E
FRANCE		
Lyon	45° 42~ N	4° 47~ E
Marseilles	43° 18~ N	5° 23~ E
Nantes	47° 15~ N	1° 34~ W
Nice	43° 42~ N	7° 16~ E
Paris	48° 49~ N	2° 29~ E
Strasbourg	48° 35~ N	7° 46~ E

	LONGITUDE	LATITUDE
FRENCH GUIANA		
Cayenne	4° 56~ N	52° 27~ W
GERMANY		
Berlin (West)	52° 27~ N	13° 18~ E
Hamburg	53° 33~ N	9° 58~ E
Hannover	52° 24~ N	9° 40~ E
Mannheim	49° 34~ N	8° 28~ E
Munich	48° 9~ N	11° 34~ E
GHANA		
Accra	5° 33~ N	0° 12~ W
GIBRALTAR		
Gibraltar	36° 9~ N	5° 22~ W
GREECE		
Athens	37° 58~ N	23° 43~ E
Thessaloniki	40° 37~ N	22° 57~ E
GREENLAND		
Narsarssuaq	61° 11~ N	45° 25~ W
GUATEMALA		
Guatemala City	14° 37~ N	90° 31~ W
GUYANA		
Georgetown	6° 50~ N	58° 12~ W

	LONGITUDE	LATITUDE
IRAN		
Abadan	30° 21~ N	48° 16~ E
Meshed	36° 17~ N	59° 36~ E
Tehran	35° 41~ N	51° 25~ E
IRAQ		
Baghdad	33° 20~ N	44° 24~ E
Mosul	36° 19~ N	43° 9~ E
IRELAND		
Dublin	53° 22~ N	6° 21~ W
Shannon	52° 41~ N	8° 55~ W
IRIAN BARAT		
Manokwari	0° 52~ S	134° 5~ E
ISRAEL		
Jerusalem	31° 47~ N	35° 13~ E
Tel Aviv	32° 6~ N	34° 47~ E
ITALY		
Milan	45° 27~ N	9° 17~ E
Naples	40° 53~ N	14° 18~ E
Rome	41° 48~ N	12° 36~ E
IVORY COAST		
Abidjan	5° 19~ N	4° 1~ W

Place	Latitude	Longitude
HAITI		
Port au Prince	18° 33~ N	72° 20~ W
HONDURAS		
Tegucigalpa	14° 6~ N	87° 13~ W
HONG KONG		
Hong Kong	22° 18~ N	114° 10~ E
HUNGARY		
Budapest	47° 31~ N	19° 2~ E
ICELAND		
Reykjavik	64° 8~ N	21° 56~ E
INDIA		
Ahmenabad	23° 2~ N	72° 35~ E
Bangalore	12° 57~ N	77° 37~ E
Bombay	18° 54~ N	72° 49~ E
Calcutta	22° 32~ N	88° 20~ E
Madras	13° 4~ N	80° 15~ E
Nagpur	21° 9~ N	79° 7~ E
New Delhi	28° 35~ N	77° 12~ E
INDONESIA		
Djakarta	6° 11~ S	106° 50~ E
Kupang	10° 10~ S	123° 34~ E
Makassar	5° 8~ S	119° 28~ E
Medan	3° 35~ N	98° 41~ E
Palembang	3° 0~ S	104° 46~ E
Surabaya	7° 13~ S	112° 43~ E
JAPAN		
Fukuoka	33° 35~ N	130° 27~ E
Sapporo	43° 4~ N	141° 21~ E
Tokyo	35° 41~ N	139° 46~ E
JORDAN		
Amman	31° 57~ N	35° 57~ E
KENYA		
Nairobi	1° 16~ S	36° 48~ E
KOREA		
Pyongyang	39° 2~ N	125° 41~ E
Seoul	37° 34~ N	126° 58~ E
LEBANON		
Beirut	33° 54~ N	35° 28~ E
LIBERIA		
Monrovia	6° 18~ N	10° 48~ W
LIBYA		
Benghazi	32° 6~ N	20° 4~ E
MADAGASCAR		
Tananarive	18° 55~ S	47° 33~ E
MALAYSIA		
Kuala Lumpur	3° 7~ N	101° 42~ E
Penang	5° 25~ N	100° 19~ E
MARTINIQUE		
Fort de France	14° 37~ N	61° 5~ W

(Continued)

	LONGITUDE	LATITUDE		LONGITUDE	LATITUDE
MEXICO			**RUSSIA**		
Guadalajara	20° 41~ N	103° 20~ W	Alma Ata	43° 14~ N	76° 53~ E
Merida	20° 58~ N	89° 38~ W	Archangel	64° 33~ N	40° 32~ E
Mexico City	19° 24~ N	99° 12~ W	Kaliningrad	54° 43~ N	20° 30~ E
Monterrey	25° 40~ N	100° 18~ W	Krasnoyarsk	56° 1~ N	92° 57~ E
Vera Cruz	19° 12~ N	96° 8~ W	Kiev	50° 27~ N	30° 30~ E
MOROCCO			Kharkov	50° 0~ N	36° 14~ E
Casablanca	33° 35~ N	7° 39~ W	Kuibyshev	53° 11~ N	50° 6~ E
NEPAL			Leningrad	59° 56~ N	30° 16~ E
Katmandu	27° 42~ N	85° 12~ E	Minsk	53° 54~ N	27° 33~ E
NETHERLANDS			Moscow	55° 46~ N	37° 40~ E
Amsterdam	52° 23~ N	4° 55~ E	Odessa	46° 29~ N	30° 44~ E
NEW ZEALAND			Petropavlovsk	52° 53~ N	158° 42~ E
Auckland	36° 51~ S	174° 46~ E	Rostov on Don	47° 13~ N	39° 43~ E
Christchurch	43° 32~ S	172° 37~ E	Sverdlovsk	56° 49~ N	60° 38~ E
Wellington	41° 17~ S	174° 46~ E	Tashkent	41° 20~ N	69° 18~ E
NICARAGUA			Tbilisi	41° 43~ N	44° 48~ E
Managua	12° 10~ N	86° 15~ W	Vladivostok	43° 7~ N	131° 55~ E
NIGERIA			Volgograd	48° 42~ N	44° 31~ E
Lagos	6° 27~ N	3° 24~ E	**SAUDI ARABIA**		
NORWAY			Dhahran	26° 17~ N	50° 9~ E
Bergen	60° 24~ N	5° 19~ E	Jedda	21° 28~ N	39° 10~ E

City	Latitude	Longitude
Oslo	59° 56~ N	10° 44~ E
PAKISTAN		
Karachi	24° 48~ N	66° 59~ E
Lahore	31° 35~ N	74° 20~ E
Peshwar	34° 1~ N	71° 35~ E
PANAMA		
Panama City	8° 58~ N	79° 33~ W
PAPUA NEW GUINEA		
Port Moresby	9° 29~ S	147° 9~ E
PARAGUAY		
Asuncion	25° 17~ S	57° 30~ W
PERU		
Lima	12° 5~ S	77° 3~ W
PHILIPPINES		
Manila	14° 35~ N	120° 59~ E
POLAND		
Krakow	50° 4~ N	19° 57~ E
Warsaw	52° 13~ N	21° 2~ E
PORTUGAL		
Lisbon	38° 43~ N	9° 8~ W
PUERTO RICO		
San Juan	18° 29~ N	66° 7~ W
RUMANIA		
Bucharest	44° 25~ N	26° 6~ E
Riyadh	24° 39~ N	46° 42~ E
SENEGAL		
Dakar	14° 42~ N	17° 29~ W
SINGAPORE		
Singapore	1° 18~ N	103° 50~ E
SOMALIA		
Mogadiscio	2° 2~ N	49° 19~ E
SOUTH AFRICA		
Cape Town	33° 56~ S	18° 29~ E
Johannesburg	26° 11~ S	28° 3~ E
Pretoria	25° 45~ S	28° 14~ E
SOUTH YEMEN		
Aden	12° 50~ N	45° 2~ E
SPAIN		
Barcelona	41° 24~ N	2° 9~ E
Madrid	40° 25~ N	3° 41~ W
Valencia	39° 28~ N	0° 23~ W
SRI LANKA		
Colombo	6° 54~ N	79° 52~ E
SUDAN		
Khartoum	15° 37~ N	32° 33~ E
SURINAM		
Paramaribo	5° 49~ N	55° 9~ W
SWEDEN		
Stockholm	59° 21~ N	18° 4~ E

(Continued)

	LONGITUDE	LATITUDE
SWITZERLAND		
Zurich	47° 23~ N	8° 33~ E
SYRIA		
Damascus	33° 30~ N	36° 20~ E
TAIWAN		
Tainan	22° 57~ N	120° 12~ E
Taipei	25° 2~ N	121° 31~ E
TANZANIA		
Dar es Salaam	6° 50~ S	39° 18~ E
THAILAND		
Bangkok	13° 44~ N	100° 30~ E
TRINIDAD		
Port of Spain	10° 40~ N	61° 31~ W
TUNISIA		
Tunis	36° 47~ N	10° 12~ E
TURKEY		
Adana	36° 59~ N	35° 18~ E
Ankara	39° 57~ N	32° 53~ E
Istanbul	40° 58~ N	28° 50~ E
Izmir	38° 26~ N	27° 10~ E
UNITED KINGDOM		
Belfast	54° 36~ N	5° 55~ W

	LONGITUDE	LATITUDE
Birmingham	52° 29~ N	1° 56~ W
Cardiff	51° 28~ N	3° 10~ W
Edinburgh	55° 55~ N	3° 11~ W
Glasgow	55° 52~ N	4° 17~ W
London	51° 29~ N	0° 0~ W
URUGUAY		
Montevideo	34° 51~ S	56° 13~ W
VENEZUELA		
Caracas	10° 30~ N	66° 56~ W
Maracaibo	10° 39~ N	71° 36~ W
VIETNAM		
Da Nang	16° 4~ N	108° 13~ E
Hanoi	21° 2~ N	105° 52~ E
Ho Chi Minh City (Saigon)	10° 47~ N	106° 42~ E
YUGOSLAVIA		
Belgrade	44° 48~ N	20° 28~ E
ZAIRE		
Kinshasa (Leopoldville)	4° 20~ S	15° 18~ E
Kisangani (Stanleyville)	0° 26~ S	15° 14~ E

PHOTOVOLTAIC SYSTEM SUPPORT
HARDWARE AND PHOTO GALLERY

The photographs and graphics of Figures B.1 to B.17 are courtesy of UNIRAC Corporation.

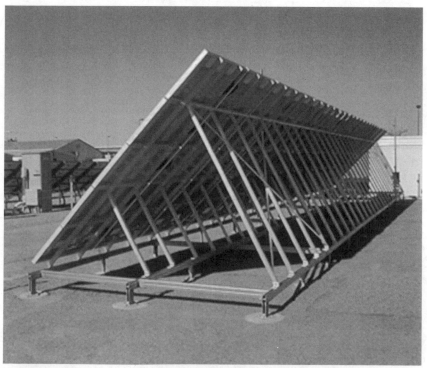

Figure B.1 Roof-mount semiadjustable tilt PV supports structure.

Figure B.2   Ground-mount semiadjustable tilt PV supports structure.

Figure B.3   Roof-mount fixed tilt PV support structure.

Figure B.4   Field-mount semiadjustable tilt PV support structure.

Figure B.5   Pipe-mounted semiadjustable tilt PV support structure.

Figure B.6   Pipe-mounted fixed tilt PV support structure.

Figure B.7   Pipe-mounted fixed tilt PV support hardware.

Figure B.8   Pipe-mounted semiadjustable tilt PV support structure.

Figure B.9   Pipe-mounted manually adjustable tilt PV support structure.

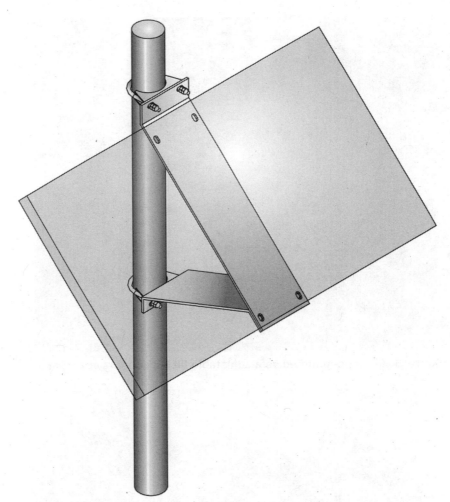

Figure B.10   Ground-mount fixed tilt PV support system.

Figure B.11   Ground-mount fixed tilt PV support system hardware detail.

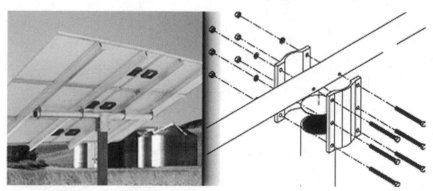

Figure B.12   Ground-mount fixed tilt PV support system graphics.

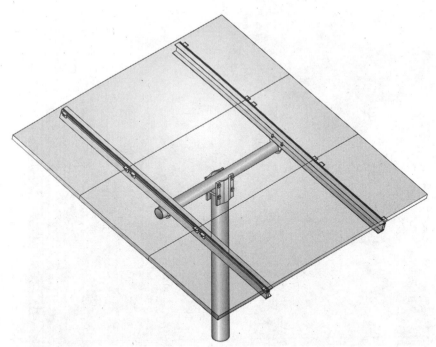

Figure B.13   Roof-mount fixed tilt PV support system using simple channel hardware.

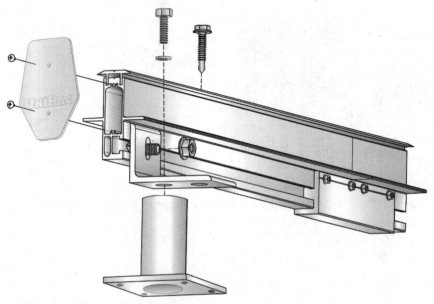

Figure B.14   Railing hardware.

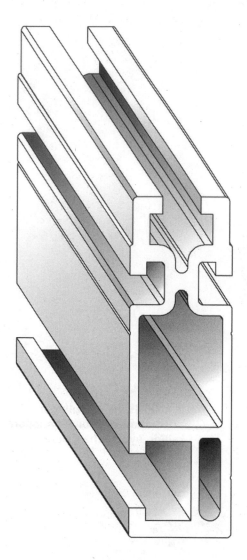

Figure B.15   Cross section of a reinforced PV support railing.

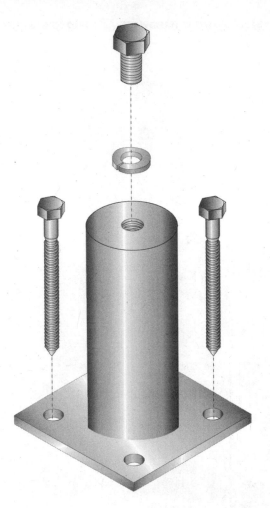

Figure B.16 Railing support stand-offs for PV support railing system hardware.

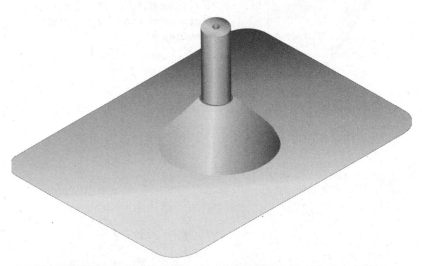

Figure B.17 Waterproof boots for PV support rail stand-offs.

Figure B.18   Desert floor mount solar power installation.

Photo courtesy of Shell Solar present SolarWorld.

Figure B.19   Solar power cogeneration for agricultural water irrigation.
Photo courtesy of WaterWorld.

Figure B.20   Solar power cogeneration for water irrigation.
Photo courtesy of WaterWorld.

Figure B.21   25 KW Solar mega concentrator power co-generation farm. *Photo courtesy of AMONIX.*

Figure B.22   30 KW Solar mega concentrator power co-generation farm. *Photo courtesy of AMONIX.*

Figure B.23   Solar power system installation blended in rock bolder.
Photo courtesy of California Green.

Figure B.24   Solar power farm installation in Mojave Desert.
Photo courtesy of Grant Electric.

Figure B.25   Solar BIPV integration in building roof structure.
Photo courtesy of Atlantis Energy Systems.

Figure B.26   Solar power roof slate installation in residential building roof structure. *Photo courtesy of Atlantis Energy Systems.*

Figure B.27 Laminated glass solar power installation in residential building roof structure. *Photo courtesy of Atlantis Energy Systems.*

Figure B.28 Laminated glass solar power installation in residential building roof structure. *Photo courtesy of Sharp Solar.*

Figure B.29   **Laminated glass BIPV solar power installation in commercial building entrance.** *Photo courtesy of Golden Solar Energy.*

Figure B.30   **BIPV solar power canopy installation.** *Photo courtesy of Atlantis Energy Systems.*

Figure B.31   BIPV solar power logia in Water and Life Museum. *Photo courtesy of Fotoworks.*

Figure B.32   10 single axis tracker Bavaria Solar Park. *Photo courtesy of Sun Power.*

Figure B.33   Roof-mount non penetrating platform solar power system. *Photo courtesy of Sun Power.*

Figure B.34   Ground-mount single axis tracker solar power system.
Photo courtesy of Sun Power.

Figure B.35   Close-up of ground-mount single axis tracker solar power system. *Photo courtesy of Sun Power.*

C

CALIFORNIA ENERGY COMMISSION
CERTIFIED EQUIPMENT

Certified Photovoltaic Modules

TABLE C.1   LIST OF ELIGIBLE PHOTOVOLTAIC MODULES CALIFORNIA ENERGY COMMISSION EMERGING

RENEWABLES PROGRAM (FEBRUARY 2005)

MANUFACTURER NAME	MODULE MODEL NUMBER	DESCRIPTION	CEC PTC* RATING	NOTES
ASE Americas, Inc.	ASE-100-ATF/17-34	100W/17V EFG Module, framed	89.7	NA
ASE Americas, Inc.	ASE-300-DGF/17-285	285W/17V EFG Module, framed	255.4	NA
ASE Americas, Inc.	ASE-300-DGF/17-300	300W/17V EFG Module, framed	269.1	NA
ASE Americas, Inc.	ASE-300-DGF/17-315	315W/17V EFG Module, framed	283	NA
ASE Americas, Inc.	ASE-300-DGF/25-145	145W/25V EFG Module, framed	128.5	NA
ASE Americas, Inc.	ASE-300-DGF/34-195	195W/34V EFG Module, framed	173.5	NA
ASE Americas, Inc.	ASE-300-DGF/42-240	240W/42V EFG Module, framed	214	NA
ASE Americas, Inc.	ASE-300-DGF/50-260	260W/50V EFG Module, framed	232.6	NA
ASE Americas, Inc.	ASE-300-DGF/50-265	265W/50V EFG Module, framed	237	NA
ASE Americas, Inc.	ASE-300-DGF/50-285	285W/50V EFG Module, framed	255.4	NA
ASE Americas, Inc.	ASE-300-DGF/50-300	300W/50V EFG Module, framed	269.1	NA
ASE Americas, Inc.	ASE-300-DGF/50-315	315W/50V EFG Module, framed	282.8	NA
AstroPower, Inc.	AP-100	100W Single-Crystal Module (was AP-1006)	88.9	NA
AstroPower, Inc.	AP-1006	100W Single-Crystal Module	88.9	NA
AstroPower, Inc.	AP-110	110W Single-Crystal Module (was AP-1106)	97.9	NA
AstroPower, Inc.	AP-1106	110W Single-Crystal Module	97.9	NA
AstroPower, Inc.	AP-120	120W Single-Crystal Module (was AP-1206)	107	NA
AstroPower, Inc.	AP-1206	120W Single-Crystal Module	107	NA

AstroPower, Inc.	AP-50-GA	50W Single-Crystal Module	44	NA
AstroPower, Inc.	AP-50-GT	50W Single-Crystal Module	44	NA
AstroPower, Inc.	AP-55-GA	55W Single-Crystal Module	48.5	NA
AstroPower, Inc.	AP-55-GT	55W Single-Crystal Module	48.5	NA
AstroPower, Inc.	AP-6105	65W Single-Crystal Module	58.8	NA
AstroPower, Inc.	AP-65	65W Single-Crystal Module (was AP-6105)	58.8	NA
AstroPower, Inc.	AP-7105	75W Single-Crystal Module	68	NA
AstroPower, Inc.	AP-75	75W Single-Crystal Module (was AP-7105)	68	NA
AstroPower, Inc.	AP6-160	160W Single-Crystal Module	141.1	NA
AstroPower, Inc.	AP6-170	170W Single-Crystal Module	150.1	NA
AstroPower, Inc.	APi-030-MNA	30W Single-Crystal Module w/o connectors (was AP-30)	26.5	NA
AstroPower, Inc.	APi-030-MNB	30W Single-Crystal Module w/o connectors (was AP-30) B	26.5	NA
AstroPower, Inc.	APi-045-MNA	45W Single-Crystal Module w/o connectors (was AP-45)	39.7	NA
AstroPower, Inc.	APi-045-MNB	45W Single-Crystal Module w/o connectors (was AP-45) B	39.7	NA
AstroPower, Inc.	APi-050-MNA	50W Single-Crystal Module w/o connectors (was AP-50)	44.2	NA
AstroPower, Inc.	APi-050-MNB	50W Single-Crystal Module w/o connectors (was AP-50) B	44.2	NA
AstroPower, Inc.	APi-055-GCA	55W Single-Crystal Module w/MC connectors (was AP-50)	48.9	NA

(Continued)

TABLE C.1 LIST OF ELIGIBLE PHOTOVOLTAIC MODULES CALIFORNIA ENERGY COMMISSION EMERGING *(Continued)*

RENEWABLES PROGRAM (FEBRUARY 2005)

MANUFACTURER NAME	MODULE MODEL NUMBER	DESCRIPTION	CEC PTC* RATING	NOTES
AstroPower, Inc.	APi-055-GCB	55W Single-Crystal Module w/MC connectors (was AP-55-GA)	48.9	NA
AstroPower, Inc.	APi-065-MNA	65W Single-Crystal Module w/o connectors (was AP-65) w/o connectors (was AP-50) B	57.7	NA
AstroPower, Inc.	APi-065-MNB	65W Single-Crystal Module w/o connectors (was AP-65) B	57.7	NA
AstroPower, Inc.	APi-070-MNA	70W Single-Crystal Module w/o connectors (was AP-70)	62.2	NA
AstroPower, Inc.	APi-070-MNB	70W Single-Crystal Module w/o connectors (was AP-70) B	62.2	NA
AstroPower, Inc.	APi-100-MCA	100W Single-Crystal Module w/MC connectors (was AP-100)	88.7	NA
AstroPower, Inc.	APi-100-MCB	100W Single-Crystal Module w/MC connectors (was AP-100) B	88.7	NA
AstroPower, Inc.	APi-100-MNA	100W Single-Crystal Module w/o connectors (was AP-100)	88.7	NA
AstroPower, Inc.	APi-100-MNB	100W Single-Crystal Module w/o connectors (was AP-100) B	88.7	NA
AstroPower, Inc.	APi-110-MNB	110W Single-Crystal Module w/o connectors (was AP-110) B	97.8	NA
AstroPower, Inc.	APi-110-MCA	110W Single-Crystal Module w/MC connectors (was AP-110)	97.8	NA

AstroPower, Inc.	APi-110-MCB	110W Single-Crystal Module w/MC connectors (was AP-110) B	97.8	NA
AstroPower, Inc.	APi-110-MNA	110W Single-Crystal Module w/o connectors (was AP-110)	97.8	NA
AstroPower, Inc.	APi-165-MCA	165W Single-Crystal Module w/MC connectors (was AP-165)	146.7	NA
AstroPower, Inc.	APi-165-MCB	165W Single-Crystal Module w/MC connectors (was AP-165) B	146.7	NA
AstroPower, Inc.	APi-173-MCA	173W Single-Crystal Module w/MC connectors (was AP-173)	154	NA
AstroPower, Inc.	APi-173-MCB	173W Single-Crystal Module w/MC connectors (was AP-173) B	154	NA
AstroPower, Inc.	APx-045-MNA	45W Apex Module w/o connectors (was APx-45)	38.6	NA
AstroPower, Inc.	APx-045-MNB	45W Apex Module w/o connectors (was APx-45) B	38.6	NA
AstroPower, Inc.	APx-050-MNA	50W Apex Module w/o connectors (was APx-50)	42.9	NA
AstroPower, Inc.	APx-065-MNA	65W Apex Module w/o connectors (was APx-65)	55.7	NA
AstroPower, Inc.	APx-065-MNB	65W Apex Module w/o connectors (was APx-65) B	55.7	NA
AstroPower, Inc.	APx-070-MNA	70W Apex Module w/o connectors (was APx-70)	60.1	NA
AstroPower, Inc.	APx-070-MNB	70W Apex Module w/o connectors (was APx-70) B	60.1	NA
AstroPower, Inc.	APx-075-MNA	75W Apex Module w/o connectors (was APx-75)	64.4	NA

(Continued)

TABLE C.1 LIST OF ELIGIBLE PHOTOVOLTAIC MODULES CALIFORNIA ENERGY COMMISSION EMERGING (Continued)

RENEWABLES PROGRAM (FEBRUARY 2005)

MANUFACTURER NAME	MODULE MODEL NUMBER	DESCRIPTION	CEC PTC* RATING	NOTES
AstroPower, Inc.	APx-075-MNB	75W Apex Module w/o connectors (was APx-75) B	64.4	NA
AstroPower, Inc.	APx-130	130W Apex Silicon Film Module	112	NA
AstroPower, Inc.	APx-130-MCA	130W Apex Silicon Film Module w/MC connectors (was APx-130)	111.6	NA
AstroPower, Inc.	APx-130-MCB	130W Apex Silicon Film Module w/MC connectors (was APx-130) B w/o connectors (was APx-65)	111.6	NA
AstroPower, Inc.	APx-130-MNA	130W Apex Silicon Film Module w/o connectors (was APx-130)	111.6	NA
AstroPower, Inc.	APx-130-MNB	130W Apex Silicon Film Module w/o connectors (was APx-130) B	111.6	NA
AstroPower, Inc.	APx-140	140W Apex Silicon Film Module	121	NA
AstroPower, Inc.	APx-140-MCA	140W Apex Silicon Film Module w/MC connectors (was APx-140)	120.4	NA
AstroPower, Inc.	APx-140-MCB	140W Apex Silicon Film Module w/MC connectors (was APx-140) B	120.4	NA
AstroPower, Inc.	APx-140-MNA	140W Apex Silicon Film Module w/o connectors (was APx-140)	120.4	NA
AstroPower, Inc.	APx-140-MNB	140W Apex Silicon Film Module w/o connectors (was APx-140) B	120.4	NA
AstroPower, Inc.	APx-45	45W Apex Module	38.8	NA
AstroPower, Inc.	APx-50	50W Apex Module	43.2	NA

Manufacturer	Model	Description		
AstroPower, Inc.	APx-65	65W Apex Module	56	NA
AstroPower, Inc.	APX-75	75W Apex Module	64.8	NA
AstroPower, Inc.	APx-75	75W Apex Module	64.8	NA
AstroPower, Inc.	APX-80	80W Apex Module	69.2	NA
AstroPower, Inc.	APx-90	90W Apex Module	77.8	NA
AstroPower, Inc.	APx050-MNB	50W Apex Module w/o connectors (was APx-50) B	42.9	NA
AstroPower, Inc.	LAP-425	425W Single-Crystal Large Area Panel, frameless	378	NA
AstroPower, Inc.	LAP-440	440W Single-Crystal Large Area Panel, frameless	391.8	NA
AstroPower, Inc.	LAP-460	460W Single-Crystal Large Area Panel, frameless	409.9	NA
AstroPower, Inc.	LAP-480	480W Single-Crystal Large Area Panel, frameless	428.2	NA
AstroPower, Inc.	LAPX-300	300W Apex Large Area Panel, Frameless	259.2	NA
Atlantis Energy, Inc.	AP-F	11.8W Shingle Module (AstroPower cells)	10.7	NA
Atlantis Energy, Inc.	AP-G	12.0W Shingle Module (AstroPower cells)	10.8	NA
Atlantis Energy, Inc.	AP-H	12.2W Shingle Module (AstroPower cells)	11	NA
Atlantis Energy, Inc.	SM-II	12.2W Shingle Module (Siemens cells)	11	NA
Atlantis Energy, Inc.	SP-A	13.3W Shingle Module (Sharp cells)	11.4	NA
Atlantis Energy, Inc.	SP-B	12.7W Shingle Module (Sharp cells)	10.9	NA
Atlantis Energy, Inc.	SP-C	11.9W Shingle Module (Sharp cells)	10.2	NA
Atlantis Energy, Inc.	SX-D	11.6W Shingle Module (Solarex cells)	10.5	NA
Atlantis Energy, Inc.	SX-E	11.0W Shingle Module (Solarex cells)	9.9	NA

(Continued)

TABLE C.1 LIST OF ELIGIBLE PHOTOVOLTAIC MODULES CALIFORNIA ENERGY COMMISSION EMERGING (Continued)

RENEWABLES PROGRAM (FEBRUARY 2005)

MANUFACTURER NAME	MODULE MODEL NUMBER	DESCRIPTION	CEC PTC* RATING	NOTES
Baoding Yingli New Energy Resources Co. Ltd.	110(17)P1447X663	110W Crystalline Silicon Solar Cells	96.6	NA
[LM2]Baoding Yingli New Energy Resources Co. Ltd.	120(17)P1447X663	120W Crystalline Silicon Solar Cells	105.6	NA
Baoding Yingli New Energy Resources Co. Ltd.	30(17)P754X350	30W Crystalline Silicon Solar Cells	26.3	NA
Baoding Yingli New Energy Resources Co. Ltd.	40(17)P516X663	40W Crystalline Silicon Solar Cells	35.1	NA
Baoding Yingli New Energy Resources Co. Ltd.	50(17)P974X453	50W Crystalline Silicon Solar Cells	43.9	NA
Baoding Yingli New Energy Resources Co. Ltd.	75(17)P1172X541	75W Crystalline Silicon Solar Cells	65.9	NA
Baoding Yingli New Energy Resources Co. Ltd.	85(17)P1172X541	85W Crystalline Silicon Solar Cells	74.9	NA
BP Solar	BP SX 140S	140W 24V Polycrystalline Module w/multicontact connector.	122.1	NA
BP Solar	BP SX 150S	150W 24V Polycrystalline Module w/multicontact conn.	131.1	NA
BP Solar	BP2140S	140W 24V Single-Crystal Module w/multicontact conn.	122.1	NA
BP Solar	BP2150S	150W 24V Single-Crystal Module w/multicontact conn.	131.1	NA
BP Solar	BP270U	70W BP Solar Single-Crystal Module (universal frame)	61	NA

BP Solar	BP270UL	70W Single-Crystal Module	61	NA
BP Solar	BP275U	75W Single-Crystal Module (universal frame)	65.5	NA
BP Solar	BP275UL	75W Single-Crystal Module	65.5	NA
BP Solar	BP3115S	115W 12V Polycrystalline Module w/multicontact conn.	101.7	NA
BP Solar	BP3115U	115W 12V Polycrystalline Module, universal frame	101.7	NA
BP Solar	BP3123XR	123W Polycrystalline Module w/multicontact conn.	108.2	NA
BP Solar	BP3125S	125W 12V Polycrystalline Module w/multicontact conn.	110.8	NA
BP Solar	BP3125U	125W 12V Polycrystalline Module, universal frame	110.8	NA
BP Solar	BP3126XR	126W Polycrystalline Module w/multicontact connector	110.8	NA
BP Solar	BP3140B	140W 24V Polycrystalline Module, new AR, multicontact; bronze frame	123.8	NA
BP Solar	BP3140S	140W 24V Polycrystalline Module, new AR w/multicontact conn.	123.8	NA
BP Solar	BP3150B	150W 24V Polycrystalline Module, new AR, multicontact; bronze frame	132.9	NA
BP Solar	BP3150S (2003+)	150W (2003 Rating) 24V Polycrystalline Module, w/multicontact conn.	133	NA
BP Solar	BP3160B (2003+)	160W (2003 Rating) 24V Polycrystalline Module, new AR, multicontact; bronze frame	142.1	NA

(Continued)

TABLE C.1 LIST OF ELIGIBLE PHOTOVOLTAIC MODULES CALIFORNIA ENERGY COMMISSION EMERGING (Continued)

RENEWABLES PROGRAM (FEBRUARY 2005)

MANUFACTURER NAME	MODULE MODEL NUMBER	DESCRIPTION	CEC PTC* RATING	NOTES
BP Solar	BP3160QS	160W 16V Polycrystalline Module w/multicontact conn.	142	NA
BP Solar	BP3160S (2003+)	160W (2003 Rating) 24V Polycrystalline Module, new AR w/multicontact conn.	142.1	NA
BP Solar	BP360U	60W 12V Polycrystalline Module, universal frame	53	NA
BP Solar	BP365U	65W 12V Polycrystalline Module, universal frame	57.6	NA
BP Solar	BP375S (2003+)	75W (2003 Rating) Polycrystalline Module (universal frame), new AR w/multicontact conn.	66.4	NA
BP Solar	BP380S (2003+)	80W (2003 Rating) Polycrystalline Module (universal frame), new AR w/multicontact conn.	71	NA
BP Solar	BP380U (2003+)	80W (2003 Rating) Polycrystalline Module (universal frame), new AR	71	NA
BP Solar	BP4150H	150W 24V Single-Crystal Module (universal frame), new AR	132.5	NA
BP Solar	BP4150S	150W 24V Single-Crystal Module, new AR w/multicontact conn.	132.6	NA
BP Solar	BP4160H	160W 24V Single-Crystal Module (universal frame), new AR	141.6	NA
BP Solar	BP4160S	160W 24V Single-Crystal Module, new AR w/multicontact conn.	141.7	NA

BP Solar	BP4165B	165W 24V Monocrystalline Module, multicontact, bronze frame	146.1	NA
BP Solar	BP4165S	165W 24V Monocrystalline Module w/multicontact conn.	146.1	NA
BP Solar	BP4170H	170W 24V Single-Crystal Module (universal frame), new AR	150.7	NA
BP Solar	BP4170S	170W 24V Single-Crystal Module (universal frame), new AR, multicontact	150.7	NA
BP Solar	BP4175B	175W 24V Monocrystalline Module, multicontact, bronze frame	155.2	NA
BP Solar	BP4175I	175W 24V Monocrystalline Module w/multicontact conn., integral frame	155.2	NA
BP Solar	BP4175S	175W 24V Monocrystalline Module w/multicontact conn.	155.2	NA
BP Solar	BP475S (2003+)	75W (2003 Rating) Single-Crystal Module (universal frame), new AR, multicontact conn.	66.3	NA
BP Solar	BP475U (2003+)	75W (2003 Rating) Single-Crystal Module (universal frame), new AR	66.3	NA
BP Solar	BP480S (2003+)	80W (2003 Rating) Single-Crystal Module (universal frame), new AR w/multicontact conn.	70.8	NA
BP Solar	BP480U (2003+)	80W (2003 Rating) Single-Crystal Module (universal frame), new AR	70.8	NA
BP Solar	BP485H	85W Single-Crystal Module (universal frame), new AR	75.3	NA
BP Solar	BP485S	85W Single-Crystal Module (universal frame), new AR, multicontact	75.3	NA

(Continued)

TABLE C.1 LIST OF ELIGIBLE PHOTOVOLTAIC MODULES CALIFORNIA ENERGY COMMISSION EMERGING (Continued)

RENEWABLES PROGRAM (FEBRUARY 2005)

MANUFACTURER NAME	MODULE MODEL NUMBER	DESCRIPTION	CEC PTC* RATING	NOTES
BP Solar	BP485U	85W Single-Crystal Module (universal frame), new AR	75.3	NA
BP Solar	BP5160S (2003+)	160W (2003 Rating) 24V Buried Grid Single-Crystal Module w/multicontact conn.	141.3	NA
BP Solar	BP5170S (2003+)	170W (2003 Rating) 24V Buried Grid Single-Crystal Module w/multicontact conn.	150.4	NA
BP Solar	BP580U (2003+)	80W (2003 Rating) Buried Grid Single Crystal Module (universal frame)	70.6	NA
BP Solar	BP585DB	85W Buried Grid Single-Crystal Module, multicontact, bronze frame	75.1	NA
BP Solar	BP585KD (2003+)	85W (2003 Rating) Buried Grid Single-Crystal Module (frameless) with special fasteners	75.1	NA
BP Solar	BP585S	85W 12V Buried Grid Single Crystal Module w/multicontact conn.	75.1	NA
BP Solar	BP585U (2003+)	85W (2003 Rating) Buried Grid Single-Crystal Module (universal frame)	75.1	NA
BP Solar	BP585UL (2003+)	85W (2003 Rating) Buried Grid Single-Crystal Module	75.1	NA
BP Solar	BP590UL (2003+)	90W (2003 Rating) Buried Grid Single-Crystal Module	79.7	NA
BP Solar	BP7170S	170W 24V Saturn Single-Crystal Module w/multicontact conn.	151.1	NA

BP Solar	BP7175S	175W 24V Saturn Single-Crystal Module w/multicontact conn.	154.9	NA
BP Solar	BP7180S	180W 24V Saturn Single-Crystal Module w/multicontact conn.	160.2	NA
BP Solar	BP7185S	185W 24V Saturn Single-Crystal Module w/multicontact conn.	164	NA
BP Solar	BP785S	85W 12V Saturn Single-Crystal Module w/multicontact conn.	75.5	NA
BP Solar	BP790DB	90W 12V Saturn Single-Crystal Module, dark frame	80	NA
BP Solar	BP790S	90W 12V Saturn Single-Crystal Module w/multicontact conn.	80	NA
BP Solar	BP790U	90W 12V Saturn Single-Crystal Module, universal frame	80	NA
BP Solar	BP845I	45W Millennia 2J a-Si Module (medium voltage, voltage, integral frame)	42.4	NA
BP Solar	BP850I	50W Millennia 2J a-Si Module (medium voltage, integral frame)	47.1	NA
BP Solar	BP855I	55W Millennia 2J a-Si Module (medium voltage, integral frame)	51.9	NA
BP Solar	BP970B	70W Thin-film CdTe Laminate w/mounting brackets	62.3	NA
BP Solar	BP970I	70W Thin-film CdTe Module (Integra frame)	62.2	NA
BP Solar	BP980B	80W Thin-film CdTe Laminate w/mounting brackets	71.3	NA
BP Solar	BP980I	80W Thin-film CdTe Module (Integra frame)	71.2	NA

(Continued)

TABLE C.1 LIST OF ELIGIBLE PHOTOVOLTAIC MODULES CALIFORNIA ENERGY COMMISSION EMERGING (Continued)

RENEWABLES PROGRAM (FEBRUARY 2005)

MANUFACTURER NAME	MODULE MODEL NUMBER	DESCRIPTION	CEC PTC* RATING	NOTES
BP Solar	BP990B	90W Thin-film CdTe Laminate w/mounting brackets	80.4	NA
BP Solar	BP990I	90W Thin-film CdTe Module (Integra frame)	80.3	NA
BP Solar	MST-43I	43W Millennia 2J a-Si Module (med. voltage, Integra frame)	40.5	NA
BP Solar	MST-43LV	43W Millennia 2J a-Si Module (low voltage, universal frame)	40.5	NA
BP Solar	MST-43MV	43W Millennia 2J a-Si Module (med. voltage, universal frame)	40.5	NA
BP Solar	MST-45LV	45W Millennia 2J a-Si Module (low voltage, universal frame)	42.4	NA
BP Solar	MST-45MV	45W Millennia 2J a-Si Module (medium voltage, universal frame)	42.4	NA
BP Solar	MST-50I	50W Millennia 2J a-Si Module (med. voltage, Integra frame)	47.1	NA
BP Solar	MST-50LV	50W Millennia 2J a-Si Module (low voltage, universal frame)	47.1	NA
BP Solar	MST-50MV	50W Millennia 2J a-Si Module (med. voltage, universal frame)	47.1	NA
BP Solar	MST-55MV	55W Millennia 2J a-Si Module (med. voltage, universal frame)	51.9	NA
BP Solar	MSX-110	110W Solarex Polycrystalline Module	95.6	NA

BP Solar	MSX-120	120W Solarex Polycrystalline Module	104.5	NA
BP Solar	MSX-240	240W Solarex Polycrystalline Module	209.1	NA
BP Solar	MSX-50	50W Solarex Polycrystalline Module	43.5	NA
BP Solar	MSX-56	56W Solarex Polycrystalline Module	48.7	NA
BP Solar	MSX-60	60W Solarex Polycrystalline Module	52.2	NA
BP Solar	MSX-64	64W Solarex Polycrystalline Module	55.8	NA
BP Solar	MSX-77	77W Solarex Polycrystalline Module	67	NA
BP Solar	MSX-80U	80W Solarex Polycrystalline Module (universal frame)	69.7	NA
BP Solar	MSX-83	83W Solarex Polycrystalline Module	72.3	NA
BP Solar	SX-110S	110W Solarex poly-Si Module (univ. frame, multicontact conn.)	95.6	NA
BP Solar	SX-110U	110W Solarex poly-Si Module (univ. frame)	95.6	NA
BP Solar	SX-120S	120W Solarex poly-Si Module (univ. frame, multicontact conn.)	104.6	NA
BP Solar	SX-120U	120W Solarex poly-Si Module (univ. frame)	104.6	NA
BP Solar	SX-160S	160W 24V Polycrystalline Module w/multicontact connection	142	NA
BP Solar	SX-40D	40W Solarex poly-Si Module (direct-mount frame)	34.8	NA
BP Solar	SX-40M	40W Solarex poly-Si Module (multimount frame)	34.8	NA
BP Solar	SX-40U	40W Solarex poly-Si Module (univ. frame)	34.8	NA
BP Solar	SX-50D	50W Solarex poly-Si Module (direct-mount frame)	43.5	NA

(Continued)

TABLE C.1 LIST OF ELIGIBLE PHOTOVOLTAIC MODULES CALIFORNIA ENERGY COMMISSION EMERGING *(Continued)*

RENEWABLES PROGRAM (FEBRUARY 2005)

MANUFACTURER NAME	MODULE MODEL NUMBER	DESCRIPTION	CEC PTC* RATING	NOTES
BP Solar	SX-50M	50W Solarex poly-Si Module (multimount frame)	43.5	NA
BP Solar	SX-50U	50W Solarex poly-Si Module (univ. frame)	43.5	NA
BP Solar	SX-55D	55W Solarex poly-Si Module (direct-mount frame)	47.8	NA
BP Solar	SX-55U	55W Solarex poly-Si Module (univ. frame)	47.8	NA
BP Solar	SX-60D	60W Solarex poly-Si Module (direct-mount frame)	52.2	NA
BP Solar	SX-60U	60W Solarex poly-Si Module (univ. frame)	52.2	NA
BP Solar	SX-65D	65W Solarex poly-Si Module (direct-mount frame)	56.7	NA
BP Solar	SX-65U	65W Solarex poly-Si Module (univ. frame)	56.7	NA
BP Solar	SX-75	75W Solarex poly-Si Module	65.2	NA
BP Solar	SX-75TS	75W Solarex poly-Si Module (125mm cells, low-profile/MC)	65.4	NA
BP Solar	SX-75TU	75W Solarex poly-Si Module (125mm cells, SPJB)	65.4	NA
BP Solar	SX-80	80W Solarex poly-Si Module	69.7	NA
BP Solar	SX-85	85W Solarex poly-Si Module	74.1	NA
BP Solar	SX140B	140W 24V Polycrystalline Module w/multicontact, bronze frame	123.8	NA
BP Solar	SX150B	150W 24V Polycrystalline Module w/multicontact, bronze frame	132.9	NA

Manufacturer	Model	Description		
BP Solar	SX160B	160W 24V Polycrystalline Module w/multicontact, bronze frame	142	NA
BP Solar	TF-80B	80W Thin-film CdTe Laminate w/mounting brackets	71.3	NA
BP Solar	TF-80I	80W Thin-film CdTe Module (Integra frame)	71.2	NA
BP Solar	TF-90B	90W Thin-film CdTe Laminate w/mounting brackets	80.4	NA
BP Solar	TF-90I	90W Thin-film CdTe Module (Integra frame)	80.3	NA
BP Solar	VLX-53	53W Solarex Value-Line poly-Si Module	46.2	NA
BP Solar	VLX-80	80W Solarex Value-Line poly-Si Module	69.7	NA
Dunasolar Inc.	DS-30	30W Unframed 2J a-Si Module	28.8	NA
Dunasolar Inc.	DS-40	40W Unframed 2J a-Si Module	38.4	NA
Energy Photovoltaics, Inc.	EPV-30	30W Unframed 2J a-Si Module	28.8	NA
Energy Photovoltaics, Inc.	EPV-40	40W Unframed 2J a-Si Module	38.4	NA
Evergreen Solar	E-25	25W String Ribbon poly-Si Module	22.2	NA
Evergreen Solar	E-28	28W String Ribbon poly-Si Module	24.9	NA
Evergreen Solar	E-30	30W String Ribbon poly-Si Module	26.7	NA
Evergreen Solar	E-50	50W String Ribbon poly-Si Module	44.4	NA
Evergreen Solar	E-56	56W String Ribbon poly-Si Module	49.8	NA
Evergreen Solar	E-60	60W String Ribbon poly-Si Module	53.4	NA
Evergreen Solar	EC-102	102W Cedar Line Module	91.2	NA
Evergreen Solar	EC-110	110W Cedar Line Module	98.4	NA
Evergreen Solar	EC-115	115W String Ribbon Cedar Line Module	103.1	NA

(Continued)

TABLE C.1 LIST OF ELIGIBLE PHOTOVOLTAIC MODULES CALIFORNIA ENERGY COMMISSION EMERGING (Continued)

RENEWABLES PROGRAM (FEBRUARY 2005)

MANUFACTURER NAME	MODULE MODEL NUMBER	DESCRIPTION	CEC PTC* RATING	NOTES
Evergreen Solar	EC-47	47W Cedar Line Module	41.9	NA
Evergreen Solar	EC-51	51W Cedar Line Module	45.6	NA
Evergreen Solar	EC-55	55W Cedar Line Module	49.2	NA
Evergreen Solar	EC-94	94W Cedar Line Module	83.9	NA
Evergreen Solar	ES-112	112W String Ribbon poly-Si AC Module (with Trace MS100)	99.7	NA
Evergreen Solar	ES-240	240W String Ribbon poly-Si AC Module (with AES MI-250)	213.8	NA
First Solar, LLC	FS-40	40W Thin-Film CdTe Laminate	38	NA
First Solar, LLC	FS-40D	40W Thin-Film CdTe Module w/D-channel mounting rails	38	NA
First Solar, LLC	FS-45	45W Thin-Film CdTe Laminate	42.8	NA
First Solar, LLC	FS-45D	45W Thin-Film CdTe Module with mounting rails	42.8	NA
First Solar, LLC	FS-50	50W/65V Thin-Film CdTe Laminate	47.6	NA
First Solar, LLC	FS-50C	50W/65V Thin-Film CdTe Laminate w/C-channel mounting rails	47.6	NA
First Solar, LLC	FS-50D	50W Thin-Film CdTe Module with mounting rails	47.6	NA
First Solar, LLC	FS-50Z	50W/65V Thin-Film CdTe Laminate w/Z-channel mounting rails	47.6	NA
First Solar, LLC	FS-55	55W Thin-Film CdTe Laminate	52.4	NA

Manufacturer	Model	Description		
First Solar, LLC	FS-55D	55W Thin-Film CdTe Module w/D-channel mounting rails	52.4	NA
First Solar, LLC	FS-60	60W Thin-Film CdTe Laminate	57.2	NA
First Solar, LLC	FS-60D	60W Thin-Film CdTe Module with mounting rails	57.2	NA
GE Energy	GEPV-030-MNA	30W Single-Crystal Module w/o connectors	26.5	NA
GE Energy	GEPV-030-MNB	30W Single-Crystal Module w/o connectors B	26.5	NA
GE Energy	GEPV-045-MNA	45W Single-Crystal Module w/o connectors	39.7	NA
GE Energy	GEPV-045-MNB	45W Single-Crystal Module w/o connectors B	39.7	NA
GE Energy	GEPV-050-MNA	50W Single-Crystal Module w/o connectors	44.2	NA
GE Energy	GEPV-050-MNB	50W Single-Crystal Module w/o connectors B	44.2	NA
GE Energy	GEPV-055-GCA	55W Single-Crystal Module w/MC connectors	48.9	NA
GE Energy	GEPV-055-GCB	55W Single-Crystal Module w/MC connectors B	48.9	NA
GE Energy	GEPV-065-MNA	65W Single-Crystal Module w/o connectors	57.7	NA
GE Energy	GEPV-065-MNB	65W Single-Crystal Module w/o connectors B	57.7	NA
GE Energy	GEPV-070-MNA	70W Single-Crystal Module w/o connectors	62.2	NA
GE Energy	GEPV-070-MNB	70W Single-Crystal Module w/o connectors B	62.2	NA
GE Energy	GEPV-100-MCA	100W Single-Crystal Module w/MC connectors	88.7	NA
GE Energy	GEPV-100-MCB	100W Single-Crystal Module w/MC connectors B	88.7	NA
GE Energy	GEPV-100-MNA	100W Single-Crystal Module w/o connectors	88.7	NA

(Continued)

TABLE C.1 LIST OF ELIGIBLE PHOTOVOLTAIC MODULES CALIFORNIA ENERGY COMMISSION EMERGING (Continued)

RENEWABLES PROGRAM (FEBRUARY 2005)

MANUFACTURER NAME	MODULE MODEL NUMBER	DESCRIPTION	CEC PTC* RATING	NOTES
GE Energy	GEPV-100-MNB	100W Single-Crystal Module w/o connectors B	88.7	NA
GE Energy	GEPV-110-MCA	110W Single-Crystal Module w/MC connectors	97.8	NA
GE Energy	GEPV-110-MCB	110W Single-Crystal Module w/MC connectors B	97.8	NA
GE Energy	GEPV-110-MNA	110W Single-Crystal Module w/o connectors	97.8	NA
GE Energy	GEPV-110-MNB	110W Single-Crystal Module w/o connectors B	97.8	NA
GE Energy	GEPV-165-MCA	165W Single-Crystal Module w/MC connectors	146.7	NA
GE Energy	GEPV-165-MCB	165W Single-Crystal Module w/MC connectors B	146.7	NA
GE Energy	GEPV-173-MCA	173W Single-Crystal Module w/MC connectors	154	NA
GE Energy	GEPV-173-MCB	173W Single-Crystal Module w/MC connectors B	154	NA
Isofoton	I-100/12	100W Monocrystalline Rail-Mounted 2 Module X	89.5	NA
Isofoton	I-100/24	100W Monocrystalline Rail-Mounted Module	89.5	NA
Isofoton	I-106/12	106W Monocrystalline Rail-Mounted Module	95	NA
Isofoton	I-106/24	106W Monocrystalline Rail-Mounted Module X2	95	NA
Isofoton	I-110-24	110W Monocrystalline Rail-Mounted 2 Module X	98.7	NA
Isofoton	I-110/12	110W Monocrystalline Rail-Mounted Module	98.7	NA
Isofoton	I-130/12	130W Monocrystalline Rail-Mounted Module	116	NA
Isofoton	I-130/24	130W Monocrystalline Rail-Mounted 2 Module X	116	NA
Isofoton	I-140 R/12	140W Monocrystalline Rail-Mounted Module	125	NA
Isofoton	I-140 R/24	140W Monocrystalline Rail-Mounted 3 Module X	125	NA
Isofoton	I-140 S/12	140W Monocrystalline Rail-Mounted 4 Module X	125	NA

Manufacturer	Model	Description		
Isofoton	I-140 S/24	140W Monocrystalline Rail-Mounted 5 Module X	125	NA
Isofoton	I-150	150W Monocrystalline Rail-Mounted 2	134.3	NA
Isofoton	I-150 S/12	150W Monocrystalline Rail-Mounted Module	134.3	NA
Isofoton	I-150 S/24	150W Monocrystalline Rail-Mounted Module	134.3	NA
Isofoton	I-159	159W Monocrystalline Rail-Mounted Module	142.6	NA
Isofoton	I-165	165W Monocrystalline Rail-Mounted Module	148.1	NA
Isofoton	I-36	36W Monocrystalline Rail-Mounted Module	32.2	NA
Isofoton	I-50	50W Monocrystalline Rail-Mounted Module	44.8	NA
Isofoton	I-53	53W Monocrystalline Rail-Mounted Module	47.5	NA
Isofoton	I-55	55W Monocrystalline Rail-Mounted Module	49.4	NA
Isofoton	I-65	65W Monocrystalline Rail-Mounted Module	58	NA
Isofoton	I-70 R	70W Monocrystalline Rail-Mounted Module	62.5	NA
Isofoton	I-70 S	70W Monocrystalline Rail-Mounted Module X2	62.5	NA
Isofoton	I-75	75W Monocrystalline Rail-Mounted Module	67.1	NA
Isofoton	I-94/12	94W Monocrystalline Rail-Mounted Module X2	84.2	NA
Isofoton	I-94/24	94W Monocrystalline Rail-Mounted Module	84.2	NA
Kaneka Corporation	CSA201	58W a-Si Module	54.1	NA
Kaneka Corporation	CSA211	58W a-Si Module (CSA)	54.1	NA
Kaneka Corporation	CSB211	58W a-Si Module (CSB)	54.1	NA
Kaneka Corporation	GSA211	60W a-Si Module	56	NA
Kaneka Corporation	LSU205	58W a-Si Module	54.1	NA
Kaneka Corporation	TSA211	116W a-Si Twin Type Module	108.2	NA
Kaneka Corporation	TSB211	116W a-Si Twin Type Module B	108.2	NA

(Continued)

TABLE C.1 LIST OF ELIGIBLE PHOTOVOLTAIC MODULES CALIFORNIA ENERGY COMMISSSION EMERGING (Continued)

RENEWABLES PROGRAM (FEBRUARY 2005)

MANUFACTURER NAME	MODULE MODEL NUMBER	DESCRIPTION	CEC PTC* RATING	NOTES
Kaneka Corporation	TSC211	120W a-Si Twin Type Module (TSC)	111.9	NA
Kaneka Corporation	TSD211	120W a-Si Twin Type Module (TSD)	111.9	NA
Kyocera Solar, Inc.	KC120-1	120W High-Efficiency Multicrystal PV Module	105.7	NA
Kyocera Solar, Inc.	KC125G	125W High-Efficiency Multicrystal PV Module	111.8	NA
Kyocera Solar, Inc.	KC158G	158W High-Efficiency Multicrystal PV Module	139.7	NA
Kyocera Solar, Inc.	KC167G	167W High-Efficiency Multicrystal PV Module	149.6	NA
Kyocera Solar, Inc.	KC187G	187W Multicrystal PV Module, Deep Blue	167.4	NA
Kyocera Solar, Inc.	KC50	50W High-Efficiency Multicrystal PV Module	43.9	NA
Kyocera Solar, Inc.	KC60	60W High-Efficiency Multicrystal PV Module	52.7	NA
Kyocera Solar, Inc.	KC70	70W High-Efficiency Multicrystal PV Module	61.6	NA
Kyocera Solar, Inc.	KC80	80W High-Efficiency Multicrystal PV Module	70.4	NA
Matrix Solar/Photowatt	PW1000-100	100W Large-Scale Dual-Voltage Multi-Si Module	90	NA
Matrix Solar/Photowatt	PW1000-105	105W Large-Scale Dual-Voltage Multi-Si Module	94.6	NA
Matrix Solar/Photowatt	PW1000-90	90W Large-Scale Dual-Voltage Multi-Si Module	80.9	NA
Matrix Solar/Photowatt	PW1000-95	95W Large-Scale Dual-Voltage Multi-Si Module	85.4	NA
Matrix Solar/Photowatt	PW1250-115	115W Large-Scale Dual-Voltage Multi-Si Module	103.7	NA
Matrix Solar/Photowatt	PW1250-125	125W Large-Scale Dual-Voltage Multi-Si Module	112.9	NA
Matrix Solar/Photowatt	PW1250-135	135W Large-Scale Dual-Voltage Multi-Si Module	122.2	NA
Matrix Solar/Photowatt	PW1650-155	155W Large-Scale Dual-Voltage Multi-Si Module	139.9	NA
Matrix Solar/Photowatt	PW1650-165	165W Large-Scale Dual-Voltage Multi-Si Module	149	NA
Matrix Solar/Photowatt	PW1650-175	175W Large-Scale Dual-Voltage Multi-Si Module	158.3	NA

Manufacturer	Model	Description		
Matrix Solar/Photowatt	PW750-70	70W Large-Scale Multi-Si Module	62.9	NA
Matrix Solar/Photowatt	PW750-75	75W Large-Scale Multi-Si Module	67.5	NA
Matrix Solar/Photowatt	PW750-80	80W Large-Scale Multi-Si Module	72	NA
Matrix Solar/Photowatt	PW750-90	90W Large-Scale Multi-Si Module	81.2	NA
MC Solar	BP970B	70W Thin-Film CdTe Laminate w/mounting brackets	62.3	NA
MC Solar	BP980B	80W Thin-Film CdTe Laminate w/mounting brackets	71.3	NA
MC Solar	BP990B	90W Thin-Film CdTe Laminate w/mounting brackets	80.4	NA
MC Solar	TF-80B	80W Thin-Film CdTe Laminate w/mounting brackets (now BP980B)	71.3	NA
MC Solar	TF-90B	90W Thin-Film CdTe Laminate w/mounting brackets (now BP990B)	80.4	NA
Midway Labs, Inc.	MLB3416-115	115W Concentrator (335x) Module	105	NA
Mitsubishi Electric Corporation	PV-MF110EC3	110W Polycrystalline Lead-free Solder Module w/o cable	98.4	NA
Mitsubishi Electric Corporation	PV-MF120EC3	120W Polycrystalline Lead-free Solder Module w/o cable	107.6	NA
Mitsubishi Electric Corporation	PV-MF125EA2LF	125W Polycrystalline Lead-free Solder Module w/MC connector	110.7	NA
Mitsubishi Electric Corporation	PV-MF130E	130W Polycrystalline Module w/multicontact connectors	115.2	NA
Mitsubishi Electric Corporation	PV-MF130EA2LF	130W Polycrystalline Lead-free Solder Module w/MC connector	115.2	NA
Mitsubishi Electric Corporation	PV-MF160EB3	160W Polycrystalline Lead-free Solder Module w/MC connector	142.4	NA

(Continued)

TABLE C.1 LIST OF ELIGIBLE PHOTOVOLTAIC MODULES CALIFORNIA ENERGY COMMISSION EMERGING (Continued)

RENEWABLES PROGRAM (FEBRUARY 2005)

MANUFACTURER NAME	MODULE MODEL NUMBER	DESCRIPTION	CEC PTC* RATING	NOTES
Mitsubishi Electric Corporation	PV-MF165EB3	165W Polycrystalline Lead-free Solder Module w/MC connector	146.9	NA
Mitsubishi Electric Corporation	PV-MF170EB3	170W Polycrystalline Lead-free Solder Module w/MC connector	152.5	NA
Pacific Solar Pty Limited	PP-USA-213-B5	150W BP Solar 2150L SunEmpower Modular Mount	131.8	NA
Pacific Solar Pty Limited	PP-USA-213-L6	160W BP Solar 5160L SunEmpower Modular Mount	140.9	NA
Pacific Solar Pty Limited	PP-USA-213-L7	170W BP Solar 5170L SunEmpower Modular Mount	149.9	NA
Pacific Solar Pty Limited	PP-USA-213-N5	150W BP Solar 4150L SunEmpower Modular Mount	131.8	NA
Pacific Solar Pty Limited	PP-USA-213-N6	160W BP Solar 4160L SunEmpower Modular Mount	140.9	NA
Pacific Solar Pty Limited	PP-USA-213-N7	170W BP Solar 4170L SunEmpower Modular Mount	149.9	NA
Pacific Solar Pty Limited	PP-USA-213-P5	150W BP Solar 3150L SunEmpower Modular Mount	131.8	NA
Pacific Solar Pty Limited	PP-USA-213-P6	160W BP Solar 3160L SunEmpower Modular Mount	140.9	NA
Pacific Solar Pty Limited	PP-USA-213-S5	150W Shell Solar SP-150-PL, -PLC SunEmpower Modular Mount	135	NA

Manufacturer	Model	Description		
Powerlight Corp.	PL-AP-120L	120W PowerGuard Roof Tile (AstroPower)	104.9	NA
Powerlight Corp.	PL-AP-130	130W PowerGuard Roof Tile (AstroPower AP-130)	115.1	NA
Powerlight Corp.	PL-AP-65	One AstroPower AP-65 laminate mounted on one PowerGuard backerboard	56.6	NA
Powerlight Corp.	PL-AP-65 Double Module	Two AstroPower AP-65 laminates mounted on one PowerGuard backerboard	113.2	NA
Powerlight Corp.	PL-AP-75 Double Module	150W PowerGuard Roof Tile (two AstroPower modules)	131.1	NA
Powerlight Corp.	PL-APx-110-SL	110W PowerGuard Roof Tile (AstroPower)	93.7	NA
Powerlight Corp.	PL-ASE-100	100W PowerGuard Roof Tile (ASE Americas)	88.6	NA
Powerlight Corp.	PL-BP-2150S	150W PowerGuard Roof Tile (BP Solar)	129.2	NA
Powerlight Corp.	PL-BP-3160L	160W PowerGuard Roof Tile (BP Solar)	139.2	NA
Powerlight Corp.	PL-BP-380L Double Module	160W PowerGuard Roof Tile (Two BP-380L modules)	139.1	NA
Powerlight Corp.	PL-BP-485L Double Module	170W PowerGuard Roof Tile (Two BP-485L modules)	148	NA
Powerlight Corp.	PL-BP-TF-80L	80W PowerGuard Roof Tile (BP Solar)	70.7	NA
Powerlight Corp.	PL-FS-415-A	50W PowerGuard Roof Tile (First Solar)	47.4	NA
Powerlight Corp.	PL-KYOC-FL120-1B	120W PowerGuard Roof Tile (Kyocera)	103.8	NA
Powerlight Corp.	PL-KYOC-FL125	125W PowerGuard Roof Tile (Kyocera)	110.8	NA
Powerlight Corp.	PL-KYOC-FL158	158W PowerGuard Roof Tile (Kyocera)	138.2	NA
Powerlight Corp.	PL-KYOC-FL167	167W PowerGuard Roof Tile (Kyocera)	148	NA
Powerlight Corp.	PL-MST-43	43W PowerGuard Roof Tile (Solarex a-Si)	40.5	NA
Powerlight Corp.	PL-MSX-120	120W PowerGuard Roof Tile (Solarex poly-Si)	103.7	NA

(Continued)

TABLE C.1 LIST OF ELIGIBLE PHOTOVOLTAIC MODULES CALIFORNIA ENERGY COMMISSION EMERGING (Continued)

RENEWABLES PROGRAM (FEBRUARY 2005)

MANUFACTURER NAME	MODULE MODEL NUMBER	DESCRIPTION	CEC PTC* RATING	NOTES
Powerlight Corp.	PL-PW-750	75–80W PowerGuard Roof Tile (Matrix Solar/Photowatt)	69.9	NA
Powerlight Corp.	PL-PW-750 Double Module	150W PowerGuard Roof Tile (two Matrix Solar Photowatt modules)	135.3	NA
Powerlight Corp.	PL-SHAR-ND-N6E1D	146W PowerGuard Roof Tile (Sharp Corp.)	125.1	NA
Powerlight Corp.	PL-SP-135	135W (Pre-2003 Rating) PowerGuard Roof Tile (Siemens)	119.1	NA
Powerlight Corp.	PL-SP-135 (2003+)	135W (2003 Rating) PowerGuard Roof Tile (Siemens)	119.8	NA
Powerlight Corp.	PL-SP-150-24L	150W PowerGuard Roof Tile (Rectangular Siemens)	133.5	NA
Powerlight Corp.	PL-SP-150-CPL	150W PowerGuard Roof Tile (Siemens)	133.6	NA
Powerlight Corp.	PL-SP-70 Double Module	Two Sharp SP-70 modules on one PowerGuard tile	124.8	NA
Powerlight Corp.	PL-SP-75	75W PowerGuard Roof Tile (Siemens)	66.8	NA
Powerlight Corp.	PL-SP-75 Double Module	150W PowerGuard Roof Tile (Two Siemens modules)	133.7	NA
Powerlight Corp.	PL-SQ75-CPL	75W PowerGuard Roof Tile (Shell)	65.1	NA
Powerlight Corp.	PL-SQ75-CPL Double Module	150W PowerGuard Roof Tile (Shell)	130.3	NA
Powerlight Corp.	PL-SQ77-CPL Double Module	154W Double Module PowerGuard Roof Tile (Shell Solar)	134.6	NA

Manufacturer	Model	Description		
Powerlight Corp.	PL-SQ85-P Double Module	170W PowerGuard Roof Tile (Two Shell modules)	152.3	NA
Powerlight Corp.	PL-SY-HIP-190BA2	190W PowerGuard Roof Tile (Sanyo)	177.5	NA
Powerlight Corp.	PL-SY-HIP-190CA2	190W PowerGuard Roof Tile (Sanyo)	177.5	NA
Powerlight Corp.	PL-SY-HIP-H552BA2	175W PowerGuard Roof Tile (Sanyo)	162.2	NA
RWE SCHOTT Solar	ASE-250DGF/17	250W/17V EFG Module, framed	223.7	NA
RWE SCHOTT Solar	ASE-250DGF/50	250W/50V EFG Module, framed	223.7	NA
RWE SCHOTT Solar	ASE-270DGF/17	270W/17V EFG Module, framed	242	NA
RWE SCHOTT Solar	ASE-270DGF/50	270W/50V EFG Module, framed	242	NA
RWE SCHOTT Solar	SAPC-175	175W Monocrystalline Silicon Module	154.4	NA
Sanyo Electric Co. Ltd.	HIP-167BA	167W HIT Hybrid a-Si/c-Si Solar Cell Module	156.7	NA
Sanyo Electric Co. Ltd.	HIP-175BA3	175W HIT Hybrid a-Si/c-Si Solar Cell Module	163.3	NA
Sanyo Electric Co. Ltd.	HIP-175BA5	175W HIT Hybrid a-Si/c-Si Solar Cell Module (5)	163.3	NA
Sanyo Electric Co. Ltd.	HIP-180BA	180W HIT Hybrid a-Si/c-Si Solar Cell Module	169.1	NA
Sanyo Electric Co. Ltd.	HIP-180BA3	180W HIT Hybrid a-Si/c-Si Solar Cell Module (3)	168	NA
Sanyo Electric Co. Ltd.	HIP-180BA5	180W HIT Hybrid a-Si/c-Si Solar Cell Module (5)	168	NA
Sanyo Electric Co. Ltd.	HIP-190BA	190W HIT Hybrid a-Si/c-Si Solar Cell Module	178.7	NA
Sanyo Electric Co. Ltd.	HIP-190BA1	190W HIT Hybrid a-Si/c-Si Solar Cell Module (std. j.b.)	178.7	NA

(Continued)

TABLE C.1 LIST OF ELIGIBLE PHOTOVOLTAIC MODULES CALIFORNIA ENERGY COMMISSION EMERGING *(Continued)*

RENEWABLES PROGRAM (FEBRUARY 2005)

MANUFACTURER NAME	MODULE MODEL NUMBER	DESCRIPTION	CEC PTC* RATING	NOTES
Sanyo Electric Co. Ltd.	HIP-190BA2	190W HIT Hybrid a-Si/c-Si Solar Cell Module (std. j.b. w/addl. wiring)	178.7	NA
Sanyo Electric Co. Ltd.	HIP-190BA3	190W HIT Hybrid a-Si/c-Si Solar Cell Module (3)	178.7	NA
Sanyo Electric Co. Ltd.	HIP-190BA5	190W HIT Hybrid a-Si/c-Si Solar Cell Module (5)	178.7	NA
Sanyo Electric Co. Ltd.	HIP-G751BA1	167W HIT Hybrid a-Si/c-Si Solar Cell Module (std. j.b.)	155.8	NA
Sanyo Electric Co. Ltd.	HIP-G751BA2	167W HIT Hybrid a-Si/c-Si Solar Cell Module (std. j.b. w/addl. wiring)	155.8	NA
Sanyo Electric Co. Ltd.	HIP-H552BA1	175W HIT Hybrid a-Si/c-Si Solar Cell Module (std. j.b.)	163.3	NA
Sanyo Electric Co. Ltd.	HIP-H552BA2	175W HIT Hybrid a-Si/c-Si Solar Cell Module (std. j.b. w/addl. wiring)	163.3	NA
Sanyo Electric Co. Ltd.	HIP-J54BA1	180W HIT Hybrid a-Si/c-Si Solar Cell Module (std. j.b.)	168.1	NA
Sanyo Electric Co. Ltd.	HIP-J54BA2	180W HIT Hybrid a-Si/c-Si Solar Cell Module (std. j.b. w/addl. wiring)	168.1	NA
Schott Applied Power Corp.	SAPC-123	123W Multisilicon Module	107.8	NA
Schott Applied Power Corp.	SAPC-165	165W Multicrystalline Silicon Module	144.8	NA
Schott Applied Power Corp.	SAPC-80	80W Multisilicon Module	70.2	NA
Schuco USA LP	S125-SP	130W Polycrystalline Module w/multicontact connectors	115.2	NA

Schuco USA LP	S158-SP	165W Polycrystalline Module w/multicontact connectors	146.9	NA
Schuco USA LP	S162-SP	170W Polycrystalline Lead-free Solder Module w/MC connector	152.5	NA
Sharp Corporation	ND-160U1Z	160 W, Multicrystalline Silicon Module	140.6	Changed Power Temp Coefficient
Sharp Corporation	ND-167U1	167W Multisilicon Module	146.9	Changed Power Temp Coefficient
Sharp Corporation	ND-167U3	167W Multisilicon Module (3)	146.9	Changed Power Temp Coefficient
Sharp Corporation	ND-70ELU	70W Multicrystalline Silicon Module (left)	61.1	Changed Power Temp Coefficient
Sharp Corporation	ND-70ERU	70W Multicrystalline Silicon Module (right)	61.1	Changed Power Temp Coefficient
Sharp Corporation	ND-L3E1U	123W Multisilicon Module	108.1	Changed Power Temp Coefficient
Sharp Corporation	ND-L3EJE	123W Multisilicon Module (w/junction box)	108.1	Changed Power Temp Coefficient
Sharp Corporation	ND-N0ECU	140W Multisilicon Residential Module	123	Changed Power Temp Coefficient
Sharp Corporation	ND-N6E1U	146W Multisilicon Module	128.3	Changed Power Temp Coefficient
Sharp Corporation	ND-Q0E2U	160W Multisilicon Module	140.6	Changed Power Temp Coefficient
Sharp Corporation	NE-165U1	165W Multisilicon Module (flat screw type, same as NE-Q5E2U)	145.2	Changed Power Temp Coefficient

(Continued)

TABLE C.1 LIST OF ELIGIBLE PHOTOVOLTAIC MODULES CALIFORNIA ENERGY COMMISSION EMERGING (Continued)

RENEWABLES PROGRAM (FEBRUARY 2005)

MANUFACTURER NAME	MODULE MODEL NUMBER	DESCRIPTION	CEC PTC* RATING	NOTES
Sharp Corporation	NE-80E1U	80W Multisilicon Module	70.4	Changed Power Temp Coefficient
Sharp Corporation	NE-80EJE	80W Multisilicon Module (w/junction box)	70.4	Changed Power Temp Coefficient
Sharp Corporation	NE-K125U1	125W Multisilicon Module (nonflat screw type, black color frame)	110	Changed Power Temp Coefficient
Sharp Corporation	NE-K125U2	125W Multisilicon Module (flat screw type)	110.1	Changed Power Temp Coefficient
Sharp Corporation	NE-Q5E1U	165W Multisilicon Module (nonflat screw type)	145.2	Changed Power Temp Coefficient
Sharp Corporation	NE-Q5E2U	165W Multisilicon Module (flat screw type)	145.2	Changed Power Temp Coefficient
Sharp Corporation	NT-175U1	175W Monocrystalline Silicon Module	154.2	Changed Power Temp Coefficient
Sharp Corporation	NT-188U1	188W Single-Crystal Silicon Module	166	Changed Power Temp Coefficient
Sharp Corporation	NT-R5E1U	175 W Multisilicon Module	154.2	Changed Power Temp Coefficient
Sharp Corporation	NT-S5E1U	185W Multisilicon Module	163.3	NA
Shell Solar Industries	SM110	110W PowerMax Module	99.2	NA
Shell Solar Industries	SP130-PC	130W PowerMax Monocrystalline Module w/cable assembly	116.6	NA
Shell Solar Industries	SP140-PC	140W PowerMax Monocrystalline Module w/cable assembly	125.8	NA

Shell Solar Industries	SP150-PC	150W PowerMax Monocrystalline Module w/cable assembly	134.9	NA
Shell Solar Industries	SQ140-P	140W PowerMax Monocrystalline Module	123.4	NA
Shell Solar Industries	SQ140-PC	140W PowerMax Monocrystalline Module w/multicontact cable assembly	123.4	NA
Shell Solar Industries	SQ150-P	150W PowerMax Monocrystalline Module	132.5	NA
Shell Solar Industries	SQ150-PC	150W PowerMax Monocrystalline Module w/multicontact cable assembly	132.5	NA
Shell Solar Industries	SQ160-P	160W PowerMax Monocrystalline Module	141.5	NA
Shell Solar Industries	SQ160-PC	160W PowerMax Monocrystalline Module w/multicontact cable assembly	141.5	NA
Shell Solar Industries	SQ165-P	165W PowerMax Monocrystalline Module	149.1	NA
Shell Solar Industries	SQ165-PC	165W PowerMax Monocrystalline Module w/multicontact cable assembly	149.1	NA
Shell Solar Industries	SQ175-P	175W PowerMax Monocrystalline Module	158.3	NA
Shell Solar Industries	SQ175-PC	175W PowerMax Monocrystalline Module w/multicontact cable assembly	158.3	NA
Shell Solar Industries	SQ70	75W PowerMax Monocrystalline Module	61.8	NA
Shell Solar Industries	SQ75	75W PowerMax Monocrystalline Module	66.3	NA
Shell Solar Industries	SQ80	80W PowerMax Monocrystalline Module	70.8	NA
Shell Solar Industries	SQ80-P	80W PowerMax Monocrystalline Module	72.3	NA
Shell Solar Industries	SQ85-P	85W PowerMax Monocrystalline Module	76.9	NA
Siemens Solar Industries	SM-110	110W PowerMax Module	99.2	NA
Siemens Solar Industries	SM10	10W PowerMax Module	9	NA
Siemens Solar Industries	SM20	20W PowerMax Module	18	NA

(Continued)

TABLE C.1 LIST OF ELIGIBLE PHOTOVOLTAIC MODULES CALIFORNIA ENERGY COMMISSION EMERGING (Continued)

RENEWABLES PROGRAM (FEBRUARY 2005)

MANUFACTURER NAME	MODULE MODEL NUMBER	DESCRIPTION	CEC PTC* RATING	NOTES
Siemens Solar Industries	SM46	46W PowerMax Module	41.5	NA
Siemens Solar Industries	SM46J	46W PowerMax Module w/conduit-ready J-box	41.5	NA
Siemens Solar Industries	SM50	50W PowerMax Module	45	NA
Siemens Solar Industries	SM50-H	50W 33 cell PowerMax Module	45.1	NA
Siemens Solar Industries	SM50-HJ	50W 33 cell PowerMax Module w/conduit-ready J-box	45.1	NA
Siemens Solar Industries	SM50-J	50W PowerMax Module w/conduit-ready J-box	45	NA
Siemens Solar Industries	SM55	55W PowerMax Module	49.6	NA
Siemens Solar Industries	SM55-J	55W PowerMax Module w/conduit-ready J-box	49.6	NA
Siemens Solar Industries	SM6	6 W PowerMax Module	5.4	NA
Siemens Solar Industries	SP130-24P	130W 24V PowerMax Module	116.7	NA
Siemens Solar Industries	SP140-24P	140W 24V PowerMax Module	125.8	NA
Siemens Solar Industries	SP150-24P	150W 24V PowerMax Module	134.9	NA

Manufacturer	Model	Description		
Siemens Solar Industries	SP18	18W 6V/12V PowerMax Module	16.2	NA
Siemens Solar Industries	SP36	36W 6V/12V PowerMax Module	32.4	NA
Siemens Solar Industries	SP65	65W 6V/12V PowerMax Module	58.4	NA
Siemens Solar Industries	SP70	70W 6V/12V PowerMax Module	62.9	NA
Siemens Solar Industries	SP75	75W 6V/12V PowerMax Module	67.5	NA
Siemens Solar Industries	SR100	100W 6V/12V PowerMax Module	89.9	NA
Siemens Solar Industries	SR50	50W 6V/12V PowerMax Module	44.9	NA
Siemens Solar Industries	SR90	90W 6V/12V PowerMax Module	80.8	NA
Siemens Solar Industries	ST36	36W 12V PowerMax Module	31.7	NA
Siemens Solar Industries	ST40	40W 12V PowerMax Module	35.3	NA
Solar Integrated Technologies	SR2001A	816W Flat-Plate Single-Ply Roofing Membrane	771.6	NA
Solar Integrated Technologies	SR2004	1488W Flat-Plate Single-Ply Roofing Membrane	1407	NA
Solar Integrated Technologies	SR2004A	744W Flat-Plate Single-Ply Roofing Membrane	703.5	NA

(Continued)

TABLE C.1 LIST OF ELIGIBLE PHOTOVOLTAIC MODULES CALIFORNIA ENERGY COMMISSION EMERGING (Continued)

RENEWABLES PROGRAM (FEBRUARY 2005)

MANUFACTURER NAME	MODULE MODEL NUMBER	DESCRIPTION	CEC PTC* RATING	NOTES
Solar Integrated Technologies	SR372	372W Flat-Plate Single-Ply Roofing Membrane	351.7	NA
Solec International, Inc.	S-055	55W Framed Crystalline Solar Electric Module	47.7	NA
Solec International, Inc.	S-100D	100W Dual-Voltage Crystalline Solar Electric Module	86.5	NA
Solec International, Inc.	SQ-080	80W Framed Crystalline Solar Electric Module	68.9	NA
Solec International, Inc.	SQ-090	90W Framed Crystalline Solar Electric Module	77.8	NA
Spire Solar Chicago	SS75	75W Rail-Mounted Monocrystalline Module	66.7	NA
Spire Solar Chicago	SSC 75	75W Rail-Mounted Monocrystalline Module x	66.7	NA
Sunpower Corporation	SPR-200	200W Monocrystalline Module	180	NA
Sunpower Corporation	SPR-210	210W Monocrystalline Module	190.9	NA
Sunpower Corporation	SPR-90	90W Monocrystalline Module	81.8	NA
SunWize Technologies, LLC	SW 100	100W Monocrystalline PV Module	87.1	NA
SunWize Technologies, LLC	SW 110	110W Monocrystalline PV Module	96	NA
SunWize Technologies, LLC	SW 115	115W Monocrystalline PV Module	100.5	NA
SunWize Technologies, LLC	SW 120	120W Monocrystalline PV Module	105	NA

SunWize Technologies, LLC	SW 150L	150W Monocrystalline PV Module	130.5	NA
SunWize Technologies, LLC	SW 155L	155W Monocrystalline PV Module	134.9	NA
SunWize Technologies, LLC	SW 160L	160W Monocrystalline PV Module	139.4	NA
SunWize Technologies, LLC	SW 165L	165W Monocrystalline PV Module	143.9	NA
SunWize Technologies, LLC	SW 75	75W Monocrystalline PV Module	65.1	NA
SunWize Technologies, LLC	SW 85	85W Monocrystalline PV Module	73.9	NA
SunWize Technologies, LLC	SW 90	90W Monocrystalline PV Module	78.4	NA
SunWize Technologies, LLC	SW 95	95W Monocrystalline PV Module	82.8	NA
United Solar Systems Corp.	ASR-120	120W Arch. Standing Seam 3J a-Si Module	110.9	NA
United Solar Systems Corp.	ASR-128	128W Arch. Standing Seam 3J a-Si Module	118.3	NA
United Solar Systems Corp.	ASR-136	136W Arch. Standing Seam 3J a-Si Module	130	NA
United Solar Systems Corp.	ASR-60	60W Arch. Standing Seam 3J a-Si Module	55.4	NA
United Solar Systems Corp.	ASR-64	64W Arch. Standing Seam 3J a-Si Module	59.1	NA
United Solar Systems Corp.	ASR-68	68W Arch. Standing Seam 3J a-Si Module	65	NA

(Continued)

TABLE C.1 LIST OF ELIGIBLE PHOTOVOLTAIC MODULES CALIFORNIA ENERGY COMMISSION EMERGING (Continued)

RENEWABLES PROGRAM (FEBRUARY 2005)

MANUFACTURER NAME	MODULE MODEL NUMBER	DESCRIPTION	CEC PTC* RATING	NOTES
United Solar Systems Corp.	ES-116	116W a-Si Module with Black Anodized Frame	109.8	NA
United Solar Systems Corp.	ES-124	124W a-Si Module with Black Anodized Frame	117.4	NA
United Solar Systems Corp.	ES-58	58W a-Si Module with Black Anodized Frame	54.9	NA
United Solar Systems Corp.	ES-62T	62W a-Si Module with Black Anodized Frame	58.7	NA
United Solar Systems Corp.	PVL-116(DM)	116W Field Applied 3J a-Si Laminate, Deck-Mounted	107.4	NA
United Solar Systems Corp.	PVL-116(PM)	116W Field Applied 3J a-Si Laminate, Purlin-Mounted	109.9	NA
United Solar Systems Corp.	PVL-124	124W Field Applied 3J a-Si Laminate	118.5	NA
United Solar Systems Corp.	PVL-128(DM)	128W Field Applied 3J a-Si Laminate, Deck-Mounted	118.4	NA
United Solar Systems Corp.	PVL-128(PM)	128W Field Applied 3J a-Si Laminate, Purlin-Mounted	121.2	NA
United Solar Systems Corp.	PVL-136	136W Field Applied 3J a-Si Laminate	130	NA
United Solar Systems Corp.	PVL-29(DM)	29W Field Applied 3J a-Si Laminate, Deck-Mounted	26.9	NA
United Solar Systems Corp.	PVL-29(PM)	29W Field Applied 3J a-Si Laminate, Purlin-Mounted	27.5	NA

Company	Model	Description		
United Solar Systems Corp.	PVL-31	31W Field Applied 3J a-Si Laminate	29.6	NA
United Solar Systems Corp.	PVL-58(DM)	58W Field Applied 3J a-Si Laminate, Deck-Mounted	53.7	NA
United Solar Systems Corp.	PVL-58(PM)	58W Field Applied 3J a-Si Laminate, Purlin-Mounted	55	NA
United Solar Systems Corp.	PVL-62	62W Field Applied 3J a-Si Laminate	59.3	NA
United Solar Systems Corp.	PVL-64(DM)	64W Field Applied 3J a-Si Laminate, Deck-Mounted	59.2	NA
United Solar Systems Corp.	PVL-64(PM)	64W Field Applied 3J a-Si Laminate, Purlin-Mounted	60.6	NA
United Solar Systems Corp.	PVL-68	68W Field Applied 3J a-Si Laminate	65	NA
United Solar Systems Corp.	PVL-87(DM)	87W Field Applied 3J a-Si Laminate, Deck-Mounted	80.6	NA
United Solar Systems Corp.	PVL-87(PM)	87W Field Applied 3J a-Si Laminate, Purlin-Mounted	82.5	NA
United Solar Systems Corp.	PVL-93	93W Field Applied 3J a-Si Laminate	88.9	NA
United Solar Systems Corp.	PVR10T	62W a-Si Roof Module	59.3	NA
United Solar Systems Corp.	PVR15T	93W a-Si Roof Module	88.9	NA
United Solar Systems Corp.	PVR20T	124W a-Si Roof Module	118.5	NA
United Solar Systems Corp.	PVR5T	31W a-Si Roof Module	29.6	NA

(Continued)

TABLE C.1 LIST OF ELIGIBLE PHOTOVOLTAIC MODULES CALIFORNIA ENERGY COMMISSION EMERGING (*Continued*)

RENEWABLES PROGRAM (FEBRUARY 2005)

MANUFACTURER NAME	MODULE MODEL NUMBER	DESCRIPTION	CEC PTC* RATING	NOTES
United Solar Systems Corp.	SFS-11L-10	10–68W Structural Standing Seam 3J a-Si Module	650	NA
United Solar Systems Corp.	SFS-11L-11	11–68W Structural Standing Seam 3J a-Si Module	715	NA
United Solar Systems Corp.	SFS-11L-12	12–68W Structural Standing Seam 3J a-Si Module	780	NA
United Solar Systems Corp.	SFS-22L-10	10–136W Structural Standing Seam 3J a-Si Module	1300	NA
United Solar Systems Corp.	SFS-22L-11	11–136W Structural Standing Seam 3J a-Si Module	1430	NA
United Solar Systems Corp.	SFS-22L-12	12–136W Structural Standing Seam 3J a-Si Module	1560	NA
United Solar Systems Corp.	SHR-15	15W Shingle 3J a-Si Module	13.9	NA
United Solar Systems Corp.	SHR-17	17W Shingle 3J a-Si Module	15.7	NA
United Solar Systems Corp.	SSR-120	120W Structural Standing Seam 3J a-Si Module, Purlin-Mounted	113.5	NA
United Solar Systems Corp.	SSR-120(DM)	120W Structural Standing Seam 3J a-Si Module, Deck-Mounted	110.9	NA
United Solar Systems Corp.	SSR-120J	120W Structural Standing Seam 3J a-Si Module w/J-box, Purlin-Mounted	113.5	NA
United Solar Systems Corp.	SSR-120J(DM)	120W Structural Standing Seam 3J a-Si Module w/J-box, Deck-Mounted	110.9	NA

Manufacturer	Model	Description		
United Solar Systems Corp.	SSR-128	128W Structural Standing Seam 3J a-Si Module, Purlin-Mounted	121.1	NA
United Solar Systems Corp.	SSR-128(DM)	128W Structural Standing Seam 3J a-Si Module, Deck-Mounted	118.3	NA
United Solar Systems Corp.	SSR-128J	128W Structural Standing Seam 3J a-Si Module w/J-box, Purlin-Mounted	121.1	NA
United Solar Systems Corp.	SSR-128J(DM)	128W Structural Standing Seam 3J a-Si Module w/J-box, Deck-Mounted	118.3	NA
United Solar Systems Corp.	SSR-136	136W Structural Standing Seam 3J a-Si Module	130	NA
United Solar Systems Corp.	SSR-60	60W Structural Standing Seam 3J a-Si Module, Purlin-Mounted	56.8	NA
United Solar Systems Corp.	SSR-60(DM)	60W Structural Standing Seam 3J a-Si Module, Deck-Mounted	55.4	NA
United Solar Systems Corp.	SSR-60J	60W Structural Standing Seam 3J a-Si Module w/J-box, Purlin-Mounted	56.8	NA
United Solar Systems Corp.	SSR-60J(DM)	60W Structural Standing Seam 3J a-Si Module w/J-box, Deck-Mounted	55.4	NA
United Solar Systems Corp.	SSR-64	64W Structural Standing Seam 3J a-Si Module, Purlin-Mounted	60.6	NA
United Solar Systems Corp.	SSR-64(DM)	64W Structural Standing Seam 3J a-Si Module, Deck-Mounted	59.1	NA
United Solar Systems Corp.	SSR-64J	64W Structural Standing Seam 3J a-Si Module w/J-box, Purlin-Mounted	60.6	NA
United Solar Systems Corp.	SSR-64J(DM)	64W Structural Standing Seam 3J a-Si Module w/J-box, Deck-Mounted	59.1	NA
United Solar Systems Corp.	SSR-68	68W Structural Standing Seam 3J a-Si Module	65	NA

(*Continued*)

TABLE C.1 LIST OF ELIGIBLE PHOTOVOLTAIC MODULES CALIFORNIA ENERGY COMMISSION EMERGING (Continued)

RENEWABLES PROGRAM (FEBRUARY 2005)

MANUFACTURER NAME	MODULE MODEL NUMBER	DESCRIPTION	CEC PTC* RATING	NOTES
United Solar Systems Corp.	US-116	116W Framed Triple-Junction a-Si Module	109.9	NA
United Solar Systems Corp.	US-32	32W Framed Triple-Junction a-Si Module	30.3	NA
United Solar Systems Corp.	US-39	39W Framed Triple-Junction a-Si Module	36.9	NA
United Solar Systems Corp.	US-42	42W Framed Triple-Junction a-Si Module	39.8	NA
United Solar Systems Corp.	US-60	60W Framed Triple-Junction a-Si Module	56.8	NA
United Solar Systems Corp.	US-64	64W Framed Triple-Junction a-Si Module	60.6	NA
Webel-SL Energy Systems	W1000-100	100W Monocrystalline PV Module	87.1	NA
Webel-SL Energy Systems	W1000-110	110W Monocrystalline PV Module	96	NA
Webel-SL Energy Systems	W1000-115	115W Monocrystalline PV Module	100.5	NA
Webel-SL Energy Systems	W1000-120	120W Monocrystalline PV Module	105	NA
Webel-SL Energy Systems	W1600-150	150W Monocrystalline PV Module	130.5	NA
Webel-SL Energy Systems	W1600-155	155W Monocrystalline PV Module	134.9	NA

Webel-SL Energy Systems	W1600-160	160W Monocrystalline PV Module	139.4	NA
Webel-SL Energy Systems	W1600-165	165W Monocrystalline PV Module	143.9	NA
Webel-SL Energy Systems	W900-75	75W Monocrystalline PV Module	65.1	NA
Webel-SL Energy Systems	W900-80	80W Monocrystalline PV Module	69.5	NA
Webel-SL Energy Systems	W900-85	85W Monocrystalline PV Module	73.9	NA
Webel-SL Energy Systems	W900-90	90W Monocrystalline PV Module	78.4	NA

*PTC stands for "PVUSA Test Conditions. The PTC watt rating is based on 1000- W/m^2 solar irradiance, 20°C ambient temperature, and 1-m/s wind speed. The PTC watt rating is lower than the Standard Test Conditions (STC), a watt-rating used by manufacturers.

TABLE C.2 CEC CERTIFIED INVERTERSIST OF ELIGIBLE INVERTERS CALIFORNIA ENERGY COMMISSION EMERGING RENEWABLES PROGRAM (FEBRUARY 2005)

MANUFACTURER NAME	INVERTER MODEL NUMBER	DESCRIPTION	POWER RATING (W)	75% LOAD EFFICIENCY	APPROVED BUILT-IN METER	NOTES
Alpha Technologies, Inc.	Solaris 3500	3.5kW, 240Vac, 96-200Vdc, NEMA-3R, Grid Interactive PV Inverter, LCD, MPPT	3,500	93	Yes	NA
Ballard Power Systems Corporation	EPC-PV-208-30kW	Utility Interactive 208V 30kW PV Power Converter System	30,000	95	No	NA
Ballard Power Systems Corporation	EPC-PV-208-75kW	Utility Interactive 75kW PV Power Converter System	75,000	93	Yes	NA
Ballard Power Systems Corporation	EPC-PV-480-30kW	Utility Interactive 480V 30kW PV Power Converter System	30,000	95	No	NA
Ballard Power Systems Corporation	EPC-PV-480-75kW	Utility Interactive 75kW PV Power Converter System	75,000	93	Yes	NA
Beacon Power Corporation	M5	+ 5kW Power Conversion System	5,000	90	No	NA
Bergey Windpower Co.	Gridtek 10	10kW, 240Vac Split-Phase, Utility Interactive Inverter	10,000	93	No	NA
Fronius USA, LLC	IG 2000	2,000W Grid-tied Units with Integrated Breakers and LCD	2,000	94	Yes	NA

Fronius USA, LLC	IG 2500-LV	2,350W Grid-tied Units with Integrated Breakers and LCD	2,350	94	Yes	NA
Fronius USA, LLC	IG 3000	2,700W Grid-tied Units with Integrated Breakers and LCD	2,700	94	Yes	NA
Fronius USA, LLC	IG 4000	4,000W Grid-tied Unit with Integrated Disconnects and Performance Meter	4,000	94	Yes	NA
Fronius USA, LLC	IG 4500-LV	4,500W Grid-tied Unit with Integrated Disconnects and Performance Meter	4,500	93	Yes	NA
Fronius USA, LLC	IG 5100	5,100W Grid-tied Unit with Integrated Disconnects and Performance Meter	5,100	94	Yes	NA
Magnetek	PVI-3000-I-OUTD-US	3kW, 150-600 VDC Utility Interactive Inverter	3,000	93	Yes	NA
Nextek Power Systems, Inc.	NPS-1000	1,000W Direct Coupling dc Rectifier	1,000	94	No	NA
OutBack Power Systems	GTFX 2524	2,500W Utility Interactive (w/battery backup) 24Vdc Inverter	2,500	91	No	NA
OutBack Power Systems	GTFX 3048	3,000W Utility Interactive (w/battery backup) 48Vdc Inverter	3,000	92	No	NA
OutBack Power Systems	GVFX 3524	3,500W Utility Interactive (w/battery backup) 24Vdc Inverter	3,500	91	No	NA

(Continued)

TABLE C.2 CEC CERTIFIED INVERTERSIST OF ELIGIBLE INVERTERS CALIFORNIA ENERGY COMMISSION EMERGING RENEWABLES PROGRAM (FEBRUARY 2005) (Continued)

MANUFACTURER NAME	INVERTER MODEL NUMBER	DESCRIPTION	POWER RATING (W)	75% LOAD EFFICIENCY	APPROVED BUILT-IN METER	NOTES
OutBack Power Systems	GVFX 3648	3,600W Utility Interactive (w/battery backup) 48Vdc Inverter	3,600	92	No	NA
Pacific Solar Pty Limited	SDEIP2-09	240W, 240V Module PV Inverter for the SunEmpower (PP-USA-213)	240	93	No	NA
PV Powered LLC	PVP1100E	1,100W Utility Interactive Inverter	1,100	95	No	NA
PV Powered LLC	PVP1800	1,800W Utility Interactive Inverter	1,800	95	Yes	NA
PV Powered LLC	PVP2800-208	2,800W (208Vac) Utility Interactive Inverter	2,800	97	Yes	NA
PV Powered LLC	PVP2800-240	2,800W (240Vac) Utility Interactive Inverter	2,800	97	Yes	NA
SatCon Power Systems Canada Ltd.	AE-100-60-PV-A	Three-phase 100kW Utility Interactive Inverter	100,000	96	No	NA
SatCon Power Systems Canada Ltd.	AE-225-60-PV-A	225kW Three-phase Inverter 480Vac	225,000	95	No	NA
SatCon Power Systems Canada Ltd.	AE-30-60-PV-E	30kW Single-phase Utility Interactive Inverter	30,000	93	No	NA
SatCon Power Systems Canada Ltd.	AE-50-60-PV-A	50kW 480Vac Three-phase Utility Interactive Inverter	50,000	94	Yes	NA
Sharp Corporation	JH-3500U	Utility Interactive Inverter 240Vac L-L, 3.5kW	3,500	92	Yes	NA

Manufacturer	Model	Description				
SMA America	SB6000U	Sunny Boy 6,000W Utility Interactive Inverter with Performance Meter	6,000	94	Yes	NA
SMA America	SC125U	125kW 3-phase 480Vac, 275-600Vdc, Utility Interactive Inverter	125,000	95	Yes	NA
SMA America	SWR1100U	1,100W, 240Vac Sunny Boy String Inverter	1,100	93	No	NA
SMA America	SWR1100U-SBD	1,100W, 240Vac Sunny Boy String Inverter with display	1,100	93	Yes	NA
SMA America	SWR1800U	1.8kW, 120Vac Sunny Boy String Inverter	1,800	93	No	NA
SMA America	SWR1800U-SBD	1.8kW, 120Vac Sunny Boy String Inverter, with display	1,800	93	Yes	NA
SMA America	SWR2500U (208V)	2.1kW, 208Vac Sunny Boy String Inverter	2,100	94	No	NA
SMA America	SWR2500U (240V)	2.5kW, 240Vac Sunny Boy String Inverter	2,500	94	No	NA
SMA America	SWR2500U-SBD (208V)	2.1kW, 208Vac Sunny Boy String Inverter, with Display	2,100	94	Yes	NA
SMA America	SWR2500U-SBD (240V)	2.5kW, 240Vac Sunny Boy String Inverter, with display	2,500	94	Yes	NA
SMA America	SWR700U	700W, 120Vac Sunny Boy String Inverter	700	93	No	NA

(Continued)

TABLE C.2 CEC CERTIFIED INVERTERSIST OF ELIGIBLE INVERTERS CALIFORNIA ENERGY COMMISSION EMERGING RENEWABLES PROGRAM (FEBRUARY 2005) (Continued)

MANUFACTURER NAME	INVERTER MODEL NUMBER	DESCRIPTION	POWER RATING (W)	75% LOAD EFFICIENCY	APPROVED BUILT-IN METER	NOTES
SMA America	SWR700U-SBD	700W, 120Vac Sunny Boy String Inverter with display	700	93	Yes	NA
Solectria Renewables, LLC	PVI 13kW	13kW 208 and 480Vac Commercial Grid-tied Solar PV Inverter	13,200	94	No	NA
Xantrex Technology, Inc.	BWT10240	10kW, 240Vac Split-phase, Utility Interactive Inverter	10,000	93	No	NA
Xantrex Technology, Inc.	GT3.0-NA-DS-240	3.0kW, 240Vac, 195-600Vdc Grid-Tied Inverter	3,000	94	Yes	NA
Xantrex Technology, Inc.	PV-100208	100kW 208Vac/3-phase Utility Interactive Inverter	100,000	95	No	NA
Xantrex Technology, Inc.	PV-100S-208	100kW 208Vac, 330-600Vdc Inverter System with automatic transformer disconnect	100,000	95	Yes	NA
Xantrex Technology, Inc.	PV-100S-480	100kW 480Vac, 330-600Vdc Inverter System with automatic transformer disconnect	100,000	95	Yes	NA
Xantrex Technology, Inc.	PV-10208	10kW 208Vac/3-phase Utility Interactive Inverter	10,000	94	Yes	NA
Xantrex Technology, Inc.	PV-15208	15kW 208Vac/3-phase Utility Interactive Inverter	15,000	95	Yes	NA

Manufacturer	Model	Description				
Xantrex Technology, Inc.	PV-20208	20kW 208Vac/3-phase Utility Interactive Inverter	20,000	94	Yes	NA
Xantrex Technology, Inc.	PV-225208	225kW 208Vac/3-phase Utility Interactive Inverter	225,000	95	No	NA
Xantrex Technology, Inc.	PV-30208	30kW 208Vac/3-phase Utility Interactive Inverter	30,000	94	Yes	NA
Xantrex Technology, Inc.	PV-45208	45kW 208Vac/3-phase Utility Interactive Inverter	45,000	94	Yes	NA
Xantrex Technology, Inc.	PV-5208	5kW, 208Vac/3-phase, Photovoltaic Utility Interactive Inverter	5,000	93	No	NA
Xantrex Technology, Inc.	STXR1000	1.0kVA, 42–85Vdc, 240Vac, Trace Engr. Sine Wave Inverter (Sunsweep MPPT)	1,000	88	No	NA
Xantrex Technology, Inc.	STXR1500	1.5kVA, 42–85Vdc, 240Vac, Trace Engr. Sine Wave Inverter (Sunsweep MPPT)	1,500	89	No	NA
Xantrex Technology, Inc.	STXR1500 v5.0	1.5kVA, 42–85Vdc, 240Vac, Trace Eng. Sine Wave Inverter (Sunsweep MPPT)	1,500	89	Yes	NA
Xantrex Technology, Inc.	STXR2000	2.0kVA, 42–85Vdc, 240Vac, Trace Engr. Sine Wave Inverter (Sunsweep MPPT)	2,000	90	No	NA
Xantrex Technology, Inc.	STXR2500	2.5kVA, 42–75Vdc, 240Vac, Trace Engr. Sine Wave Inverter (Sunsweep MPPT)	2,500	90	No	NA
Xantrex Technology, Inc.	STXR2500 v5.0	2.5kVA, 42–75Vdc, 240Vac, Trace Eng. Sine Wave Inverter (Sunsweep MPPT)	2,500	90	Yes	NA

(Continued)

TABLE C.2 CEC CERTIFIED INVERTERSIST OF ELIGIBLE INVERTERS CALIFORNIA ENERGY COMMISSION EMERGING RENEWABLES PROGRAM (FEBRUARY 2005) (Continued)

MANUFACTURER NAME	INVERTER MODEL NUMBER	DESCRIPTION	POWER RATING (W)	75% LOAD EFFICIENCY	APPROVED BUILT-IN METER	NOTES
Xantrex Technology, Inc.	SW4024 (w GTI)	4.0kVA, 24Vdc, 120Vac, Trace Engr. batt. bkp., Sine Wave Inverter	4,000	88	No	NA
Xantrex Technology, Inc.	SW4048 (w GTI)	4.0kVA, 48Vdc, 120Vac, Trace Engr. batt. bkp., Sine Wave Inverter	4,000	88	No	NA
Xantrex Technology, Inc.	SW5548 (w GTI)	5.5kVA, 48Vdc, 120Vac, Trace Engr. batt. bkp., Sine Wave Inverter	5,500	89	No	NA

TABLE C.3 LIST OF ELIGIBLE SYSTEM PERFORMANCE METERS CALIFORNIA ENERGY COMMISSION EMERGING RENEWABLES PROGRAM (FEBRUARY 2005)

MANUFACTURER NAME	MODEL NUMBER	DISPLAY TYPE	NOTES
ABB/Elster	1S	LCD	NA
ABB/Elster	2S	LCD	NA
ABB/Elster	3S	LCD	NA
ABB/Elster	A3 Alpha	LCD	NA
ABB/Elster	AB1	Cyclometer	NA
ABB/Elster	ABS	Cyclometer	NA
ABB/Elster	Alpha	LCD	NA
ABB/Elster	Alpha Plus	LCD	NA
ABB/Elster	REX	LCD	NA
Astropower	APM2 SunChoice	LCD	NA
BP Solar	HSSM-1	LCD	NA
Brand Electronic	20-1850	LCD	NA
Brand Electronic	20-1850CI	LCD	NA
Brand Electronic	20-CTR	LCD	NA
Brand Electronic	21-1850CI	LCD	NA
Brand Electronic	4-1850	LCD	NA
Brand Electronic	ONE Meter	LCD	NA
Draker Solar Design	PVDAQ Basic	Computer monitor	NA
Draker Solar Design	PVDAQ Commercial	Computer monitor	NA
E-MON	D-MON 208100 KIT	LCD	NA
E-MON	D-MON 208100C KIT	LCD	NA

(Continued)

TABLE C.3 LIST OF ELIGIBLE SYSTEM PERFORMANCE METERS CALIFORNIA ENERGY COMMISSION EMERGING RENEWABLES PROGRAM (FEBRUARY 2005) (Continued)

MANUFACTURER NAME	MODEL NUMBER	DISPLAY TYPE	NOTES
E-MON	D-MON 208100D KIT	LCD	NA
E-MON	D-MON 2081600 KIT	LCD	NA
E-MON	D-MON 2081600C KIT	LCD	NA
E-MON	D-MON 2081600D KIT	LCD	NA
E-MON	D-MON 208200 KIT	LCD	NA
E-MON	D-MON 208200C KIT	LCD	NA
E-MON	D-MON 208200D KIT	LCD	NA
E-MON	D-MON 20825 KIT	LCD	NA
E-MON	D-MON 20825C KIT	LCD	NA
E-MON	D-MON 20825D KIT	LCD	NA
E-MON	D-MON 208320 KIT	LCD	NA
E-MON	D-MON 208320C KIT	LCD	NA
E-MON	D-MON 208320D KIT	LCD	NA
E-MON	D-MON 208400 KIT	LCD	NA
E-MON	D-MON 208400C KIT	LCD	NA
E-MON	D-MON 208400D KIT	LCD	NA
E-MON	D-MON 20850 KIT	LCD	NA
E-MON	D-MON 20850C KIT	LCD	NA
E-MON	D-MON 20850D KIT	LCD	NA
E-MON	D-MON 208800 KIT	LCD	NA
E-MON	D-MON 208800C KIT	LCD	NA

E-MON	D-MON 208800D KIT	LCD	NA
E-MON	D-MON 480100 KIT	LCD	NA
E-MON	D-MON 480100C KIT	LCD	NA
E-MON	D-MON 480100D KIT	LCD	NA
E-MON	D-MON 4801600 KIT	LCD	NA
E-MON	D-MON 4801600 KIT	LCD	NA
E-MON	D-MON 4801600C KIT	LCD	NA
E-MON	D-MON 4801600C KIT	LCD	NA
E-MON	D-MON 4801600D KIT	LCD	NA
E-MON	D-MON 4801600D KIT	LCD	NA
E-MON	D-MON 480200 KIT	LCD	NA
E-MON	D-MON 480200C KIT	LCD	NA
E-MON	D-MON 480200D KIT	LCD	NA
E-MON	D-MON 48025 KIT	LCD	NA
E-MON	D-MON 48025C KIT	LCD	NA
E-MON	D-MON 48025D KIT	LCD	NA
E-MON	D-MON 4803200 KIT	LCD	NA
E-MON	D-MON 4803200C KIT	LCD	NA
E-MON	D-MON 4803200D KIT	LCD	NA
E-MON	D-MON 480400 KIT	LCD	NA
E-MON	D-MON 480400C KIT	LCD	NA
E-MON	D-MON 480400D KIT	LCD	NA
E-MON	D-MON 48050 KIT	LCD	NA
E-MON	D-MON 48050C KIT	LCD	NA

(Continued)

TABLE C.3   LIST OF ELIGIBLE SYSTEM PERFORMANCE METERS CALIFORNIA ENERGY COMMISSION EMERGING RENEWABLES PROGRAM (FEBRUARY 2005) (Continued)

MANUFACTURER NAME	MODEL NUMBER	DISPLAY TYPE	NOTES
E-MON	D-MON 48050D KIT	LCD	NA
E-MON	D-MON 480800 KIT	LCD	NA
E-MON	D-MON 480800C KIT	LCD	NA
E-MON	D-MON 480800D KIT	LCD	NA
E-MON	E-CON 2120100-SA KIT	LCD	NA
E-MON	E-CON 2120200-SA KIT	LCD	NA
E-MON	E-CON 212025-SA KIT	LCD	NA
E-MON	E-CON 212050-SA KIT	LCD	NA
E-MON	E-CON 2277100-SA KIT	LCD	NA
E-MON	E-CON 22772000-SA KIT	LCD	NA
E-MON	E-CON 227725-SA KIT	LCD	NA
E-MON	E-CON 227750-SA KIT	LCD	NA
E-MON	E-CON 3208100-SA KIT	LCD	NA
E-MON	E-CON 3208200-SA KIT	LCD	NA
E-MON	E-CON 320825-SA KIT	LCD	NA
E-MON	E-CON 320850-SA KIT	LCD	NA
Fat Spaniel Technologies, Inc.	PV2Web	LCD and other digital display types	(PC-based for SMA-America inverters)
General Electric	I70S	Electromechanical or cyclometer register	NA
General Electric	KV	LCD	NA

General Electric	KV2	LCD	NA
Global Power Products	ENER-COMM ECE-100	LCD	NA
Global Power Products	ENER-COMM ECE-200	LCD	NA
Global Power Products	ENER-COMM ECED-100	LCD	NA
Home Energy Systems, Inc.	100 A	LCD	NA
Home Energy Systems, Inc.	25 A	LCD	NA
Home Energy Systems, Inc.	50 A	LCD	NA
Integrated Metering Systems	1101201	LCD	NA
Integrated Metering Systems	1101201-T	LCD	NA
Integrated Metering Systems	1101202	LCD	NA
Integrated Metering Systems	1101202-T	LCD	NA
Integrated Metering Systems	1102401	LCD	NA
Integrated Metering Systems	1102401-T	LCD	NA
Integrated Metering Systems	1102402	LCD	NA
Integrated Metering Systems	1102402-T	LCD	NA
Integrated Metering Systems	1102771	LCD	NA
Integrated Metering Systems	1102771-T	LCD	NA
Integrated Metering Systems	1102772	LCD	NA
Integrated Metering Systems	1102772-T	LCD	NA
Integrated Metering Systems	1103471	LCD	NA
Integrated Metering Systems	1103471-T	LCD	NA
Integrated Metering Systems	1103472	LCD	NA
Integrated Metering Systems	1103472-T	LCD	NA
Integrated Metering Systems	1201201	LCD	NA

(Continued)

TABLE C.3 LIST OF ELIGIBLE SYSTEM PERFORMANCE METERS CALIFORNIA ENERGY COMMISSION EMERGING RENEWABLES PROGRAM (FEBRUARY 2005) (Continued)

MANUFACTURER NAME	MODEL NUMBER	DISPLAY TYPE	NOTES
Integrated Metering Systems	1201201-T	LCD	NA
Integrated Metering Systems	1201202	LCD	NA
Integrated Metering Systems	1201202-T	LCD	NA
Integrated Metering Systems	1202401	LCD	NA
Integrated Metering Systems	1202401-T	LCD	NA
Integrated Metering Systems	1202402	LCD	NA
Integrated Metering Systems	1202402-T	LCD	NA
Integrated Metering Systems	1202771	LCD	NA
Integrated Metering Systems	1202771-T	LCD	NA
Integrated Metering Systems	1202772	LCD	NA
Integrated Metering Systems	1202772-T	LCD	NA
Integrated Metering Systems	1203471	LCD	NA
Integrated Metering Systems	1203471-T	LCD	NA
Integrated Metering Systems	1203472	LCD	NA
Integrated Metering Systems	1203472-T	LCD	NA
Integrated Metering Systems	1301201	LCD	NA
Integrated Metering Systems	1301201-T	LCD	NA
Integrated Metering Systems	1301202	LCD	NA
Integrated Metering Systems	1301202-T	LCD	NA
Integrated Metering Systems	1302401	LCD	NA
Integrated Metering Systems	1302401-T	LCD	NA

Integrated Metering Systems	1302402	LCD	NA
Integrated Metering Systems	1302402-T	LCD	NA
Integrated Metering Systems	1302771	LCD	NA
Integrated Metering Systems	1302771-T	LCD	NA
Integrated Metering Systems	1302772	LCD	NA
Integrated Metering Systems	1302772-T	LCD	NA
Integrated Metering Systems	1303471	LCD	NA
Integrated Metering Systems	1303471-T	LCD	NA
Integrated Metering Systems	1303472	LCD	NA
Integrated Metering Systems	1303472-T	LCD	NA
Integrated Metering Systems	2111201	LCD	NA
Integrated Metering Systems	2111201-T	LCD	NA
Integrated Metering Systems	2111202	LCD	NA
Integrated Metering Systems	2111202-T	LCD	NA
Integrated Metering Systems	2112401	LCD	NA
Integrated Metering Systems	2112401-T	LCD	NA
Integrated Metering Systems	2112402	LCD	NA
Integrated Metering Systems	2112402-T	LCD	NA
Integrated Metering Systems	2112771	LCD	NA
Integrated Metering Systems	2112771-T	LCD	NA
Integrated Metering Systems	2112772	LCD	NA
Integrated Metering Systems	2112772-T	LCD	NA
Integrated Metering Systems	2113471	LCD	NA
Integrated Metering Systems	2113471-T	LCD	NA

(Continued)

TABLE C.3 LIST OF ELIGIBLE SYSTEM PERFORMANCE METERS CALIFORNIA ENERGY COMMISSION EMERGING RENEWABLES PROGRAM (FEBRUARY 2005) (Continued)

MANUFACTURER NAME	MODEL NUMBER	DISPLAY TYPE	NOTES
Integrated Metering Systems	2113472	LCD	NA
Integrated Metering Systems	2113472-T	LCD	NA
Integrated Metering Systems	2221201	LCD	NA
Integrated Metering Systems	2221201-T	LCD	NA
Integrated Metering Systems	2221202	LCD	NA
Integrated Metering Systems	2221202-T	LCD	NA
Integrated Metering Systems	2222401	LCD	NA
Integrated Metering Systems	2222401-T	LCD	NA
Integrated Metering Systems	2222402	LCD	NA
Integrated Metering Systems	2222402-T	LCD	NA
Integrated Metering Systems	2222771	LCD	NA
Integrated Metering Systems	2222771-T	LCD	NA
Integrated Metering Systems	2222772	LCD	NA
Integrated Metering Systems	2222772-T	LCD	NA
Integrated Metering Systems	2223471	LCD	NA
Integrated Metering Systems	2223471-T	LCD	NA
Integrated Metering Systems	2223472	LCD	NA
Integrated Metering Systems	2223472-T	LCD	NA
iSYS Systems	PVM-Net	Computer monitor	NA
Landis + Gyr Inc.	AL Altimus 2S	LCD	NA
OutBack Power Systems	MX60	LCD	NA

Manufacturer	Product	Display	Notes
Pacific Solar	Sunlogger	LCD	SunEmpower product only
Poobah Industries	SB-001	LCD	Software for PalmOne for Sunny Boy inverters
Power Measurement	ION 6200	LCD	NA
Righthand Engineering, LLC	WinVerter-Monitor	Computer monitor	PC-based for Xantrex Trace SW series inverters
Schlumberger/Sangamo	Centron C1S	LCD or cyclometer	NA
Schlumberger/Sangamo	J4S	LCD or cyclometer	NA
Schlumberger/Sangamo	J5S	LCD or cyclometer	NA
SMA America	Sunny Boy Control	LCD	NA
SMA America	Sunny Boy Control Light	LCD	NA
SMA America	Sunny Boy Control Plus	LCD	NA
SMA America	Sunny Boy Control Plus-485	LCD	NA
SMA America	Sunny Boy Control-485	LCD	NA
SMA America	Sunny Data	Computer monitor	NA
SMA America	Sunny Data Control	Computer monitor	NA
SMA America	Sunny Data Control	Computer monitor	NA
SMA America	SWR LCD	LCD	NA
SolarQuest	rMeter Monitor	LCD or computer	NA

HISTORICAL TIME LINE
OF SOLAR ENERGY

This appendix is an adaptation of the "Solar History Timeline," courtesy of the U.S. Department of Energy.

Seventh Century BC. A magnifying glass is used to concentrate the sun's rays on a fuel and light a fire for light, warmth, and cooking.

Third Century BC. Greeks and Romans use mirrors to light torches for religious purposes.

Second Century BC. As early as 212 BC, the Greek scientist Archimedes makes use of the reflective properties of bronze shields to focus sunlight and set fire to Rome's wooden ships, which were besieging. Although there is no proof that this actually happened, the Greek navy recreated the experiment in 1973 and successfully set fire to a wooden boat 50 m away.

AD 20. The Chinese report using mirrors to light torches for religious purposes.

First to Fourth Centuries. In the first to the fourth centuries, Roman bathhouses are built with large, south-facing windows to let in the sun's warmth.

Sixth Century. Sunrooms on houses and public buildings are so common that the Justinian Code establishes "sun rights" to ensure that a building has access to the sun.

Thirteenth Century. In North America, the ancestors of Pueblo people known as the *Anasazi* built south-facing cliff dwellings that captured the warmth of the winter sun.

1767. Swiss scientist Horace de Saussure is credited with building the world's first solar collector, later used by Sir John Herschel to cook food during his South African expedition in the 1830s.

1816. On September 27, 1816, Robert Stirling applies for a patent for his *economiser*, a solar thermal electric technology that concentrates the sun's thermal energy to produce electric power.

1839. French scientist Edmond Becquerel discovers the photovoltaic effect while experimenting with an electrolytic cell made up of two metal electrodes placed in an electricity-conducting solution; the electricity generation increases when exposed to light.

1860s. French mathematician August Mouchet proposes an idea for solar-powered steam engines. In the next two decades, he and his assistant, Abel Pifre, will construct the first solar-powered engines for a variety of uses. The engines are the predecessors of modern parabolic dish collectors.

1873. Willoughby Smith discovers the photoconductivity of selenium.

1876. William Grylls Adams and Richard Evans Day discover that selenium produces electricity when exposed to light. Although selenium solar cells fail to convert enough sunlight to power electrical equipment, they prove that a solid material can change light into electricity without heat or moving parts.

1880. Samuel P. Langley invents the bolometer, used to measure light from the faintest stars and the sun's heat rays. It consists of a fine wire connected to an electric circuit. When radiation falls on the wire, it becomes warmer, and this increases the electrical resistance of the wire.

1883. American inventor Charles Fritts describes the first solar cells made of selenium wafers.

1887. Heinrich Hertz discovers that ultraviolet light alters the lowest voltage capable of causing a spark to jump between two metal electrodes.

1891. Baltimore inventor Clarence Kemp patents the first commercial solar water heater.

1904. Wilhelm Hallwachs discovers that a combination of copper and cuprous oxide is photosensitive.

1905. Albert Einstein publishes his paper on the photoelectric effect, along with a paper on his theory of relativity.

1908. William J. Bailey of the Carnegie Steel Company invents a solar collector with copper coils and an insulated box, which is roughly the same collector design used today.

1914. The existence of a barrier layer in photovoltaic devices is noted.

1916. Robert Millikan provides experimental proof of the photoelectric effect.

1918. Polish scientist Jan Czochralski develops a way to grow single-crystal silicon.

1921. Albert Einstein wins the Nobel Prize for his theories explaining the photoelectric effect; for details, see his 1904 technical paper on the subject.

1932. Audobert and Stora discover the photovoltaic effect in cadmium sulfide.

1947. Because energy had become scarce during the long Second World War, passive solar buildings in the United States are in demand; Libbey-Owens-Ford Glass Company publishes a book titled, *Your Solar House*, which profiles 49 of the nation's greatest solar architects.

1953. Dr. Dan Trivich of Wayne State University makes the first theoretical calculations of the efficiencies of various materials of different band-gap widths based on the spectrum of the sun.

1954. Photovoltaic technology is born in the United States when Daryl Chapin, Calvin Fuller, and Gerald Pearson develop the silicon photovoltaic or PV cell at Bell Labs, which is the first solar cell capable of generating enough power from the sun to run everyday electrical equipment. Bell Laboratories then produces a solar cell with 6 percent efficiency, which is later augmented to 11 percent.

1955. Western Electric begins to sell commercial licenses for silicon photovoltaic technologies. Early successful products include PV-powered dollar changers and devices that decode computer punch cards and tape.

1950s. Architect Frank Bridgers designs the world's first commercial office building featuring solar water heating and design. The solar system has operated continuously since then; the Bridgers-Paxton Building is listed in the National Historic Register as the world's first solar-heated office building.

1956. William Cherry of the U.S. Signal Corps Laboratories approaches RCA Labs' Paul Rappaport and Joseph Loferski about developing a photovoltaic cell for proposed Earth-orbiting satellites.

1957. Hoffman Electronics achieves 8 percent efficient photovoltaic cells.

1958. T. Mandelkorn of U.S. Signal Corps Laboratories fabricates n-on-p (negative layer on positive layer) silicon photovoltaic cells, making them more resistant to radiation; this is critically important for cells used in space.

Hoffman Electronics achieves 9 percent solar cell efficiency.

A small array (less than 1 W) on the *Vanguard I* space satellite powers its radios. Later that year, *Explorer III, Vanguard II,* and *Sputnik-3* will be launched with PV-powered systems on board. Silicon solar cells become the most widely used energy source for space applications, and remain so today.

1959. Hoffman Electronics achieves a 10 percent efficient, commercially available cell. Hoffman also learns to use a grid contact, significantly reducing the series resistance.

On August 7, the *Explorer VI* satellite is launched with a PV array of 9600 solar cells, each measuring 1 cm × 2 cm. On October 13 *Explorer VII* is launched.

1960. Hoffman Electronics achieves 14 percent efficient photovoltaic cells.

Silicon Sensors, Inc., of Dodgeville, Wisconsin, is founded and begins producing selenium photovoltaic cells.

1962. Bell Telephone Laboratories launches Telstar, the first telecommunication satellite; its initial power is 14 W.

1963. Sharp Corporation succeeds in producing silicon PV modules.

Japan installs a 242-W photovoltaic array, the world's largest to date, on a lighthouse.

1965. Peter Glaser conceives the idea of the satellite solar power station.

1966. NASA launches the first orbiting astronomical observatory powered by a 1-kW photovoltaic array; it provides astronomical data in the ultraviolet and x-ray wavelengths filtered out by Earth's atmosphere.

1969. A "solar furnace" is constructed in Odeillo, France; it features an eight-story parabolic mirror.

1970. With help from Exxon corporation. Dr. Elliot Berman designs a significantly less costly solar cell, bringing the price down from $100/W to $20/W. Solar cells begin powering navigation warning lights and horns on offshore gas and oil rigs, lighthouses, and railroad crossings. Domestic solar applications are considered good alternatives in remote areas where utility-grid connections are too expensive.

1972. French workers install a cadmium sulfide photovoltaic system at a village school in Niger.

The Institute of Energy Conversion is established at the University of Delaware to do research and development on thin-film photovoltaic and solar thermal systems, becoming the world's first laboratory dedicated to PV research and development.

1973. The University of Delaware builds "Solar One," a PV/thermal hybrid system. Roof-integrated arrays feed surplus power through a special meter to the utility during the day; power is purchased from the utility at night. In addition to providing electricity, the arrays are like flat-plate thermal collectors; fans blow warm air from over the array to heat storage bins.

1976. The NASA Lewis Research Center starts installing the first of 83 photovoltaic power systems in every continent except Australia. They provide power for vaccine refrigeration room lighting, medical clinic lighting, telecommunications, water pumping, grain milling, and television. The project takes place from 1976 to 1985 and then from 1992 to completion in 1995. David and Christopher Wronski of RCA Laboratories produce the first amorphous silicon photovoltaic cells, which could be less expensive to manufacture than crystalline silicon devices.

1977. In July, the U.S. Energy Research and Development Administration, a predecessor of the U.S. Department of Energy, launches the Solar Energy Research Institute [today's National Renewable Energy Laboratory (NREL)], a federal facility dedicated to energy finding and improving ways to harness and use energy from the sun. Total photovoltaic manufacturing production exceeds 500 kW; 1 kW is enough power to light about ten 100-W lightbulbs.

1978. NASA's Lewis Research Center installs a 3.5-kW photovoltaic system on the Indian Reservation in southern Arizona—the world's first village system. It provides power for water pumping and residential electricity in 15 homes until 1983, when grid power reaches the village. The PV system is then dedicated to pumping water from a community well.

1980. ARCO Solar becomes the first company to produce more than 1 MW (1000 kW) of photovoltaic modules in 1 year.

At the University of Delaware, the first thin-film solar cell exceeds 10 percent efficiency; it's made of copper sulfide and cadmium sulfide.

1981. Paul MacCready builds the first solar-powered aircraft—the Solar Challenger—and flies it from France to England across the English Channel. The aircraft has more than 16,000 wing-mounted solar cells producing 3000 W of power.

1982. The first megawatt-scale PV power station goes on line in Hesperia, California. The 1-MW-capacity system, developed by ARCO Solar, has modules on 108 dual-axis trackers.

Australian Hans Tholstrup drives the first solar-powered car—the Quiet Achiever—almost 2800 mi between Sydney and in 20 days—10 days faster than the first gasoline-powered car to do so.

1983. ARCO Solar dedicates a 6-MW photovoltaic substation in central California. The 120-acre, unmanned facility supplies the Pacific Gas Electric Company's utility grid with enough power for up to 2500 homes. Solar Design Associates completes a home powered by an integrated, stand-alone, 4-kW photovoltaic system in the Hudson River Valley. Worldwide, photovoltaic production exceeds 21.3 MW, and sales top $250 million.

1984. The Sacramento Municipal Utility District commissions its first 1-MW photovoltaic electricity-generating facility.

1985. Researchers at the University of South Wales break the 20 percent efficiency barrier for silicon solar cells.

1986. The world's largest solar thermal facility is commissioned in Kramer Junction, California. The solar field contains rows of mirrors that concentrate the sun's energy onto a system of pipes circulating a heat transfer fluid. The heat transfer fluid is used to produce steam, which powers a conventional turbine to generate electricity.

1988. Dr. Alvin Marks receives patents for two solar power technologies: Lepcon and Lumeloid. Lepcon consists of glass panels covered with millions of aluminum or copper strips, each less than a thousandth of a millimeter wide. As sunlight hits the metal strips, light energy is transferred to electrons in the metal, which escape at one end in the form of electricity. Lumeloid is similar but substitutes cheaper, filmlike sheets of plastic for the glass panels and covers the plastic with conductive polymers.

1991. President George Bush announces that the U.S. Department of Energy's Solar Energy Research Institute has been designated the National Renewable Energy Laboratory.

1992. Researchers at the University of South Florida develop a 15.9 percent efficient thin-film photovoltaic cell made of cadmium telluride, breaking the 15 percent barrier for this technology.

A 7.5-kW prototype dish system that includes an advanced membrane concentrator begins operating.

1993. Pacific Gas & Electric installs the first grid-supported photovoltaic system in Kerman, California. The 500-kW system is the first "distributed power" PV installation.

The National Renewable Energy Laboratory (formerly the Solar Energy Research Institute) completes construction of its Solar Energy Research Facility; it will be recognized as the most energy-efficient of all U.S. government buildings in the world.

1994. The first solar dish generator to use a free-piston engine is hooked up to a utility grid.

The National Renewable Energy Laboratory develops a solar cell made of gallium indium phosphide and gallium arsenide; it's the first one of its kind to exceed 30 percent conversion efficiency.

1996. The world's most advanced solar-powered airplane, the *Icare*, flies over Germany. Its wings and tail surfaces are covered by 3000 superefficient solar cells, for a total area of 21 m². The U.S. Department of Energy and an industry consortium begin operating Solar Two—an upgrade of the Solar One concentrating solar power tower. Until the project's end in 1999, Solar Two demonstrates how solar energy can be stored efficiently and economically so power is produced even when the sun isn't shining; it also spurs commercial interest in power towers.

1998. On August 6, a remote-controlled, solar-powered aircraft, *Pathfinder*, sets an altitude record of 80,000 ft on its thirty-ninth consecutive flight in Mojave, California—higher than any prop-driven aircraft to date.

Subhendu Guha, a scientist noted for pioneering work in amorphous silicon, leads the invention of flexible solar shingles, a roofing material and state-of-the-art technology for converting sunlight to electricity on buildings.

1999. Construction is completed on 4 Times Square in New York, the tallest skyscraper built in the city in the 1990s. It has more energy-efficient features than any other commercial skyscraper and includes building-integrated photovoltaic (BIPV) panels on the thirty-seventh through forty-third floors on the south- and west-facing facades to produce part of the building's power.

Spectrolab, Inc., and the National Renewable Energy Laboratory develop a 32.3 percent efficient solar cell. The high efficiency results from combining three layers of photovoltaic materials into a single cell, which is most efficient and practical in devices with lenses or mirrors to concentrate the sunlight. The concentrator systems are mounted on trackers to keep them pointed toward the sun.

Researchers at the National Renewable Energy Laboratory develop a breaking prototype solar cell that measures 18.8 percent efficient, topping the previous record for thin-film cells by more than 1 percent. Cumulative installed photovoltaic capacity reaches 1000 MW, worldwide.

2000. First Solar begins production at the Perrysburg, Ohio, photovoltaic manufacturing plant, the world's largest at the time; estimates indicate that it can produce enough solar panels each year to generate 100 MW of power. At the International Space Station, astronauts begin installing solar panels on what will be the largest solar power array deployed in space, each wing consisting of an array of 2800 solar cell modules.

Industry Researchers develop a new inverter for solar electric systems that increases safety during power outages. Inverters convert the dc electric output of solar systems to alternating current—the standard for household wiring as well as for power lines to homes.

Two new thin-film solar modules developed by BP Solarex break previous performance records. The company's 0.5-m^2 module has a 10.8 percent conversion efficiency—the highest in the world for similar thin-film modules. Its 0.9-m^2 module achieves 10.6 percent efficiency and a power output of 91.5 W—the highest in the world for a thin-film module.

The 12-kW solar electric system of a Morrison, Colorado, family is the largest residential installation in the United States to be registered with the U.S. Department of Energy's Million Solar Roofs program. The system provides most of the electricity for the family of eight's 6000-ft^2 home.

2001. Home Depot begins selling residential solar power systems in three stores in California. A year later it expands sales to 61 stores nationwide.

NASA's solar-powered aircraft, *Helios*, sets a new world altitude record for non-rocket-powered craft: 96,863 ft (more than 18 mi up).

2002. ATS Automation Tooling Systems, Inc., in Canada begins commercializing spheral solar technology. Employing tiny silicon beads between two sheets of aluminum foil, this solar-cell technology uses much less silicon than conventional multicrystalline silicon solar cells, thus potentially reducing costs. The technology was first championed in the early 1990s by Texas Instruments, but TI later discontinued work on it. For more, see the DOE Photovoltaic Manufacturing Technologies Web site.

The largest solar power facility in the Northwest—the 38.7-kW system White Bluffs Solar Station—goes on line in Richland, Washington.

PowerLight Corporation installs the largest rooftop solar power system in the United States—a 1.18-MW system at Santa Rita Jail, in Dublin, California.

LIST OF SUSTAINABLE ENERGY EQUIPMENT SUPPLIERS AND CONSULTANTS

An updated listing of this roster can be accessed through www.pvpower.com/pvinteg.html. The author does not endorse the listed companies or assume responsibility for inadvertent errors. Photovoltaic (PV) design and installation companies who wish to be included in the list can register by e-mail under the Web site.

A & M Energy Solutions

Business type: solar power contractors
2118 Wilshire Blvd., #718, Santa Monica, CA 90403
Phone: 1-310-445-9888

Abraham Solar Equipment

Business type: PV systems installation, distribution
Product types: system installation, distribution
124 Creekside Pl, Pagosa Springs, CO 81147
Phone: (800) 222-7242

ABS Alaskan, Inc.

Business type: systems design, integration
Product types: systems design, integration, PV and other small power systems
2130 Van Horn Rd, Fairbanks, AK 99701
Phone: Fairbanks (907) 451-7145; Anchorage (907) 562-4949
Toll Free: (800) 478-7145
U.S. Toll Free: (800) 235-0689
Fax: Fairbanks (907) 451-1949
E-mail: abs@absak.com
URL: www.absak.com

Advanced Energy Systems, Inc.

Business type: manufacture and distribute PV systems and lighting packages
Product types: PV power packages and lighting systems and energy-efficient lighting
9 Cardinal Dr., Longwood, FL 32779
Phone: (407) 333-3325
Fax: (407) 333-4341
E-mail: magicpwr@magicnet.net
URL: www.advancednrg.com

AeroVironment, Inc.

Business type: PV systems design
Product type: systems design
222 E Huntington Dr, Monrovia, CA 91016
Phone: (818) 357-9983
Fax: (818) 359-9628
E-mail: avgill@aol.com

AES Alternative Energy Systems, Inc.

Business type: PV, wind and micro-hydrosystems design, integration
Product types: systems design, integration, charge control/load centers
Contact: J. Fernando Lamadrid B.
9 E. 78th St., New York, NY 10021
Phone: (212) 517-9326
Fax: (212) 517-5326
E-mail: aes@altenergysys.com
URL: www.altenergysys.com

Alpha Real, A.G.

Business type: PV systems design, installation
Product types: power electronics and systems engineering
Feldeggstrasse 89, CH-8008 Zurich, Switzerland
Phone: 01-383-02-08
Fax: 01-383-18-95

Altair Energy, LLC

Business type: PV systems design, installation, service
Product types: turnkey PV systems, services, warranty and maintenance contracts,
    financing
600 Corporate Cir, Suite M, Golden, CO 80401
Phone: (303) 277-0025; (800) 836-8951
Fax: (303) 277-0029
E-mail: info@altairenergy.com

ALTEN srl

Business type: PV systems design, engineering, installation, BOS, and module
    manufacture
Product types: system design, engineering, installation, modules, BOS
Via della Tecnica 57/B4, 40068 S. Lazzaro, Bologna, Italy
Phone: 39 051-6258396; 39 051-6258624
Fax: 39 051-6258398
E-mail: alten@tin.it

Alternatif Enerji Sistemleri Sanayi Ticaret Ltd. Sti.

Business type: systems design, integration product sales
Product types: systems design, integration, product sales
Nispetiye Cad. No: 18/A Blok D.6 1.Levent Istanbul, Turkey
Phone: 90 (212) 283 74 45 pbx
Fax: ~90 (212) 264 00 87
E-mail: info@alternatifenerji.com

Alternative-Energie-Technik GmbH

Business type: PV systems design, engineering, installation, distribution
Product types: system design, engineering, installation, distribution
Industriestraße, 12D-66280, Sulzbach-Neuweiler, Germany
Phone: 06897-54337
Fax: 06897-54359
E-mail: info@aet.de
URL: www.aet.de

Alternative Energy Engineering

Business type: PV systems design, installation, distribution
Product types: system design, installation, distribution
P.O. Box 339-PV, Redway, CA 95560
Phone/order line: (800) 777-6609
Phone/techline: (707) 923-7216
URL: www.alt-energy.com

Alternative Energy Store

Business type: on-line component and system sales
Product types: solar and wind
4 Swan St., Lawrence, MA 01841
Fax/voice: (877) 242-6718
Phone orders: (877) 242-6718; (207) 469-7026
URL: www.AltEnergyStore.com

Alternative Power, Inc.

Business type: PV systems design, installation, distribution
Product types: system design, installation, distribution
160 Fifth Ave, Suite 711, New York, NY 10010-7003
Phone: (212) 206-0022
Fax: (212) 206-0893
E-mail: dbuckner@altpower.com
URL: www.altpower.com

Alternative Power Systems

Business type: systems design, integration, sales
Product types: systems design, integration
Sales contact: James Hart, San Diego, CA
Phone: (877) 946-3786 (877-WindSun)
Fax: (760) 434-3407
E-mail: engineering@aapspower.com
URL: www.aapspower.com

Alternative Solar Products

Business type: PV systems design, installation, distribution
Product types: system design, installation, distribution
Contact: Greg Weidhaas
27420 Jefferson Ave., Suite 104 B, Temecula, CA 92590-2668
Phone: (909) 308-2366
Fax: (909) 308-2388

American Photovoltaic Homes and Farms, Inc.

Business type: PV systems integration
Product type: design and construct PV-integrated homes
5951 Riverdale Ave, Riverdale, NY 10471
Phone: (718) 548-0428

Applied Power Corporation

Business type: PV systems design
Product type: system design
1210 Homann Dr, SE Lacey, WA 98503
Phone: (360) 438-2110
Fax: (360) 438-2115
E-mail: info@appliedpower.com
URL: www.appliedpower.com

Arabian Solar Energy & Technology (ASET)

Business type: PV and dc systems design, manufacture, and sales
Product types: PV systems, and dc systems
11 Sherif St., Cairo, Egypt
Phone: 20 2 393 6463; 20 2 395 3996
Fax : 20 2 392 9744
E-mail: aset@asetegypt.com
URL: www.asetegypt.com

AriStar Solar Electric

Business type: PV systems sales, integration
Product types: systems sales and integration
3101 W Melinda Ln., Phoenix, AZ 85027
Phone: (623) 879-8085; (888) 878-6786
Fax: (623) 879-8096
E-mail: aristar@uswest.net
URL: www.azsolar.com

Ascension Technology, Inc.

Business type: PV systems design, integration, BOS manufacturer
Product types: systems design and BOS
PO Box 6314, Lincoln, MA 01773
Phone: (781) 890-8844
Fax: (781) 890-2050

Atlantic Solar Products, Inc.

Business type: PV systems design and integration
Product types: systems design and integration
PO Box 70060, Baltimore, MD 21237
Phone: (410) 686-2500
Fax: (410) 686-6221
E-mail: mail@atlanticsolar.com
URL: www.atlanticsolar.com

Atlantis Energy Systems

Business type: PV system designer, manufacturer
Product types: PV systems, modules
9275 Beatty Dr., Sacramento, CA 95820
Phone: (916) 438-2930
E-mail: jomo13@atlantisenergy.com

B.C. Solar

Business type: PV system design, installation, training
Product types: systems design, installation, and training
PO Box 1102, Post Falls, ID 82854
Phone: (208) 667-9608
Phone: (208) 263-4290

Big Frog Mountain Corp.

Business type: PV and wind energy systems design, sales, integration
Product types: systems design, installation, distribution
Contact: Thomas Tripp
100 Cherokee Blvd, Suite 2109, Chattanooga, TN 37405
Phone: (423) 265-0307
Fax: (423) 265-9030
E-mail: sales@bigfrogmountain.com
URL: www.bigfrogmountain.com

Burdick Technologies Unlimited (BTU)

Business type: PV systems design, installation
Product types: systems design, installation; roofing systems
701 Harlan St., #64, Lakewood, CO 80214
Phone: (303) 274-4358

C-RAN Corporation

Business type: PV systems design, packaging
Product types: systems design (water purification, lighting, security)
666 4th St., Largo, FL 34640
Phone: (813) 585-3850
Fax: (813) 586-1777

CEM Design

Business type: PV systems design
Product types: systems design, architecture
520 Anderson Ave, Rockville, MD 20850
Phone: (301) 294-0682
Fax: (301) 762-3128

CI Solar Supplies Co.

Business type: PV systems
Product type: systems
PO Box 2805, Chino, CA 91710
Phone: (800) 276-5278 (800-2SOLAR8)
E-mail: jclothi@ibm.net
URL: www.cisolar.com

California Solar

Business type: PV systems design
Product type: systems design
627 Greenwich Dr., Thousand Oaks, CA 91360
Phone: (805) 379-3113
Fax: (805) 379-3027

CANROM Photovoltaics, Inc.

Business type: systems integrator
Product types: systems design, installation
108 Aikman Ave, Hamilton, ON, Canada L8M 1P9
Phone: (905) 526-7634
Fax: (905) 526-9341
URL: www.canrom.com

Creative Energy Technologies

Business type: PV systems and products
Product types: systems and efficient appliances
10 Main St., Summit, NY 12175
Phone: (888) 305-0278
Fax: (518) 287-1459
E-mail: info@cetsolar.com
URL: www.cetsolar.com

Currin Corporation

Business type: PV systems design, installation
Product types: systems design, installation
PO Box 1191, Midland, MI 48641-1191
Phone: (517) 835-7387
Fax: (517) 835-7395

DCFX Solar Systems P/L

Business type: PV systems design, integration
Product types: systems design, integration, pyramid power system, transportable
    hybrid package systems
Mt Darragh Rd., PO Box 264, Pambula 2549 N.S.W., Australia
Phone: 612.64956922
Fax: 612.64956922
E-mail: dcfx@acr.net.au

Dankoff Solar Products, Inc.

Business type: PV systems design, installation, distribution
Product types: systems design, installation, distribution
2810 Industrial Rd., Santa Fe, NM 87505
Phone: (505) 473-3800
Fax: (505) 473-3830
E-mail: pumps@danksolar.com

Delivered Solutions

Business type: PV product distributor
Product type: PV products
PO Box 891240, Temecula, CA 92589
Phone: (800) 429-7650 (24 hours a day)
Phone: (909) 694-3820
Fax: (909) 699-6215

Design Engineering Consultants

Business type: sustainable energy systems electromechanical consultants
Contact: Dr. Peter Gevorkian
10850 Riverside Dr., Suite 509, Toluca Lake, CA 91602
Phone: 1-818-980-7583
Fax: 1-818-475-5079
E-mail: Peter@decleed.com

Direct Gain, LLC

Business type: PV systems design, installation
Product types: system design, installation
23 Coxing Rd, Cottekill, NY 12419
Phone: (914) 687-2406
Fax: (914) 687-2408
E-mail: RLewand@Worldnet.att.net

Direct Power and Water Corporation

Business type: PV systems design, installation
Product types: systems design, installation
3455-A Princeton NE, Albuquerque, NM 87107
Phone: (505) 889-3585
Fax: (505) 889-3548
E-mail: dirpowdd@directpower.com

Diversified Technologies

Business type: PV systems design, installation
Product types: systems design, installation
35 Wiggins Ave., Bedford, MA 01730-2345
Phone: (617) 466-9444

Eclectic Electric

Business type: PV systems design, installation
Product types: systems design, installation, training
127 Avenida del Monte, Sandia Park, NM 87047
Phone: (505) 281-9538

EcoEnergies, Inc.

Business type: PV systems design, integration, distribution
Product types: PV and other renewables
171 Commercial St., Sunnyvale, CA 94086
Contact: Thomas Alexander
Phone: (408) 731-1228
Fax: (408) 746-3890
URL: www.ecoenergies.com

Ecotech (HK) Ltd.

Business type: PV systems and DHW design, integration, and distribution
Product types: PV and DHW
Room 608, 6/F, Yue Fung Industrial Building 35–45, Chai Wan Kok St., Tsuen Wan, N.T., Hong Kong
Phone: (852) 2833 1252; (852) 2405 2252
Fax: (852) 2405 3252
E-mail: inquiry@ecotech.com.hk
URL: www.ecotech.com.hk

ECS Solar Energy Systems

Business type: PV systems packaging
Product types: system packaging modular power stations
6120 SW 13th St., Gainesville, FL 32608
Phone: (904) 377-8866; (904) 338-0056

Ehlert Electric and Construction

Business type: PV systems design, installation
Product types: pumping systems design, installation
HCR 62, Box 70, Cotulla, TX 78014-9708
Phone: (210) 879-2205
Fax: (210) 965-3010

Electro Solar Products, Inc.

Business type: PV systems design, installation
Product types: system design, manufacture (traffic control, lights, pumping)
502 Ives Pl, Pensacola, FL 32514
Phone: (904) 479-2191
Fax: (904) 857-0070

Electron Connection

Business type: PV systems design, installation, distribution
Product types: system design, installation, distribution
PO Box 203, Hornbrook, CA 96044
Phone/Fax: (916) 475-3401
Phone: (800) 945-7587
E-mail: econnect@snowcrest.net

Electronics Trade and Technology Development Corp. Ltd.

Business type: PV distribution and export
Product types: PV distribution and export
Contact: M. H. Rao, General Manager
3001 Redhill Ave, Bldg. 5–103, Costa Mesa, CA 92626
Phone: (714) 557-2703
Fax: (714) 545-2723
E-mail: mhrao@pacbell.net

EMI

Business type: PV systems distribution
Product types: system and products distribution
Phone: (888) 677-6527

Energy Outfitters

Business type: PV systems design, integration
Product types: systems design, integration
136 S Redwood Hwy, PO Box 1888, Cave Junction, OR 87523
Phone: (800) 467-6527 (800-GO-SOLAR)
Office phone: (541) 592-6903
Fax: (503) 592-6747
E-mail: nrgoutfit@cdsnet.net

Energy Products and Services, Inc.

Business type: PV systems design, integration
Product types: systems design, integration, training
321 Little Grove LN., Fort Myers, FL 33917-3928
Phone: (941) 997-7669
Fax: (941) 997-8828

Enertron Consultants

Business type: PV systems design, integration
Product types: systems design, building integration
418 Benvenue Ave, Los Altos, CA 94024
Phone: (415) 949-5719
Fax: (415) 948-3442

Enn Cee Enterprises

Business type: PV systems design, thermal systems, integration
Product types: solar lanterns, indoor lighting, street lighting, garden lighting, water heating, dryers
Contact: K.S. Chaugule, Managing Director; Vipul K. Chaugule, Director
#542, First Stage, C M H Rd, Indiranagar, Bangalore, Karnataka, India
Phone: 91 (080) 525 9858 (Time Zone GMT 5:30)
Fax: 91 (080) 525 9858

EnviroTech Financial, Inc.

Business type: equipment trade and finance
Contact: Mr. Gene Beck
Orange, California
Phone: 1-714-532-2731
URL: www.etfinancial.com

EV Solar Products, Inc.

Business type: PV systems design, installation
Product types: systems design, installation
Contact: Ben Mancini
2655 N Hwy 89, Chino Valley, AZ 86323
Phone: (520) 636-2201
Fax: (520) 636-1664
E-mail: evsolar@primenet.com
URL: www.evsolar.com

Feather River Solar Electric

Business type: PV systems design, integration
Product types: systems design, integration
4291 Nelson St., Taylorsville, CA 95983
Phone: (916) 284-7849

Flack & Kurtz Consulting Engineers

Business type: PV consulting engineers
Product types: systems engineering
Contact: Daniel H. Nall, AIA, PE
475 Fifth Ave., New York, NY 10017
Phone: (212) 951-2691
Fax: (212) 689-7489
E-mail: nall@ny.fk.com

Fran-Mar

Business type: PV systems design, integration
Product types: systems design, integration
9245 Babcock Rd, Camden, NY 13316
Phone: (315) 245-3916
Fax: (315) 245-3916

Gebrüder Laumans GmbH & Co. KG

Business type: tile company with installation license for BMC PV tiles in Germany
    and Benelux
Product types: roof and facade PV tile and slate installation
Stiegstrasse 88, D-41379 Brüggen, Germany
Phone: (0 21 57) 14 13 30
Fax: (0 21 57) 14 13 39

Generation Solar Renewable Energy Systems, Inc.

Business type: PV/wind systems design, integration, installation, distribution
Product types: systems design, integration, installation, distribution
Contact: Richard Heslett
340 George St. N, Suite 405, Peterborough, ON K9H 7E8
Canada phone: (705) 741-1700
E-mail: gensolar@nexicom.net
URL: www.generationsolar.com

GenSun, Incorporated

Business type: integrated PV system builder
Product types: unitized, self-contained, portable, zero installation
10760 Kendall Rd., PO Box 2000, Lucerne Valley, CA 92356
Phone: (760) 248-2689; (800) 429-3777
Fax: (760) 248-2424
E-mail: solar@gensun.com
URL: www.gensun.com

Geosolar Energy Systems, Inc.

Business type: PV systems design, integration
Product types: systems design, integration
3401 N Federal Hwy, Boca Raton, FL 33431 USA
Phone: (407) 393-7127
Fax: (407) 393-7165

Glidden Construction

Business type: PV systems design, integration
Product types: systems design, integration
3727-4 Greggory Way, Santa Barbara, CA 93105
Phone: (805) 966-5555
Fax: (805) 563-1878

Global Resource Options, LLP

Business type: PV systems design, manufacture, sales and consulting
Product type: commercial- and residential-scale PV systems, consulting, design, installation
P.O. Box 51, Strafford, VT 05072
Phone: 802-765-4632
Fax: 802-765-9983
E-mail: global@sover.net
URL: www.globalresourceoptions.com

GO Solar Company

Business type: PV systems sales and integration
Product types: systems, components, integration
12439 Magnolia Blvd 132, North Hollywood, CA 91607
Phone: (818) 566-6870
Fax: (818) 566-6879
E-mail: solarexpert@solarexpert.com
URL: www.solarexpert.com

Grant Electric

Business type: Solar power contractor
Contact: Bruce Grant
16461 Sherman Way, Suite 175, Van Nuys, California 91406
Phone: 1-818-375-1977

Great Northern Solar

Business type: PV systems design, integration, distribution
Product types: Systems design, integration, distributor
Rte 1, Box 71, Port Wing, WI 54865
Phone: (715) 774-3374

Great Plains Power

Business type: PV systems design, integration
Product types: systems design, integration
1221 Welch St., Golden, CO 80401
Phone: (303) 239-9963
Fax: (303) 233-0410
E-mail: solar@bewellnet.com

Green Dragon Energy

Business type: systems integrator
Product type: PV and wind systems
2 Llwynglas, Bont-Dolgadfan, Llanbrynmair, Powys SY19 7AR, Wales, UK
Phone: 44 (0) 1650 521 589
Mobile: 0780 386 0003
E-mail: dragonrg@globalnet.co.uk

Heinz Solar

Business type: PV lighting systems
Product types: lighting systems design, integration
16575 Via Corto East, Desert Hot Springs, CA 92240
Phone: (619) 251-6886
Fax: (619) 251-6886

Henzhen Topway Solar Co., Ltd/Shenzhen BMEC

Business type: assembles and manufactures components and packages
Product types: lanterns, lamps, systems, BOS
RM8-202, Hualian Huayuan, Nanshan Dadao, Nanshan, Shenzhen, P.R. China, Post Code: 518052
Phone: 86 755 6402765; 6647045; 6650787
Fax: 86 755 6402722
E-mail: info@bangtai.com
URL: www.bangtai.com

High Resolution Solar

Business type: PV system design, integration, distribution
Product types: systems design, integration, distributor
Contact: Jim Mixan
7209 S 39th St., Omaha, NE 68147
Phone: (402) 738-1538
E-mail: jmixansolar@worldnet.att.net

Hitney Solar Products, Inc.

Business type: PV systems design, integration
Product types: systems design, integration
2655 N Hwy 89, Chino Valley, AZ 86323
Phone: (520) 636-1001
Fax: (520) 636-1664

Horizon Industries

Business type: PV systems and product distribution, service
Product types: systems and product distributor, service
2120 LW Mission Rd, Escondido, CA 92029
Phone: (888) 765-2766 (888-SOLAR NOW)
Fax: (619) 480-8322

Hutton Communications

Business type: PV systems design, integration
Product types: systems design, integration
5470 Oakbrook Pkwy, #G, Norcross, GA 30093
Phone: (770) 729-9413
Fax: (770) 729-9567

I.E.I., Intercon Enterprises, Inc.

Business type: North American distributor, Helios Technology Srl
Product type: PV modules
Contact: Gilbert Stepanian
12140 Hidden Brook Terr, N Potomac, MD 20878
Phone: (301) 926-6097
Fax: (301) 926-9367
E-mail: gilberts@erols.com

Independent Power and Light

Business type: PV systems design, integration, distribution
Product types: systems design, integration, distributor
RR 1, Box 3054, Hyde Park, VT 05655
Phone: (802) 888-7194

Innovative Design

Business type: architecture and design services
Product types: systems design and integration
850 West Morgan St., Raleigh, NC 27603
Phone: (919) 832-6303
Fax: (919) 832-3339
E-mail: innovativedesign@mindspring.com
URL: www.innovativedesign.net
Nevada Office: 8275 S Eastern Suite 220, Las Vegas, NV 89123
Phone: (702) 990-8413
Fax: (702) 938-1017

Integrated Power Corporation

Business type: PV systems design, integration
Product types: systems design, integration
7618 Hayward Rd, Frederick, MD 21702
Phone: (301) 663-8279
Fax: (301) 631-5199
E-mail: sales@integrated-power.com

Integrated Solar, Ltd.

Business type: PV system design, integration, distribution, installation, service
Product type: design, integrator, distributor, catalog
1331 Conant St., Suite 107, Maumee, OH 43537
Phone: (419) 893-8565
Fax: (419) 893-0006
E-mail: ISL 11@ix.netcom.com

Inter-Island Solar Supply

Business type: PV systems design, integration, distribution
Product types: systems design, integration, distribution
761 Ahua St., Honolulu, HI 96819
Phone: (808) 523-0711
Fax: (808) 536-5586
URL: www.solarsupply.com

ITALCOEL s.r.l., Electronic & Energy Control Systems

Business type: system integrator and BOS manufacturer
Product types: PV systems, PV inverters, design
66, Loc. Crognaleto, I-65010 Villanova (PE), Italy EU
Phone: 39.85.4440.1
Fax: 39.85.4440.240
E-mail: dayafter@iol.it

Jade Mountain

Business type: catalog sales
Product types: PV system components and loads
PO Box 4616, Boulder, CO 80306-4616
Phone: (800) 442-1972; (303) 449-6601
Fax: 303-449-8266
E-mail: jade-mtn@indra.com
URL: www.jademountain.com

Johnson Electric Ltd.

Business type: PV systems design, integration, distribution
Product types: systems design, integration, distributor
2210 Industrial Dr., PO Box 673, Montrose, CO 81402
Phone: (970) 249-0840
Fax: (970) 249-1248

Kyocera Solar, Inc.

Business type: manufacturer and distributor
Product types: PV modules and systems
7812 E Acoma, Scottsdale, AZ 85260
Phone: (800) 223-9580; (480) 948-8003
Fax: (480) 483-2986
E-mail: info@kyocerasolar.com
URL: www.kyocerasolar.com

L and P Enterprise Solar Systems

Business type: PV systems design, integration
Product types: systems design, integration
PO Box 305, Lihue, HI 96766
Phone: (808) 246-9111
Fax: (808) 246-3450

Light Energy Systems

Business type: PV systems design, contracting, consulting
Product types: systems design, integration
965 D Detroit Ave, Concord, CA 94518
Phone: (510) 680-4343
E-mail: solar@lightenergysystems.com
URL: www.lightenergysystems.com

Lotus Energy Pvt. Ltd.

Business type: PV systems design, integration; BOS manufacture; training
Product types: systems design, integration; BOS; training
Contact: Jeevan Goff, Managing Director
PO Box 9219, Kathmandu, Nepal
Phone: 977 (1) 418 203 (Time Zone GMT 5:45)
Fax: 977 (1) 412 924
E-mail: Jeevan@lotusnrg.com.np
URL: www.southasia.com/Nepaliug/lotus

Moonlight Solar

Business type: PV systems design, integration
Product types: design, contracting, repair
3451 Cameo LN, Blacksburg, VA 24060
Phone/Fax: (540) 953-1046
E-mail: moonlightsolar@moonlightsolar.com
E-mail: URL: www.moonlightsolar.com

Mytron Systems Ind.

Business type: PV systems
Product type: solar cookers, lantern, and PV systems
161, Vidyut Nagar B, Ajmer Rd. Jaipur 302021, India
Phone/Fax: 91-141-351434
E-mail: yogeshc@jpl.dot.net.in

Nekolux: Solar, Wind & Water Systems

Business type: PV, wind and micro-hydrosystems design and installation
Product types: systems design, integration, distributor
Contact: Vladimir Nekola
1433 W. Chicago Ave, Chicago, IL 60622
Phone: 312-738-3776
E-mail: vladimir@nekolux.com
URL: www.nekolux.com

New England Solar Electric (formerly Fowler Solar Electric)

Business type: PV systems design, integration
Product types: systems design, integration, book
401 Huntington Rd, PO Box 435, Worthington, MA 01098
Phone: (800) 914-4131
URL: www.newenglandsolar.com

Nextek Power Systems, Inc.

Business type: lighting integration
Product types: dc lighting for commercial applications using fluorescent or HID lighting
992 S Second St., Ronkonkoma, NY 11779
Phone: (631) 585-1005
Fax: (631) 585-8643
E-mail: davem@nextekpower.com
URL: www.nextekpower.com
West Coast Office: 921 Eleventh St., Suite 501, Sacramento, CA 95814
Phone: (916) 492-2445
Fax: (916) 492-2176
E-mail: patrickm@nextekpower.com

Ning Tong High-Tech

Business type: BOS manufacturer
Product types: solar garden light, solar traffic light, batteries, solar tracker, solar modules, portable solar systems, solar simulator and tester
Room 404, 383 Panyu Rd., Shanghai, P.R. China 200052
Phone: 86 21 62803172
Fax: 86 21 62803172
E-mail: songchao38@21cn.com
URL: www.ningtong-tech.com

North Coast Power

Business Type: PV systems dealer
Product Type: PV systems dealer
PO Box 151, Cazadero, CA 95421
Phone: (800) 799-1122
Fax: (877) 393-3955
E-mail: mmiller@utilityfree.com
URL: www.utilityfree.com

Northern Arizona Wind & Sun

Business type: PV systems integration, products distribution
Product types: systems and products integrator and distributor
PO Box 125, Tolleson, AZ 85353
Phone: (888) 881-6464 or (623) 877-2317
Fax: (623) 872-9215
E-mail: Windsun@Windsun.com
URL: www.solar-electric.com, www.windsun.com
Flagstaff Office: 2725 E Lakin Dr, #2, Flagstaff, AZ 86004
Phone: (800) 383-0195; (928) 526-8017
Fax: (928) 527-0729

Northern Power Systems

Business type: power systems design, integration, installation
Product types: controllers, systems design, integration, installation
182 Mad River Park, Waitsfield, VT 05401
Phone: (802) 496-2955, X266
Fax: (802) 879-8600
E-mail: rmack@northernpower.com
URL: www.northernpower.com

Northwest Energy Storage

Business type: PV systems design, integration, distribution
Product types: systems design, integration, distributor
10418 Hwy 95 N, Sandpoint, ID
Phone: (800) 718-8816; (208) 263-6142

Occidental Power

Business type: PV systems design, integration, installation
Product types: systems design, integration, installation
3629 Taraval St., San Francisco, CA 94116
Phone: (415) 681-8861
Fax: (415) 681-9911
E-mail: solar@oxypower.com
URL: www.oxypower.com

Off Line Independent Energy Systems

Business type: PV systems design, integration
Product types: systems design, integration
PO Box 231, North Fork, CA 93643
Phone: (209) 877-7080
E-mail: ofln@aol.com

Oman Solar Systems Company, L.L.C. (Division of AJAY Group of Companies)

Business type: systems, design, integration, installation, consulting
Product types: PV systems, wind generators, water pumps, solar hot water systems
Contact: N.R. Rao, PO Box 1922, RUWI 112, Oman
Phone: 00968-592807; 595756; 591692
Fax: 00968-591122; 7715490
E-mail: oss.marketing@ajaygroup.com

Phasor Energy Company

Business type: PV systems design, integration
Product types: systems design, integration
4202 E Evans Dr, Phoenix, AZ 85032-5469
Phone: (602) 788-7619
Fax: (602) 404-1765

Photovoltaic Services Network, LLC (PSN)

Business type: PV systems design, integration, package grid-tied systems
Product types: systems design, integration
215 Union Blvd., Suite 620, Lakewood, CO 80228
Phone: (303) 985-0717; (800) 836-8951
Fax: (303) 980-1030
E-mail: tschuyler@neosdenver.com

Planetary Systems

Business type: PV systems design, integration, distributor
Product types: systems design, integration, distributor
PO Box 9876, 2400 Shooting Iron Ranch Rd, Jackson, WY 83001
Phone/Fax: (307) 734-8947

Positive Energy, Inc.

Business type: PV systems design, integration
Product types: systems design, integration
3900 Paseo del Sol, #201, Santa Fe, NM 87505
Phone: (505) 424-1112
Fax: (505) 424-1113
E-mail: info@positivenergy.com
URL: www.positivenergy.com

PowerPod Corp.

Business type: PV systems design, integration
Product types: modular PV systems for village electrification
PO Box 321, Placerville, CO 81430
Phone: (970) 728-3159
Fax: (970) 728-3159
E-mail: solar@rmi.com
URL: www.powerpod.com

Rainbow Power Company Ltd.

Business type: system design, integration, maintenance, repair
Product types: systems integration, systems design, maintenance and repair
1 Alternative Way, PO Box 240, Nimbin, NSW, Australia, 2480
Phone: (066) 89 1430
Fax: (066) 89 1109
International phone: 61 66 89 1088
International fax: 61 66 89 1109
E-mail: rpcltd@nor.com.au
URL: www.rpc.com.au

Real Goods Trading Company

Business type: PV systems design, integration, distribution, catalog
Product types: systems design, integration, distributor, catalog
966 Mazzoni St., Ukiah, CA 95482
Phone: (800) 762-7325
E-mail: realgood@realgoods.com
URL: www.realgoods.com

Remote Power, Inc.

Business type: PV systems design, integration
Product types: systems design, integration
12301 N Grant St., #230 Denver, CO 80241-3130
Phone: (800) 284-6978
Fax: (303) 452-9519
E-mail: RPILen@aol.com

Renewable Energy Concepts, Inc.

Business type: PV system design, installation, sales
Product types: PV panels, wind turbines, inverters, batteries
1545 Higuera St., San Luis Obispo, CA 93401
Phone: (805) 545-9700; (800) 549-7053
Fax: (805) 547-0496
E-mail: info@reconcepts.com
URL: www.reconcepts.com

Renewable Energy Services, Inc., of Hawaii

Business type: PV systems design, integration
Product types: systems design, integration
PO Box 278, Paauilo, HI 96776
Phone: (808) 775-8052
Fax: (808) 7775-0852

Resources & Protection Technology

Business type: PV / BIPV system design, integration, distribution
Product types: PV, solar thermal, heat pump
4A, Block 2, Dragon Centre, 25 Wun Sha St. Tai Hang, Hong Kong
Phone: (852) 8207 0801
Fax: (852) 8207 0802
E-mail: info@rpt.com.hk
URL: www.rpt.com.hk

RGA, Inc.

Business type: PV lighting systems
Product types: lighting systems
454 Southlake Blvd, Richmond, VA 233236
Phone: (804) 794-1592
Fax: (804) 3779-1016

RMS Electric

Business type: PV systems design, integration
Product types: systems design, integration
2560 28th St., Boulder, CO 80301
Phone: (303) 444-5909
Fax: (303) 444-1615
E-mail: info@rmse.com
URL: www.rmse.com

Roger Preston Partners

Business type: system design, integration, engineering
Product types: systems integration, systems design, energy engineering
1050 Crown Point Pkwy, Suite 1100, Atlanta, GA 30338
Phone: (770) 394-7175
Fax: (770) 394-0733
E-mail: rpreston@atl.mindspring.com

Roseville Solar Electric

Business type: PV systems design, integration, distribution, and installation
Product type: grid-tie, battery backup, residential, commercial
Contact: Kevin Hahner
PO Box 38590, Sacramento, CO 38590
Phone: (916) 772-6977; (916) 240-6977
E-mail: khahner@juno.com

SBT Designs

Business type: system sales and installation
Product types: system sales and installation
25840 IH-10 West #1, Boerne, TX 78006
Phone: (210) 698-7109
Fax: (210) 698-7147
E-mail: sbtdesigns@bigplanet.com

S C Solar

Business type: PV and solar thermal systems sales, integration
Product types: systems design, sales, and distribution
7073 Henry Harris Rd, Lancaster, SC 29720
Phone/Fax: (803) 802-5522
E-mail: dwhigham@scsolar.com
URL: www.scsolar.com

SEPCO—Solar Electric Power Co.

Business type: manufacturer of PV lighting systems and OEM PV systems
Product types: PV lighting and power systems
Contact: Steven Robbins
7984 Jack James Dr., Stuart, FL 34997
Phone: (561) 220-6615
Fax: (561) 220-8616
E-mail: sepco@tcol.net

Siam Solar & Electronics Co., Ltd.

Business type: Solarex distributor
Product types: laminator of custom-size PV modules, sine wave inverters,
    12-V dc ballasts
Contact: Mr. Viwat Sri-on (Managing Director)
62/16-25 Krungthep-Nontaburi Rd, Nontaburi, 11000, Thailand
Phone: 66-2-5260578
Fax: 66-2-5260579
E-mail: sattaya@loxinfo.co.th

Sierra Solar Systems

Business type: PV systems design, integration
Product types: systems design, integration
109 Argall Way, Nevada City, CA 95959
Phone: (800) 517-6527
Fax: (916) 265-6151
E-mail: solarjon@oro.net
URL: www.sierrasolar.com

Solar Age Namibia Pty. Ltd.

Business type: PV systems design, integration
Product types: systems design, integration, village lighting
PO Box 9987, Windhoek, Namibia
Phone: 264-61-215809
Fax: 264-61-215793
E-mail: solarage@iafrica.com.na

Solar Century

Business type: PV systems design, integration
Product types: systems design and installation
91–94 Lower Marsh, London SE 1 7AB, UK
Phone: 44 (0)207 803 0100
Fax: 44 (0)207 803 0101
URL: www.solarcentury.com

Solar Creations

Business type: PV systems design, integration
Product types: systems design, integration
2189 SR 511S, Perrysville, OH 44864
Phone: (419) 368-4252

Solar Depot

Business type: PV systems design, integration
Product types: systems design, integration
61 Paul Dr., San Rafael, CA 94903
Phone: (415) 499-1333
Fax: (415) 499-0316
URL: www.solardepot.com

Solar Design Associates

Business type: PV systems design, building integration, architecture
Product types: systems design, building integration, architecture
PO Box 242, Harvard, MA 01451
Phone: (978) 456-6855
Fax: (978) 456-3030
E-mail: sda@solardesign.com
URL: www.solardesign.com

Solar Dynamics, Inc.

Business type: manufacture portable PV system package
Product type: portable PV system
152 Simsbury Rd, Building 9, Avon, CT 06001
Phone: (877) 527-6461 (877-JASMINI); (860) 409-2500
Fax: (860) 409-9144
E-mail: info@solar-dynamics.com
URL: www.solar-dynamics.com

Solar Electric, Inc.

Business type: PV systems design, integration, distribution
Product types: systems design, integration, distributor
5555 Santa Fe St., #J San Diego, CA 92109
Phone: (800) 842-5678; (619) 581-0051
Fax: (619) 581-6440
E-mail: solar@cts.com
URL: www.solarelectricinc.com

Solar Electric Engineering, Inc.

Business type: PV systems design, integration, distribution
Product types: systems design, integration, distributor
116 4th St., Santa Rosa, CA 95401
Phone: (800) 832-1986

Solar Electric Light Company (SELCO)

Business type: PV systems design, integration
Product types: systems design, integration
35 Wisconsin Cir., Chevy Chase, MD 20815
Phone: (301) 657-1161
Fax: (301) 657-1165
URL: www.selco-intl.com
India URL: www.selco-india.com
Vietnam URL: www.selco-vietnam.com
Sri Lanka URL: www.selco-srilanka.com

Solar Electric Light Fund

Business type: PV systems design, integration
Product types: systems design, integration
1734 20th St., NW., Washington, DC 20009
Phone: (202) 234-7265
Fax: (202) 328-9512
URL: www.self.org

Solar Electric Specialties Co.

Business type: PV systems design, integration
Product types: systems design, integration
PO Box 537, Willits, CA 95490
Phone: (800) 344-2003
Fax: (707) 459-5132
E-mail: seswillits@aol.com
URL: www.solarelectric.com

Solar Electric Systems of Kansas City

Business type: PV lighting systems
Product types: lighting systems
13700 W 108th St., Lenexa, KS 66215
Phone: (913) 338-1939
Fax: (913) 469-5522
E-mail: solarelectric@compuserve.com

Solar Electrical Systems

Business type: PV systems design, integration, distribution
Product types: systems design, integration, distributor
2746 W Appalachian Ct., Westlake Village, CA 91362
Phone: (805) 373-9433, (310) 202-7882
Fax: (805) 497-7121, (310) 202-1399
E-mail: ses@pacificnet.net

Solar Energy Systems of Jacksonville

Business type: PV systems design, integration
Product types: systems design, integration
4533 Sunbeam Rd, #302, Jacksonville, FL 32257
Phone: (904) 731-2549
Fax: (904) 731-1847

Solar Energy Systems Ltd.

Business type: PV systems design, integration
Product types: systems design, integration
Unit 3, 81 Guthrie St., Osborne Park, Western Australia 6017
Phone: ~61 (0)8.9204 1521
Fax: ~61 (0)8.9204 1519
E-mail: amaslin@sesltd.com.au
URL: www.sesltd.com.au

Solar Engineering and Contracting

Business type: PV systems design, integration
Product types: systems design, integration
PO Box 690, Lawai, HI 96765
Phone: (808) 332-8890
Fax: (808) 332-8629

The Solar Exchange

Business type: PV systems design, integration
Product types: water pumping and home systems
PO Box 1338, Taylor, AZ 85939
Phone: (520) 536-2029; (520) 521-0929
E-mail: solarexchange@cybertrails.com

Solar Grid

Business type: catalog sales
Product types: PV system components
2965 Staunton Rd, Huntington, WV 25702
Order line: (800) 697-4295
Tech line: (304) 697-1477
Fax: (304) 697-2531
E-mail: sales@solarg.com

Solar Integrated Technologies

Business type: manufacturer of flexible solar power mats
1837 E. Martin Luther King Jr. Blvd, Los Angeles, CA 90058
Phone: 323-231-0411
Fax: 323-231-0517

Solar Online Australia

Business type: PV and wind products, design, supply, integration
Product types: components, systems, design, integration
48 Hilldale Dr., Cameron Park NSW 2285, Australia
Phone: 61 2 4958 6771
E-mail: info@solaronline.com.au
URL: www.solaronline.com.au

Solar Outdoor Lighting, Inc. (SOL)

Business type: PV street lighting systems
Product types: street lighting systems
3131 SE Waaler St., Stuart, FL 34997
Phone: (407) 286-9461
Fax: (407) 286-9616
E-mail: lightsolar@aol.com
URL: www.solarlighting.com

Solar Quest, Becker Electric

Business type: PV systems design, integration, distribution
Product types: systems design, integration, distributor
28706 New School Rd., Nevada City, CA 95959
Phone: (800) 959-6354; (916) 292-1725
Fax: (916) 292-1321

Solar Sales Pty. Ltd.

Business type: PV systems design, integration
Product types: systems design, integration
97 Kew St., PO Box 190, Welshpool 6986, Western Australia
Phone: 618.03622111
Fax: 618.94721965
E-mail: solar@ois.com.au

Solar Sense

Business type: PV systems integration
Product types: small portable solar power systems and battery chargers
Contact: Lindsay Hardie
7725 Lougheed Hwy, Burnaby, BC, Canada V5A 4V8
Phone: (800) 648-8110; (604) 656-2132
Fax: (604) 420-1591
E-mail: info@solarsense.com
URL: www.solarsense.com

Solar-Tec Systems

Business type: PV systems sales, integration
Product types: systems sales, integration
33971-A Silver Lantern, Dana Point, CA 92629
Phone: (949) 248-9728
Fax: (949) 248-9729
URL: www.solar-tec.com

Solartronic

Business type: PV systems design, integration, sales
Product types: systems design, integration; product distributor
Morelos Sur No. 90 62070 Col. Chipitlán Cuernavaca, Mor., Mexico
Phone: 52 (73)18-9714
Fax: 52 (73)18-8609
E-mail: info@solartronic.com
URL: www.solartronic.com

Solartrope Supply Corporation

Business type: wholesale supply house
Product types: systems components
Phone: (800) 515-1617

Solar Utility Company, Inc.

Business type: PV systems design, integration
Product types: systems design, integration
Contact: Steve McKenery
6160 Bristol Pkwy, Culver City, CA 90230
Phone: (310) 410-3934
Fax: (310) 410-4185

Solar Village Institute, Inc.

Business type: PV systems design, integration
Product types: systems design, integration
PO Box 14, Saxapawhaw, NC 27340
Phone: (910) 376-9530

Solar Works!

Business type: PV systems design, integration
Product types: systems design, integration
Contact: Daniel S. Durgin
PO Box 6264, 525 Lotus Blossom Ln, Ocean View, HI 96737
Phone: (808) 929-9820
Fax: (808) 929-9831
E-mail: ddurgin@aloha.net
URL: www.solarworks.com

Solar Works, Inc.

Business type: PV systems design, integration, distribution
Product types: systems design, integration, distributor
64 Main St., Montpelier, VT 05602
Phone: (802) 223-7804
E-mail: LSeddon@solar-works.com
URL: www.solar-works.com

Sollatek

Business type: systems design, installation, BOS manufacturer
Product types: systems design and installation
Unit 4/5, Trident Industrial Estate, Blackthorne Rd, Poyle Slough, SL3 0AX,
    United Kingdom
Phone: 44 1753 6883000
Fax: 44 1753 685306
E-mail: sollatek@msn.com

Soler Energie S.A. (Total Energie Group)

Business type: PV systems design, integration
Product types: systems design, integration
BP 4100, 98713 Papeete, French Polynesia
Phone: 689 43 02 00
Fax: 689 43 46 00
E-mail: soler@mail.pf
URL: www.total-energie.fr

Solo Power

Business type: PV systems design, integration
Product types: systems design, integration
1011-B Sawmill Rd, NW Albuquerque, NM 87104
Phone: (505) 242-8340
Fax: (505) 243-5187

Soltek Solar Energy Ltd.

Business type: PV systems design, integration, distribution, catalog
Product types: systems design, integration
2-745 Vanalman Ave., Victoria, BC V8Z 3B6 Canada
E-mail: soltek@pinc.com

SOLutions in Solar Electricity

Business type: PV systems design, sales, installation, consulting, training
Product types: system design, installation, consulting, training
Contact: Joel Davidson
PO Box 5089, Culver City, CA 90231
Phone: (310) 202-7882
Fax: (310) 202-1399
E-mail: joeldavidson@earthlink.net
URL: www.solarsolar.com

Soluz, Inc.

Business type: PV systems design, integration
Product types: international systems, design and distribution
Contact: Steve Cunningham
55 Middlesex St., Suite 221, North Chelmsford, MA 01863-1561
Phone: (508) 251-5290
Fax: (508) 251-5291
E-mail: soluz@igc.apc.org

Southwest Photovoltaic Systems, Inc.

Business type: PV systems design, integration
Product types: systems design, integration
212 E Main St., Tomball, TX 77375
Phone: (713) 351-0031
Fax: (713) 351-8356
E-mail: SWPV@aol.com

Sovran Energy, Inc.

Business type: PV systems design, integration, distribution
Product types: systems design, integration, distributor
13187 Trewhitt Rd., Oyama, BC, Canada V4V 2B17
Phone: (250) 548-3642
Fax: (250) 548-3610
E-mail: sovran@sovran.ca
URL: www.sovran.ca

Star Power International Limited

Business type: PV systems integration
Product types: systems design, integration
912 Worldwide Industrial Center, 43 Shan Mei St., Fotan, Hong Kong
Phone: (852) 26885555
Fax: (852) 26056466
E-mail: starpwr@hkstar.com

Stellar Sun

Business type: PV systems integration
Product types: systems design, integration
2121 Watt St., Little Rock, AR 72227
Phone: (501) 225-0700
Fax: (501) 225-2920
E-mail: bill@stellarsun.com
URL: http://stellarsun.com

Strong Plant & Supplies FZE

Business type: PV system design, integration, distribution, consulting services
Product types: PV systems, modules, charge controllers, power centers, inverters, lighting, and pumping products
Contact: Toufic E. Kadri
PO Box 61017, Dubai, United Arab Emirates
Phone: 971 4 835 531
Fax: 971 4 835 914
E-mail: strongtk@emirates.net.ae

Sudimara Solar/PT Sudimara Energi Surya

Business type: PV systems design, integration, distribution
Product types: systems design, integration, distributor
Jl. Banyumas No. 4, Jakarta, 10310, Indonesia
Phone: 3904071-3
Fax: 361639

Sun, Wind and Fire

Business type: PV systems design, integration
Product types: systems design, integration
7637 SW 33rd Ave, Portland, OR 97219-1860
Phone: (503) 245-2661
Fax: (503) 245-0414

SunAmp Power Company

Business type: PV systems design, integration, distribution
Product types: systems design, integration, distributor
7825 E Evans, #400, Scottsdale, AZ 85260
Phone: (800) 677-6527 (800-MR SOLAR)
E-mail: sunamp@sunamp.com
URL: www.sunamp.com

Sundance Solar Designs

Business type: PV systems design, integration
Product types: systems design, integration
PO Box 321, Placerville, CO 81430
Phone: (970) 728-3159
Fax: (970) 728-3159
E-mail: solar@rmi.com

Sunelco

Business type: PV systems design, integration
Product types: systems design, integration
PO Box 1499, 100 Skeels St., Hamilton, MT 59840
Phone: (800) 338-6844; (406) 363-6924
Fax: (406) 363-6046
E-mail: sunelco@montana.com
URL: www.sunelco.com

Sunergy Systems

Business type: PV equipment and systems
Product types: equipment and systems
PO Box 70, Cremona, AB T0M 0R0 Canada
Phone: (403) 637-3973

Sunmotor International Ltd.

Business type: manufacturer and systems installation for PV water pumping
Product type: solar water pumping systems
104, 5037-50th St., Olds, AB T4H 1R8 Canada
Phone: (403) 556-8755
Fax: (403) 556-7799
URL: www.sunpump.com

Sunnyside Solar, Inc.

Business type: PV systems design, integration, distribution
Product types: systems design, integration, distributor, lighting
RD 4, Box 808, Green River Rd., Brattleboro, VT 05301
Phone: (802) 257-1482

Sunpower Co.

Business type: PV systems design, integration, distribution
Product types: systems design, integration, distributor, pumping
Contact: Leigh and Pat Westwell
RR3, Tweed, ON K0K 3J0 Canada
Phone: (613) 478-5555
E-mail: sunpower@blvl.igs.net

SunWize Technologies, Inc.

Business type: PV systems design, integration, distribution, manufacture of portable
PV systems
Product types: systems design, integration, distributor, portable PV systems
1155 Flatbush Rd., Kingston, NY 12401
Contact: Bruce Gould, VP, Sales
Phone: (800) 817-6527; (845) 336-0146
Fax: (845) 336-0457
E-mail: sunwize@besicorp.com
URL: www.sunwize.com

Superior Solar Systems, Inc.

Business type: PV systems design, integration
Product types: systems design, integration
1302 Bennett Dr., Longwood, FL 32750
Phone: (800) 478-7656 (800-4PVsolar)
Fax: (407) 331-0305

Talmage Solar Engineering, Inc.

Business type: PV systems design, integration
Product types: systems design, integration
18 Stone Rd., Kennebunkport, ME 04046
Phone: (888) 967-5945
Fax: (207) 967-5754
E-mail: tse@talmagesolar.com
URL: www.talmagesolar.com

Technical Supplies Center Ltd. (TSC)

Business type: BOS distributor and system integrator
Product type: PV, wind, batteries, charge controllers, inverters
South 60th St., East Awqaff Complex, PO Box 7186, Sana'a, Republic of Yemen
Phone: 967 1 269 500
Fax : 967 1 267 067
E-mail: ZABARAH@y.net.ye

Thomas Solarworks

Business type: PV systems design, integration
Product types: systems design, integration
PO Box 171, Wilmington, IL 60481
Phone: (815) 476-9208
Fax: (815) 476-2689

Total Energie

Business type: PV systems design, integration
Product types: systems design, integration
7, chemin du Plateau, 69570 Lyon-Dardilly, France
Phone: 33 (0)4 72 52 13 20
Fax: 33 (0)4 78 64 91 00
E-mail: infos@total-energie.fr
URL: www.total-energie.fr

Utility Power Group

Business type: PV manufacture, systems design, integration
Product types: manufacture, systems design, integration
9410-G DeSoto Ave, Chatsworth, CA 91311
Phone: (818) 700-1995
Fax: (818) 700-2518
E-mail: 71263.444@compuserve.com

Vector Delta Design Group, Inc.

Product types: turnkey electric and solar power design and integration
Contact: Dr. Peter Gevorkian
2325 Bonita Dr., Glendale, CA 91208
Phone: (818) 241-7479
Fax: (818) 243-5223
E-mail: vectordeltadesign@charter.net
URL: www.vectordelta.com

Vermont Solar Engineering

Business type: PV systems design, integration, distribution
Product types: systems design, integration, distributor
PO Box 697, Burlington, VT 05402
Phone: (800) 286-1252; (802) 863-1202
Fax: (802) 863-7908
E-mail: vtsolar1@together.net
URL: www.vtsolar.com

Whole Builders Cooperative

Business type: PV systems design, integration
Product types: systems design, integration
2928 Fifth Ave, S. Minneapolis, MN 55408-2412
Phone: (612) 824-6567
Fax: (612) 824-9387

Wind and Sun

Business type: PV systems design, integration, distribution
Product types: systems design, integration, distributor
The Howe, Watlington, Oxford OX9 5EX UK
Phone: (44) 1491-613859
Fax: (44) 1491-614164

WINSUND (Division of Hugh Jennings Ltd.)

Business type: PV and wind systems design, installation, distribution
Product types: systems design, installation, distribution
Tatham St., Sunderland SR1 2AG, England, UK
Phone: 44 191 514 7050
Fax: 44 191 564 1096
E-mail: info@winsund.com
URL: www.winsund.com

Woodland Energy

Business type: portable PV systems design, manufacture, and sales
Product type: portable PV systems
PO Box 247, Ashburnham, MA 01430
Phone: (978) 827-3311
E-mail: info@woodland-energy.com
URL: www.woodland-energy.com

WorldWater & Power Corporation

Type of business: solar power irrigation and water pumping
Pennington Business Park, 55 Route 31 South, Pennington, NJ 08534
Phone: 1-609-818-0700
E-mail: pump@waterworld.com

Zot's Watts

Business type: PV systems design, integration, distribution
Product types: systems design, integration, distributor
Contact: Zot Szurgot
1701 NE 75th St., Gainesville, FL 32641
Phone: (352) 373-1944
E-mail: roselle@gnv.fdt.net

GLOSSARY OF RENEWABLE
ENERGY POWER SYSTEMS

All those technical terms can make renewable energy systems difficult for many people to understand. This glossary aims to cover all the most commonly used terms, as well as a few of the more specific terms.

alternating current (ac): Electric current that continually reverses direction. The frequency at which it reverses is measured in cycles per second, or hertz (Hz). The magnitude of the current itself is measured in amperes (A).

alternator: A device for producing ac electricity. Usually driven by a motor, but can also be driven by other means, including water and wind power.

ammeter: An electric or electronic device used to measure current flowing in a circuit.

amorphous silicon: A noncrystalline form of silicon used to make photovoltaic modules (commonly referred to as solar panels).

ampere (A): The unit of measurement of electric current.

ampere-hour (Ah): A measurement of electric charge. One ampere-hour of charge would be removed from a battery if a current of 1 A flowed out of it for 1 hour. The ampere-hour rating of a battery is the maximum charge that it can hold.

anemometer: A device used to measure wind speed.

anode: The positive electrode in a battery, diode, or other electric device.

axial flow turbine: A turbine in which the flow of water is in the same direction as the axis of the turbine.

battery: A device, made up of a collection of cells, used for storing electricity, which can be either rechargeable or nonrechargeable. Batteries come in many forms, and include flooded cell, sealed, and dry cell.

battery charger: A device used to charge a battery by converting (usually) ac alternating voltage and current to a dc voltage and current suitable for the battery. Chargers often incorporate some form of a regulator to prevent overcharging and damage to the battery.

beta limit: The maximum power (theoretically) that can be captured by a wind turbine from the wind, which equals 59.3 percent of the wind energy.

blade: The part of a turbine that water or air reacts against to cause the turbine to spin, which is sometimes incorrectly referred to as the propeller. Most electricity-producing wind turbines will have two or three blades, whereas water-pumping wind turbines will usually have up to 20 or more.

capacitor: An electronic component used for the temporary storage of electricity, as well as for removing unwanted noise in circuits. A capacitor will block direct current but will pass alternating current.

cathode: The negative electrode in a battery, diode, or other electric device.

cell: The most basic, self-contained unit that contains the appropriate materials, such as plates and electrolyte, to produce electricity.

circuit breaker: An electric device used to interrupt an electric supply in the event of excess current flow. It can be activated either magnetically, thermally, or by a combination of both, and can be manually reset.

compact fluorescent lamp: A form of fluorescent lighting that has its tube "folded" into a "U" or other more compact shape, so as to reduce the space required for the tube.

conductor: A material used to transfer or conduct electricity, often in the form of wires.

conduit: A pipe or elongated box used to house and protect electric cables.

converter: An electronic device that converts electricity from one dc voltage level to another.

cross-flow turbine: A turbine where the flow of water is at right angles to the axis of rotation of the turbine.

current: The rate of flow of electricity, measured in amperes. Analogous to the rate of flow of water measured in liters per second, which is also measured in amperes.

Darrius rotor: A form of vertical-axis wind turbine that uses thin blades.

diode: A semiconductor device that allows current to flow in one direction, while blocking it in the other.

direct current (dc): Electric current that flows in one direction only, although it may vary in magnitude.

dry cell battery: A battery that uses a solid paste for an electrolyte. Common usage refers to these as small cylindrical "torch" cells.

earth (or ground): Refers to physically connecting a part of an electric system to the ground, done as a safety measure, by means of a conductor embedded in suitable soil.

earth-leakage circuit breaker (ELCB): A device used to prevent electrical shock hazards in mains voltage power systems; it includes independent power systems, which are also known as residual current devices (RCDs).

electricity: The movement of electrons (a subatomic particle) produced by a voltage through a conductor.

electrode: An electrically conductive material, forming part of an electric device, often used to lead current into or out of a liquid or gas. In a battery, the electrodes are also known as plates.

electrolysis: A chemical reaction caused by the passage of electricity from one electrode to another.

electrolyte: The connecting medium, often a fluid, that allows electrolysis to occur. All common batteries contain an electrolyte, such as the sulfuric acid used in lead-acid batteries.

energy: The abstract notion that makes things happen or that has the potential or ability to do work. It can be stored and converted between many different forms, such as heat, light, electricity, and motion. It is never created or destroyed but does become unavailable to us when it ends up as low-temperature heat. It is measured in joules (J) or watt-hours (Wh) but more usually megajoules (MJ) or kilowatt-hours (kWh).

equalizing charge: A flooded lead-acid battery will normally be charged in boost mode until the battery reaches 2.45 to 2.5 V per cell, at which time the connected regulator should switch into "float" mode, where the battery will be maintained at 2.3 to 2.4 V per cell. During an equalizing charge, the cells are overcharged at 2.5 to 2.6 V per cell to ensure that all cells have an equal (full charge). This is normally achieved by charging from a battery charger, though some regulators will perform this charge when energy use is low, such as when the users are not at home.

float charge: A way of charging a battery by varying the charging current so that its terminal voltage, the voltage measured directly across its terminals, "floats" at a specific voltage level.

flooded cell battery: A form of rechargeable battery where the plates are completely immersed in a liquid electrolyte. The starter battery in most cars is of the flooded cell type. Flooded cell batteries are the most commonly used type for independent and remote area power supplies.

fluorescent light: A form of lighting that uses long thin tubes of glass that contain mercury vapor and various phosphor powders (chemicals based on phosphorus) to produce white light that is generally considered to be the most efficient form of home lighting. See also *compact fluorescent lamp.*

furling: A method of preventing damage to horizontal-axis wind turbines by automatically turning them out of the wind using a spring-loaded tail or other device.

fuse: An electric device used to interrupt an electric supply in the event of excess current flow. Often consists of a wire that melts when excess current flows through it.

gel-cell battery: A form of lead-acid battery where the electrolyte is in the form of a gel or paste. Usually used for mobile installations and when batteries will be subject to high levels of shock or vibration.

generator: A mechanical device used to produce dc electricity. Coils of wire passing through magnetic fields inside the generator produce power. See also *alternator*. Most ac-generating sets are also referred to as generators.

gigawatt (GW): A measurement of power equal to a thousand million watts.

gigawatt-hour (GWh): A measurement of energy. One gigawatt-hour is equal to 1 GW being used for a period of 1 hour or 1 MW being used for 1000 hours.

halogen lamp: A special type of incandescent globe made of quartz glass and a tungsten filament, which also contains a small amount of a halogen gas (hence the name), enabling it to run at a much higher temperature than a conventional incandescent globe. Efficiency is better than a normal incandescent, but not as good as a fluorescent light.

head: The vertical distance that water will fall from the inlet of the collection pipe to the water turbine in a hydropower system.

hertz (Hz): Unit of measurement for frequency. It is equivalent to cycles per second (refer to *alternating current*). Common household mains power is normally 60 Hz.

horizontal-axis wind turbine: The most common form of wind turbine, consisting of two or three airfoil-style blades attached to a central hub, which drives a generator. The axis or main shaft of the machine is horizontal or parallel to Earth's surface.

incandescent globe: This is the most common form of light globe in the home. It usually consists of a glass globe inside of which is a wire filament that glows when electricity is passed through it. They are the least efficient of all electric lighting systems.

independent power system: A power generation system that is independent of the tile mains grid.

insolation: The level of intensity of energy from the sun that strikes earth. Usually given as watts per square meter (W/m^2). A common level in Australia in the summer is about 1000 W/m^2.

insulation: A material used to prevent the flow of electricity used in electric wires in order to prevent electric shock. Typical materials used include plastics, such as PVC and polypropylene; ceramics; and minerals, such as mica.

inverter: An electronic device used to convert dc electricity into ac electricity, usually with an increase in voltage. There are several different basic types of inverters, including sine wave and square-wave inverters.

junction box: An insulating box, usually made from plastics, such as PVC, used to protect the connection point of two or more cables.

kilowatt (kW): A measurement of power equal to 1000 W.

kilowatt-hour (kWh): A measurement of energy. One kilowatt-hour is equal to 1 kW being used for a period of 1 hour.

lead-acid battery: A type of battery that consists of plates made of lead and lead oxide, surrounded by a sulfuric acid electrolyte. It is the most common type of battery used in RAPS systems.

light-emitting diode (LED): A semiconductor device that produces light of a single color or very narrow band of colors. Light-emitting diodes are used for indicator lights, as well as for low-level lighting. They are readily available in red, green, blue, yellow, and amber. The lights have a minimum life of 100,000 hours of use.

load: The collective appliances and other devices connected to a power source. When used with a shunt regulator, a "dummy" load is often used to absorb any excess power being generated.

megawatt (MW): A measurement of power equal to 1 million W.

megawatt-hour (MWh): A measurement of power with respect to time (energy). One megawatt-hour is equal to 1 MW being used for a period of 1 hour, or 1 kW being used for 1000 hours.

meters per second (m/s): A speed measurement system often used to measure wind speed. One meter per second is equal to 2.2 mi/h or 3.6 km/h.

micro-hydrosystem: A generation system that uses water to produce electricity. Types of water turbines include Pelton, Turgo, cross-flow, overshot, and undershot waterwheels.

modified square wave: A type of waveform produced by some inverters. This type of waveform is better than a square wave but not as suitable for some appliances as a sine wave.

monocrystalline solar cell: A form of solar cell made from a thin slice of a single large crystal of silicon.

nacelle: That part of a wind generator that houses the generator, gearbox, and so forth at the top of the tower.

nickel-cadmium battery (NICAD): A form of rechargeable battery, having higher storage densities than that of lead-acid batteries. NICADs use a mixture of nickel hydroxide and nickel oxide for the anode and cadmium metal for the cathode. The electrolyte is potassium hydroxide. They are very common in small rechargeable appliances, but rarely found in independent power systems, due to their high initial cost.

noise: Unwanted electrical signals produced by electric motors and other machines that can cause circuits and appliances to malfunction.

ohm: The unit of measurement of electrical resistance. The symbol used is the uppercase Greek letter omega. A resistance of 1 ohm will allow 1 A of current to pass through it at a voltage drop of 1 V.

Ohm's law: A simple mathematical formula that allows voltage, current, or resistance to be calculated when the other two values are known. The formula is

$$V = IR$$

where V is the voltage, I is the current, and R is the resistance.

Pelton wheel: A water turbine in which specially shaped buckets attached to the periphery of a wheel are struck by a jet of water from a narrow nozzle.

photovoltaic effect: The effect that causes a voltage to be developed across the junction of two different materials when they are exposed to light.

pitch: Loosely defined as the angle of the blades of a wind or water turbine with respect to the flow of the wind or water.

plates: The electrodes in a battery. They usually take the form of flat metal plates. The plates often participate in the chemical reaction of a battery, but sometimes just provide a surface for the migration of electrons through the electrolyte.

polycrystalline silicon: Silicon used to manufacture photovoltaic panels, which is made up of multiple crystals clumped together to form a solid mass.

power: The rate of doing work or, more generally, the rate of converting energy from one form to another (measured in watts).

PVC (polyvinyl chloride): A plastic used as an insulator on electric cables, as well as for conduits. Contains highly toxic chemicals and is slowly being replaced with safer alternatives.

quasi sine wave: A description of the type of waveform produced by some inverters. See *modified square wave*.

ram pump: A water-pumping device that is powered by falling water. These devices work by using the energy of a large amount of water falling a small height to lift a small amount of water to a much greater height. In this way, water from a spring or stream in a valley can be pumped to a village or irrigation scheme on a hillside. Wherever a fall of water can be sustained, the ram pump can be used as a comparatively cheap, simple, and reliable means of raising water to considerable heights.

RAPS (remote area power supply): A power generation system used to provide electricity to remote and rural homes, usually incorporating power generated from renewable sources such as solar panels and wind generators, as well as nonrenewable sources, such as petroleum-powered generators.

rechargeable battery: A type of battery that uses a reversible chemical reaction to produce electricity, allowing it to be reused many times. Forcing electricity through the battery in the opposite direction to normal discharge reverses the chemical reaction.

regulator: A device used to limit the current and voltage in a circuit, normally to allow the correct charging of batteries from power sources, such as photovoltaic arrays and wind generators.

renewable energy: Energy that is produced from a renewable source, such as sunlight.

residual current device (RCD): See *earth-leakage circuit breaker*.

resistance: A material's ability to restrict the flow of electric current through itself (measured in ohms).

resistor: An electronic component used to restrict the flow of current in a circuit, also used specifically to produce heat, such as in a water heater element.

sealed lead-acid battery: A form of lead-acid battery where the electrolyte is contained in an absorbent fiber separator or gel between the battery's plates. The battery is sealed so that no electrolyte can escape, and thus can be used in any position, even inverted.

semiconductor: A material that only partially conducts electricity that is neither an insulator nor a true conductor. Transistors and other electronic devices are made from semiconducting materials and are often called semiconductors.

shunt: A low-value resistance, connected in series with a conductor that allows measurements of currents flowing in the conductor by measurement of voltage across the shunt, which is often used with larger devices, such as inverters to allow monitoring of the power used.

sine wave: A sinusoidal-shaped electrical waveform. Mains power is a sine wave, as is the power produced by some inverters. The sine wave is the most ideal form of electricity for running more sensitive appliances, such as radios, TVs, and computers.

solar cell: A single photovoltaic circuit usually made of silicon that converts light into electricity.

solar module: A device used to convert light from the sun directly into dc electricity by using the photovoltaic effect. Usually made of multiple silicon solar cells bonded between glass and a backing material.

solar power: Electricity generated by conversion of sunlight, either directly through the use of photovoltaic panels, or indirectly through solar-thermal processes.

solar thermal: A form of power generation using concentrated sunlight to heat water or other fluid that is then used to drive a motor or turbine.

square wave: A type of waveform produced by some inverters. The square wave is the least desirable form of electricity for running most appliances. Simple resistors, such as incandescent globes and heating elements, work well on a square wave.

storage density: The capacity of a battery compared to its weight (measured in watt-hours per kilogram).

surge: An unexpected flow of excessive current, usually caused by a high voltage that can damage appliances and other electric equipment. Also, an excessive amount of power drawn by an appliance when it is first switched on.

switch mode: A form of converting one form of electricity to another by rapidly switching it on and off and feeding it through a transformer to effect a voltage change.

tip-speed ratio: The ratio of blade tip speed to wind speed for a wind turbine.

transformer: A device consisting of two or more insulated coils of wire wound around a magnetic material, such as iron, used to convert one ac voltage to another or to electrically isolate the individual circuits.

transistor: A semiconducting device used to switch or otherwise control the flow of electricity.

turbulence: Airflow that rapidly and violently varies in speed and direction that can cause damage to wind turbines. It is often caused by objects, such as trees or buildings.

vertical-axis wind turbine: A wind turbine with the axis or main shaft mounted vertically, or perpendicular to Earth's surface. This type of turbine does not have to be turned to face the wind—it always does.

voltage: The electric pressure that can force an electric current to flow through a closed circuit (measured in volts).

voltage drop: The voltage lost along a length of wire or conductor due to the resistance of that conductor. This also applies to resistors. The voltage drop is calculated using Ohm's law.

voltmeter: An electric or electronic device used to measure voltage.

water turbine: A device that converts the motion of the flow of water into rotational motion, which is often used to drive generators or pumps. See *micro-hydrosystem*.

waterwheel: A simple water turbine, often consisting of a series of paddles or boards attached to a central wheel or hub that is connected to a generator to produce electricity or a pump to move water.

watt (W): A measurement of power commonly used to define the rate of electricity consumption of an appliance.

watt-hour (Wh): A measurement of power with respect to time (energy). One watt-hour is equal to 1 W being used for a period of 1 hour.

wind farm: A group of wind generators that usually feed power into the mains grid.

wind generator: A mechanical device used to produce electricity from the wind. Typically a form of wind turbine connected to a generator.

wind turbine: A device that converts the motion of the wind into rotational motion used to drive generators or pumps. Wind generator, wind turbine, windmill, and other terms are commonly used interchangeably to describe complete wind-powered electricity-generating machines.

yaw: The orientation of a horizontal-axis wind turbine.

zener diode: A diode often used for voltage regulation or protection of other components.

Meteorological Terms

altitude: The angle up from the horizon.

angle of incidence: The angle between the normal to a surface and the direction of the sun. Therefore, the sun will be perpendicular to the surface if the angle of incidence is zero.

azimuth: The angle from north measured on the horizon, in the order of N, E, S, and W. Thus, north is 0 degrees, and east is 90 degrees.

civil twilight: Defined as beginning in the morning and ending in the evening when the center of the sun is geometrically 6 degrees below the horizon.

horizon: The apparent intersection of the sky with Earth's surface. For rise and set computations, the observer is assumed to be at sea level, so that the horizon is geometrically 90 degrees from the local vertical direction. The inclination surface tilt is expressed as an angle to the horizontal plane. Horizontal is 0 degrees; vertical is 90 degrees.

horizontal shadow angle (HSA): The angle between the orientation of a surface and the solar azimuth.

local civil time (LCT): It is a locally agreed upon time scale. It is the time given out on the radio or television, and the time by which we usually set our clocks. Local civil time depends on the time of year and your position on Earth. It can be defined as the time at the Greenwich meridian plus the time zone and the daylight savings corrections.

orientation: The angle of a structure or surface plane relative to north in the order of N, E, S, and W. Thus, north is 0 degrees, and east is 90 degrees.

shadow angles: Shadow angles refer to the azimuth and altitude of the sun, taken relative to the orientation of a particular surface.

solar noon: The time when the sun crosses the observer's meridian. The sun has its greatest elevation at solar noon.

sunrise and sunset: Times when the upper edge of the disk of the sun is on the horizon. It is assumed that the observer is at sea level and that there are no obstructions to the horizon.

twilight: The intervals of time before sunrise and after sunset when there is natural light provided by the upper atmosphere.

vertical shadow angle (VSA): The angle between the HSA and the solar altitude, measured as a normal to the surface plane.

INDEX